Monolayers

Monolayers

E. D. Goddard, Editor

Union Carbide Corp.

A memorial symposium to N. K. Adam sponsored by the Division of Colloid and Surface Chemistry at the 168th Meeting of the American Chemical Society, Atlantic City, N. J., Sept. 11–13, 1974.

ADVANCES IN CHEMISTRY SERIES **144**

AMERICAN CHEMICAL SOCIETY

WASHINGTON, D. C. 1975

Library of Congress CIP Data

Monolayers
(Advances in chemistry series; 144)

Includes bibliographical references and index.

1. Surface chemistry—Congresses. 2. Adam, Neil Kensington. I. Adam, Neil Kensington. II. Goddard, Errol Desmond, 1926- III. American Chemical Society. Division of Colloid and Surface Chemistry. IV. Series: Advances in chemistry series; 144.

QD1.A355 no. 144 [QD506.A1] 540'.8s [541'.3453]
75-25757
ISBN 0-8412-0220-6 ADCSAJ 144 1-372 (1975)

PRINTED IN THE UNITED STATES OF AMERICA

Advances in Chemistry Series

Robert F. Gould, *Editor*

FOREWORD

ADVANCES IN CHEMISTRY SERIES was founded in 1949 by the American Chemical Society as an outlet for symposia and collections of data in special areas of topical interest that could not be accommodated in the Society's journals. It provides a medium for symposia that would otherwise be fragmented, their papers distributed among several journals or not published at all. Papers are refereed critically according to ACS editorial standards and receive the careful attention and processing characteristic of ACS publications. Papers published in ADVANCES IN CHEMISTRY SERIES are original contributions not published elsewhere in whole or major part and include reports of research as well as reviews since symposia may embrace both types of presentation.

CONTENTS

N. K. Adam

PREFACE

Periodically an individual, by dint of outstanding contributions, achieves a unique and dominating position in a particular field of science. Such was the case with N. K. Adam in the field generally referred to as "Insoluble Monolayers." He established this position of authority in the 1920's and 1930's through a series of classical papers on the monolayer properties of a wide range of long chain materials. His studies were characterized by thoroughness and an unusual degree of experimental rigor in an area demanding the use of high purity materials. Indeed, by today's standards of purity, set by the availability of sophisticated instrumental analytical methods, the compounds Adam used had an evident degree of purity which is a source of continuing wonder.

A second major contribution of Adam was the writing of his book "The Physics and Chemistry of Surfaces," first published in 1930 and brought out in two subsequent editions in 1938 and 1941. The last edition is now available in paperback reprint form. For at least three decades this was the standard text on surface science. It has the further merit of documenting in orderly fashion most of Adam's own work.

The decision of the Division of Colloid and Surface Chemistry of the American Chemical Society to hold a symposium honoring the memory of N. K. Adam provided the opportunity, both of publicly recognizing his pioneering contributions to the science of "monolayers" and of collecting together a series of papers representing current directions in monolayer research. Some 25 such papers were assembled. Very fittingly, almost half of them have come from overseas, adding the contributions of English, French, Dutch, Italian, Belgian, and Japanese schools to those of various well-known American surface scientists.

The papers fit into five broad groupings. The first, comprising single long chain component monolayers, is concerned with theory and experiment, thermodynamics, and equations of state, as applied to ionized monolayers, counterion phenomena, surface potential, and the influence of unsaturation in fatty acid monolayers. The second section deals with the thermodynamics of mixed and "penetrated" monolayers. This is follower by several papers on systems of biological interest, including lipids, proteins, glycosides, enzymes, and their interactions. Thereafter there is a section involving techniques other than standard surface pressure and potential measurements. Included are studies of surface viscosity, an oscillatory compression technique yielding a novel method of determining equations of state, and an electron microscope investigation of transferred films. This section also includes three papers on the technique of spin-labeling as a tool for obtaining information on molecular motions in monomolecular arrays. The final two papers are concerned with the properties of monolayers of synthetic polymers.

N. K. Adam would have enjoyed this symposium if for no reason other than to recognize the remarkable progress and developments that it signifies. An isolated illustration is the evident impact Adam's own work on monolayers has had in the development of theories of cell and membrane structure. It is indeed fitting that we have an introductory account of Adam's early period by a former student of his, James Danielli, whose name is firmly linked to the bimolecular leaflet theory of membrane structure. An account of Adam's activities at Southampton up to the time of his death in 1973 by Michael Phillips concludes the introductory material.

I would like to end these prefatory remarks on a personal note. As a young graduate student arriving at the Department of Colloid Science in Cambridge in 1948, I well remember the answer to my question regarding a good source of reading on surface chemistry. My supervisor and, subsequently, valued friend and mentor, the late A. E. Alexander, replied unhesitatingly "why, N. K. Adam's book of course." Although other excellent books on surface science have since appeared, it appears to me that Alexander's advice is as sound today as it was then.

E. D. Goddard

Union Carbide Corp.
Tarrytown, N.Y.
May 1975

1

Reminiscences of N. K. Adam

JAMES F. DANIELLI

The Department of Life Sciences, Worcester Polytechnic Institute, Worcester, Mass. 01609 and The Center for Theoretical Biology, State University of New York, Amherst, N. Y. 14226

N. K. Adam was unlike anyone else I have ever met. Most people can be said to resemble a number of others, and by making such comparisons one obtains useful insights. No such comparison could be made with N. K. which had validity or value.

His scientific work lay almost entirely in surface chemistry, essentially in the study of monolayers. He took up the study of monolayers, following the remarkable contribution made by Irving Langmuir. The field at that time needed a first class experimentalist who could work with rigour and dispassionate exactness. N. K. was the ideal man for this.

Judging from his behaviour, his work and his writing, he was convinced that, provided the correct experiments were done, the phenomena of nature can be shown to arise from a few simple principles. This ascetic attitude was shown in his writing, in which, by the accurate use of simple words and simple sentences, he conveyed meaning with unusual clarity. All his writings on surface chemistry had this characteristic.

Although in many ways a sober and austere person, he also had a capacity for humour and gaiety which delighted his companions, and which endeared him to his students. My undergraduate year composed a song about the faculty which was of noteworthy lewdness in some cases: the verse about N. K. was a mild good humoured comment about his use of a windy equation to which I shall refer later. I shall never forget his telling me of his own involvement, when he was an undergraduate at Cambridge, in a competition to see who could sit longest bare-bottomed on a block of ice.

I had the good fortune to attend his lectures on surface chemistry during my freshman year at University College London, in 1929. Questioning the validity of some data resulted in an invitation to spend the Summer of 1930 in his laboratory. This splendid opportunity resulted in my working for my doctoral thesis under his guidance, with occasional help from G. S. Hartley. One of the first principles of research he instilled in me was that, when one's experimental findings are not quite those

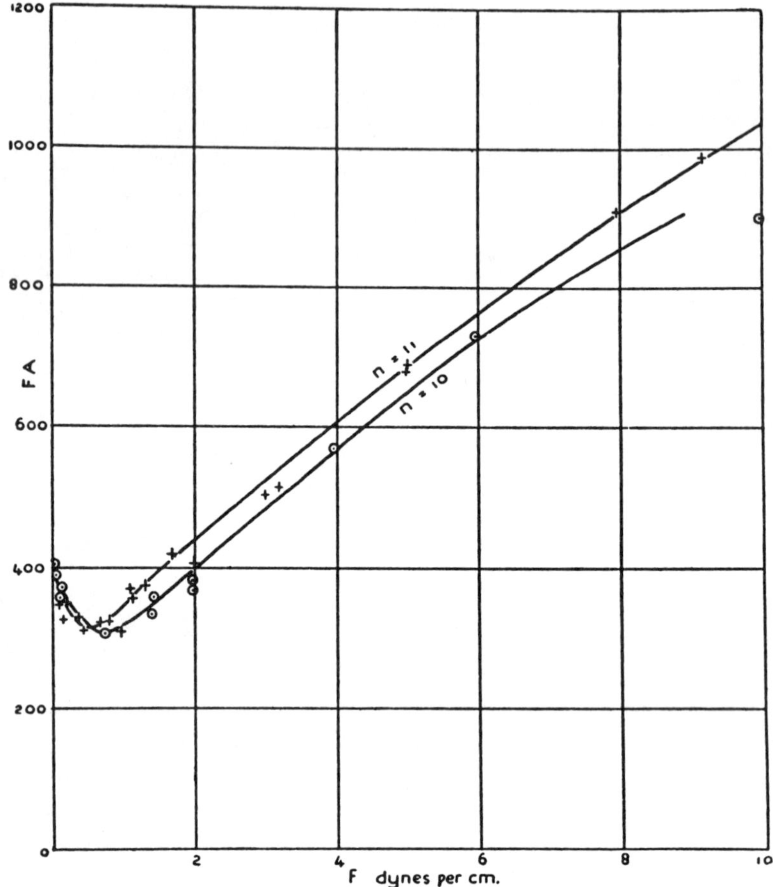

"The Physics and Chemistry of Surfaces"

Figure 1. Behaviour of the diesters of αω dibasic esters (1)

anticipated, the proper course is not to minimise, but to maximise, the discrepancies, because by so doing new discoveries are made.

This opportunity to work under his guidance enabled me to appreciate his excellence in the laboratory. As can be seen by looking at the first edition of his book "The Physics and Chemistry of Surfaces", his initial endeavour as a surface chemist was to survey the range of phenomena exhibited by monolayers, using the surface balance which George Jessop and he invented. These results he then endeavoured to explain in terms of the molecular structures of the substances concerned. One of the tools he found useful was the evidence on the shapes and sizes of organic molecules made available by Lawrence Bragg's X-ray diffraction data. This was reflected in a chapter on X-ray diffraction studies which was included in the first edition. By the time the second edition was prepared

X-ray diffraction methods were very well known and the chapter could be omitted.

The first paper on the film balance was published from Cambridge. Soon thereafter he moved to the Department of Chemistry at University College London, the head of which was F. G. Donnan, of "Donnan Equilibrium" fame. Here he enjoyed a period of intensive research, in good company, always with several research students.

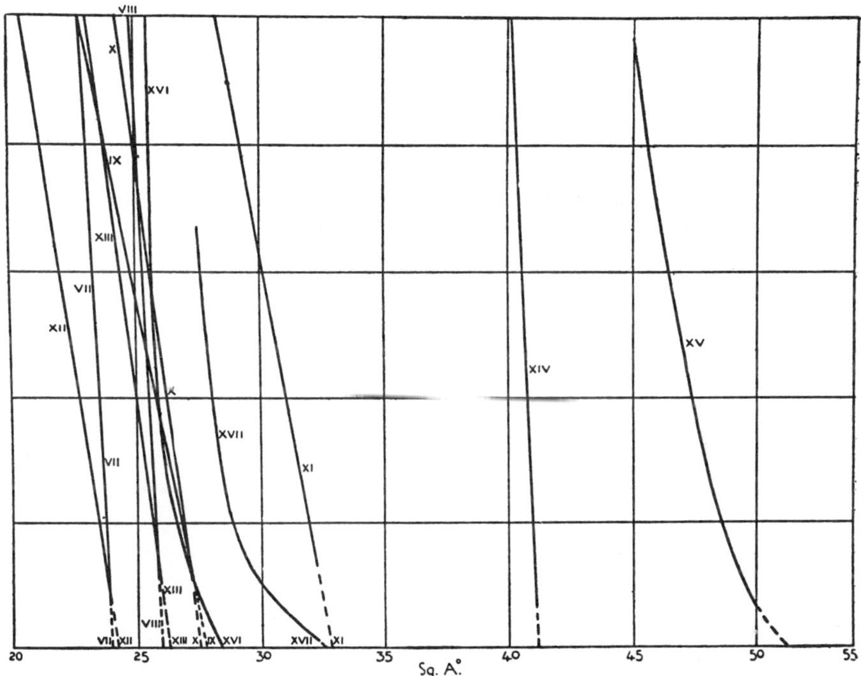

"The Physics and Chemistry of Surfaces"

Figure 2. Molecules with only one polar area at one end of a hydrocarbon chain (1)

The results he obtained with the film balance confirmed and greatly extended the results obtained by Langmuir. Several of the more interesting sets of data are set out in Figures 1–5.

Figure 1 shows the behaviour of the diesters of $\alpha\omega$ dibasic esters. These molecules he supposed would lie flat at the air–water interface, and, therefore, there would be little lateral cohesion between molecules. Thus, he argued that the molecules should obey the gas equation:

$$FA = RT.$$

This equation was received with joy and hilarity by successive genera-

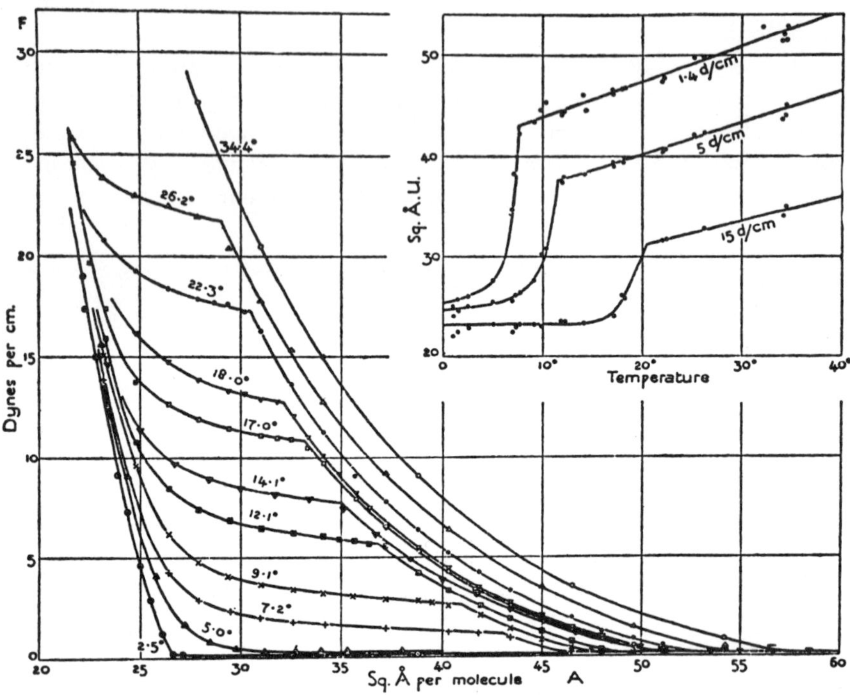

"The Physics and Chemistry of Surfaces"

Figure 3. Behaviour of myristic acid at different temperatures (1)

tions of ribald students, and eventually was written as:

$$FA = kT,$$

which diminished the noise levels in his lectures. As can be seen from Figure 1, at infinite dilution the value of *FA* does tend closely to the value of 400 postulated for a perfect gas.

In Figure 2 is shown the data for molecules which have only one polar area, at one end of a hydrocarbon chain. In monolayers of such molecules the chains lie more or less perpendicular to the plane of the surface, with a cross-sectional area primarily determined by the non-polar chain, secondarily modified by the nature of the polar groups. Such films are coherent liquids or solids when the chain length is sufficiently great. With shorter chains he found interesting transitions. Figure 3 shows the behaviour of myristic acid [$CH_3(CH_2)_{12}COOH$] at different temperatures; these films consist of two phases, one stable at low surface pressure, and one at high pressure, with a transition region. At very low pressures these films approximate to a gaseous state, as shown in Figure 4. The behaviour of molecules in monolayers shown in Figures 3 and 4 involved the existence of coherent phases in which the area per molecule was much greater than the minimum cross-section of a hydrocarbon chain. He, therefore, called these "expanded films". Their nature has interested many surface chemists since Adam's studies were made: yet no very satisfactory theory of these films has yet emerged.

The general conclusion, which emerged from these studies, was that, for substances containing one polar group, at sufficiently low temperature and sufficiently high surface pressure, condensed films were formed in which the cross-sectional area per molecule agreed well with that predicted from molecular models and from X-ray diffraction data. Molecules with two polar groups, spaced widely apart formed gaseous films. Thus, the principle emerged from these early studies that, provided the physical state of a film was understood, approximate molecular cross-sectional areas could be obtained from monolayer data.

When I became one of N. K.'s research students, he suggested that a study of steroids should be made. He had made a few preliminary studies, finding that monolayers of cholesterol and a few other steroids were stable and suitable for serious study. This was an interesting period

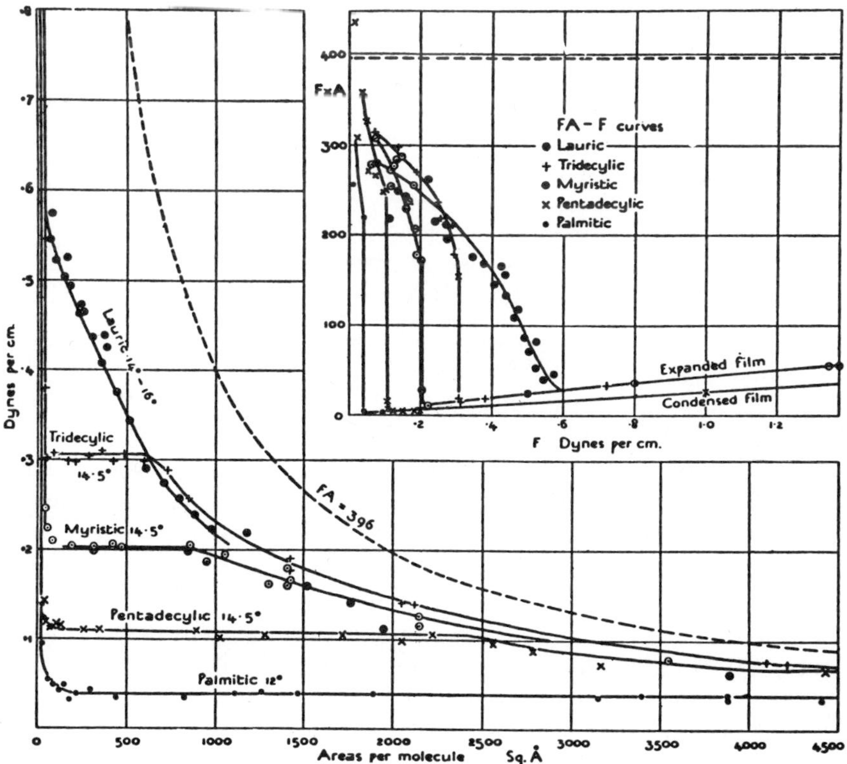

"The Physics and Chemistry of Surfaces"

Figure 4. Myristic acid films approximating a gaseous state at very low pressure (1)

in steroid chemistry. The structure of the hydrocarbon nucleus was in dispute. Work on the structure of vitamin D indicated a close relationship of this substance to the steroids. In the work we did on sterols, we used the principles outlined above to find molecular cross-sectional areas (*see* Figure 5). This enabled us to confirm one of several proposed structures

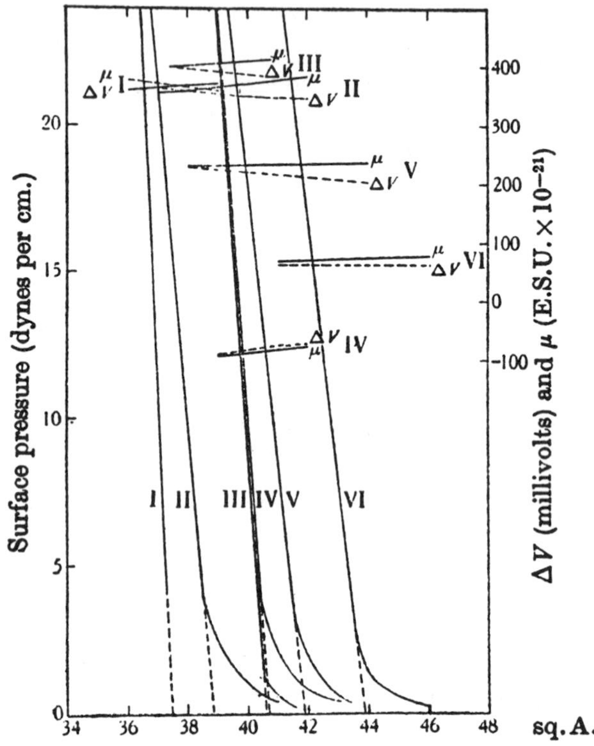

"The Physics and Chemistry of Surfaces"

Figure 5. Molecular cross-sectional areas of sterols (1)

for the steroid nucleus, and to show that the orientation of steroids in films was a function of the position of the polar groups. At this point we examined films of oestrogens, then recently isolated but of unknown structure. Figure 6 shows typical results obtained with oestrogens in monolayers. From results such as these we were able to measure the dimensions of the oestrogens, and show that they were probably steroids. Simultaneously J. D. Bernal made X-ray diffraction studies on the same molecules, and came to the same conclusions.

N. K. was a generous supervisor, ensuring that his students received recognition for their achievements. Work with him involved a good deal of fun. One unforgettable incident arose when he was asked to look into the blocking of ships' salt-water circulation pipes by growth of mussels, etc. within the pipes. Several tins of mussels attached to pipes, well sealed, were stored in his laboratory for some months. Finally, he got around to opening one. The stench which emerged emptied the top floor of the laboratory.

The last occasion on which I worked with him was in 1936, when he was asked by the Royal Society to study oil pollution, already a menance to sea birds. Tar on the beaches could conveniently be studied by examining the feet and bottoms of holiday makers. But oil on the water required more strenuous exercise. N. K. persuaded me to row him round Plymouth bay and Plymouth harbour: whilst I got the exercise, he was very contentedly measuring the surface tension of the water! In 1937, N. K. moved to the University College of Southampton, and his subsequent work is reviewed in the article by M. C. Phillips in this volume.

The work done by Adam & Langmuir on monolayers has had many repercussions. Not least was that it was from these two great men that I drew the concepts which enabled me to develop the basic principles of cell membrane structure in the 1930's. These defined cell membranes has having a lipid bilayer as their continuous phase, sandwiched between two protein layers, and penetrated by hydrophobic proteins. There are now some thousands of biologists studying these membranes, and the source of concepts remains in the work of Adam and Langmuir. I wish we knew as much about the protein components as we do about the lipids.

Looking back on the last sixty years of monolayer studies, the greatest impact was made by Langmuir and Adam. Langmuir's grasp of the

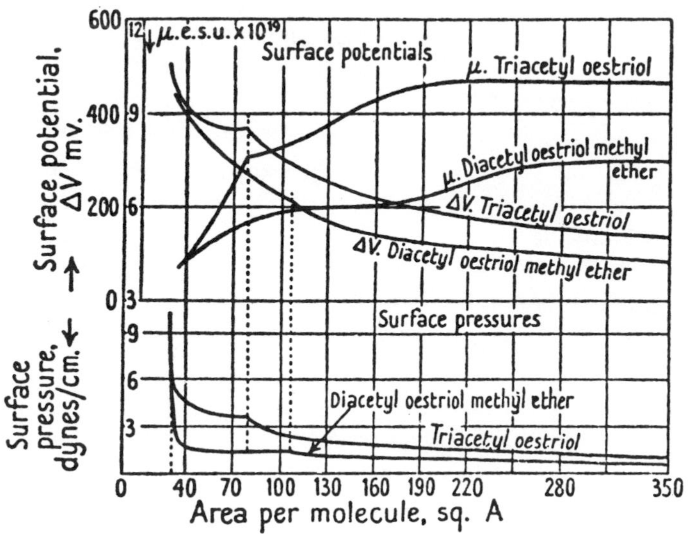

"The Physics and Chemistry of Surfaces"

Figure 6. Typical results obtained with oestrogens in monolayers (1)

properties of matter led to development of necessary theory. Adam's precision in experimentation revealed the boundaries of monolayer studies and provided reliable data which set standards in the field. Adam was not comfortable with theory, and relied upon others, particularly Hartley and Guggenheim, in evaluating theory.

Though in general a tolerant and kindly person, he was intolerant of humbug and was of great intellectual honesty. To his close associates his intellectual honesty was perhaps the most valued of his attributes. Only a fool tried to put anything over on N. K. All of us knew that, asked to read a draft paper, he would examine it with the keenest eye, and what passed his eye would pass any editor. He set a high ethical standard, shared his knowledge, and was incapable of the rather shady practices shown in *e.g.* "The Double Helix".

I do not pretend that he was devoid of human weaknesses. When tired or overworked, he was often reluctant to consider new ideas. But in my experience, N. K. was one of the easiest men to get along with, given intellectual honesty. He was never concerned to *prove* something—rather to find out whether it was right or wrong. One of the principles he instilled into his students was not to gloss over deviations from expected results. Rather to find out how to magnify the deviations, since this is the way new discoveries are made.

I wish he were still with us.

Literature Cited

1. Adam, N. K., "The Physics and Chemistry of Surfaces," The Clarendon Press, Oxford, England, 1930.

RECEIVED January 24, 1975.

2

N. K. Adam—Southampton Period

M. C. PHILLIPS

Unilever Research Laboratory Colworth/Welwyn, The Frythe, Welwyn, Herts., England

In 1937, N. K. Adam accepted the Chair and Headship of the Department of Chemistry at the University College (later University) of Southampton. He held this position for 20 years, was one of the founding fathers of the university, and prepared it well for the advent of the present large and important School of Chemistry there.

When Adam went to Southampton he joined a small, struggling, impoverished institution of uncertain future. Although the college had existed in some form since 1862, its career had been checkered. It had become a University College in 1902, and its students took the External London degrees although a number of them sat only for the two year teachers' training qualification and did not graduate.

From 1937 until the end of World War II in 1945, the total academic staff of the chemistry department (including Adam) varied between five and six. Although the total number of fulltime day degree students was only about 40 before the war, the department also provided many evening classes for an additional 40 or more so-called technical students since the University College was then acting as the Technical College for Southampton; this arrangement continued well into the 1950's as far as chemistry was concerned.

In 1937, the chemistry department occupied two single-story brick buildings; one contained the main general teaching laboratory plus stores, balance room, fume room, combustion room, the professor's laboratory/room, and one other small research laboratory; the other contained the physical chemistry teaching laboratory, two research rooms, a darkroom, and a fair-sized departmental library room. The chemistry department also had two wooden army huts—one (grandly called the research hut) was only barely suitable as an apparatus store; the other, called the inter. hut, was used for elementary laboratory teaching of Intermediate B.Sc. standard, and its decrepitude almost defies description. The benches leaned this way and that on the crazy, hummocked wooden floor, and gas, water, and electric services were crude and unpredictable—a source of constant anxiety to the lecturer-in-charge and to the one storekeeper–chief steward of the department and his youthful laboratory attendant. In addition to the laboratories, there were two raked lecture rooms and a small preparation room a short way distant on the first floor of the main college building.

This situation when Adam took over made it imperative that the new incumbent of the Chemistry Chair should devote much of his time and energy to planning a new chemistry building and generally to furthering the development and recovery of the college during a period of post-depression, financial restriction. Moreover, since Adam had been virtually a fulltime research worker for most of his career until then, he had to prepare and to develop from scratch all the advanced lectures on physical chemistry (of which he gave the greater part). Also, as a scientist of international standing, he could not entirely escape all the time-consuming chores associated with such eminence. As a result, N. K.'s life had to take an almost entirely new direction, and so his activities were largely drawn away from personal research into the other fields indicated. By the time he had coped with these changes, the outbreak of World War II once again imposed new restraints and new activities. The move to Southampton allowed him to pursue his interest in sailing and from 1937–39 he owned a small yacht in which his wife and son accompanied him on weekend and holiday sailing in the Solent and nearby waters until the war made this impossible.

During the war, purely academic research almost ceased, but in collaboration with the Chemical Defence Experimental Establishment at Porton Down, an extensive study of the vapor pressures of chemical warfare agents was undertaken in case these substances should be used by the enemy. By 1945 a number of measuring devices able to cover the range of pressures from 10^{-6} to 1000 mm of mercury were in operation at Southampton. Adam arranged, superintended, and coordinated this work, but the design, construction, and operation of these instruments was mainly the work of E. W. Balson, a former student of the college and then a member of the chemistry staff. N. K.'s skill as a mechanic and glass blower as well as his extensive knowledge of classical physical chemistry was invaluable in this work, much of which was published (*1*) after the war. His experimental skill was also invaluable in other directions. When the Southampton gas works was hit during the air raids, he obtained a huge cylinder of Propagas and connected it to the main laboratory gas supply. He modified bunsen burners by covering the lower jets with hard wax, piercing a much smaller hole with a very fine sewing needle and replacing the barrels. Although the resulting flames were only moderate, they were reasonably hot and smoke free and enabled practical work to continue more effectively than with spirit lamps.

The college was very fortunate not to suffer any major damage during the heavy air raids on Southampton, though cynics believed that the Luftwaffe, seeing below them the collection of near-derelict huts, thought the place had been blitzed already.

After the war ended in 1945, the College expanded rapidly, and new and greatly enlarged plans were drawn up by N. K. for the permanent chemistry department building. It was providential that his earlier plans had to be shelved since they would have been inadequate for the ensuing rapid growth of the department which continues today. The first stage of the new plan—a large single-story laboratory to replace the inter. hut—was opened on an entirely new site in late September 1948; a considerable second stage, whose building was supervised by Adam, followed in September 1952. These have since been overshadowed by a huge seven-story glass and concrete tower block designed and completed after Adam's time in 1962.

N. K. enjoyed teaching, and during most of his time as professor he carried a lecture load which was very heavy by modern standards; it also included advanced lectures to evening or part-time day students at various periods as well as to fulltime day students. He was impatient when students were inattentive or lazy during lectures. If he made a mistake on the blackboard and discovered it before any of the class members, he would give them a piece of his mind in straight terms, ending up "If you don't correct my mistakes before I notice them myself, I shall start making them on purpose." He was not pompous in his dealings with students as evidenced by the fact that he said one day to an unusually deferential student "You may call me Prof. Adam, N. K., or old cock, but my name is *not* 'Please Sir'." N. K. was concerned with education, not with preparation for examinations. He described some of the candidates for Southampton University Scholarship examinations in his later years as "little parrots," and any of their masters who requested copies of past papers to "spot" possible questions were likely to receive a very dusty answer.

In 1947 Adam was elected Dean of the Faculty of Science, a post of heavy administrative responsibilities that he held for 10 years until his retirement in 1957. Two of his heaviest duties were to select from among ever-increasing numbers of applicants those for admission to courses in the Science Faculty and to help draw up regulations for the faculty and for the now dawning University of Southampton, whose charter was granted in April 1952. This followed a three-year period of "special relationship" with the University of London, during which boards of examiners, which included some Southampton academic members and some from London, set the final examination papers in some subjects for the

Southampton students who, however, were still awarded London degrees, as before.

As perhaps befits a man born on Guy Fawkes day (November 5) N. K. had an explosive personality. However, his eruptions were usually short-lived, and peace was soon restored with a twinkle of the eye and an infectious chuckle; sometimes he would make peace with erstwhile savaged colleagues by bringing them a bag of cherries from the splendid tree in his garden or even a useful bucket of compost or some other little offering. On occasion, his patience was exhausted by the University administration. Told by the Principal of the old University College to be quiet in a Senate meeting, N. K. promptly replied: "I will *not* be quiet, Mr. Principal, I am a professor of this College and *you* are a mere administrator." History is silent on the sequel. He might also leave an unnecessarily lengthy meeting with the announcement that Toy Town (a popular children's radio program) was on at 5:15 p.m., and he was off home to listen. However, he always had the good of the university or his department at heart, and he would attack vigorously any proposal he believed to be detrimental to these ends.

N. K. and his wife were passionately fond of ducks. This interest spilled over into his everyday behavior. Thus, on passing a friendly colleague, he would nod gravely and say "Quack," or sometimes he might give a silent quack by apt use of thumb and forefinger. He also used this technique during long-winded or boring meetings when he would look across the table to a sympathetic colleague and give the silent quack. N. K. did not continue his classical work on monomolecular films in Southampton but concentrated his personal research on detergency (*2*) and wetting (*3*). In view of the well-known saying "like water off a duck's back," it is intriguing that he should have concerned himself with water repellency. As part of a study (*4*) of the penetration or liquids into solids he concluded that: "the superlative water-repellence of ducks feathers is due, not to any miraculously water-repellent wax, but to the well-designed structure in which the barbules are kept apart at nearly the theoretically ideal distance." Needless to say, he vented his displeasure in no uncertain way on the representatives of a detergent company who wanted to demonstrate the efficiency of their material by showing that a duck sank into a solution of the product.

During the years before his retirement he prepared a large textbook (*5*), "Physical Chemistry," which was published in 1956. After his retirement in September 1957, N. K. continued his interest in surface science and still worked on projected revisions of his books as well as reviewing publications in the surface field and giving advice on surface problems. He spent much time also on a careful study of Christian Science and published several articles (*6*) on this subject.

Professor Adam and his wife (who died one month after he did) were cremated at Southampton after Christian Science services. A large congregation attended a Memorial Service for him at Highfield Church, Southampton on October 10th, 1973. There is no doubt that he greatly loved his last University, and he is remembered with great affection and respect in Southampton.

Acknowledgment

I am indebted to G. J. Hills for giving me access to an appreciation of N. K. Adam which was recently prepared by members of the Chemistry Department at Southampton University.

Literature Cited

1. Balson, E. W., Adam, N. K., "The Vapour Pressure of Lewisite," *Trans. Faraday Soc.* (1951) **47,** 417.
2. Adam, N. K., Stevenson, D. G., "Detergent Action," *Endeavour* (1953) **12,** 25.
3. Adam, N. K., "Principles of Water-Repellency," in "Waterproofing and Water-Repellency," J. L. Moilliet, Ed., p. 1, Elsevier, Amsterdam, 1963.
4. Adam, N. K., "Principles of Penetration of Liquids into Solids," *Disc. Faraday Soc.* (1948) **3,** 5.
5. Adam, N. K., "Physical Chemistry," Clarendon Press, Oxford, 1956.
6. Adam, N. K., "A Christian Scientist's Approach to the Study of Natural Science," *Christian Sci. J.* (1962) **80,** 225.

RECEIVED January 24, 1975.

3

Entropies of Compression of Charged Monolayers at Aqueous Interfaces

J. MINGINS, N. F. OWENS, J. A. G. TAYLOR, J. H. BROOKS,[1] and B. A. PETHICA

Unilever Research, Port Sunlight Laboratory, Port Sunlight, Wirral Merseyside, England

Entropies of compression calculated from surface pressure–area isotherms for anionic or cationic monolayers of the same chain length (18 carbons) that are spread at n-hep*tane/water (O/W) or air/water (A/W) interfaces or for their equimolar mixture at the O/W interface are negative. When plotted against the logarithm of the area they give two lines—one for the A/W interface and a steeper one for the O/W interface. As well as being independent of head group, the entropies do not depend on ionic strength. A model that treats the monolayer as an ensemble of discs that interact only through hard sphere and coulombic forces explains the size of the entropies at A/W but not at O/W where an extra chain-conformation term is needed.*

The inherent asymmetry of an interface and the requirement that resident amphipathic molecules take up certain orientations that fulfill the energy requirements of the system coupled with the ability of a researcher to vary the surface number density and hence the intermolecular separation make insoluble monolayers eminent model systems for studying ionic, dipolar, and dispersion interactions. Most studies on monolayers deal with the properties of solid-condensed or liquid-condensed films at the air/water (A/W) interface. Vapor-expanded or

[1] Present address: C.S.I.R.O., Division of Textile Physics, Ryde, Sydney, N.S.W. Australia.

gaseous insoluble films at high dilutions were initially the province of one or two adventurous surface chemists (*1, 2, 3, 4*), and the classic work of Adam and Jessop (*5*) epitomizes the high level of accuracy required to measure the surface pressure (Π) to make any meaningful statement on the behavior of dilute monolayers. Since Adam's early work, the required extreme sensitivity of the surface balance and the immaculately clean conditions needed for the working interface have discouraged researchers from making a steady compilation of reliable data on very dilute monolayers and apart from one example (*6*), such data are restricted to the A/W interface. Furthermore, the data, with few exceptions (*6, 7, 8, 9, 10*), have been limited to uncharged monolayers.

These published results at high area per molecule (A) were obtained primarily to check surface equations of state; in a companion paper (*11*) the results and conclusions from charged monolayer systems are criticized in the light of our recent experiments. Because the published isotherms for charged monolayers are available for only one temperature, the investigation of monolayer structure by analyzing the entropies of compression was hitherto precluded. In this paper we compile our accurate isotherms for charged monolayers spread at moderate to high A at 5° and 20°C. Except for mixed films, the approximate range of A includes all the monolayers studied so that meaningful comparisons can be made on the basis of interface type, salt concentration, and head group. Regrettably, no data on charged monolayers spread at either interface are available to test the chain length dependence of the entropy of compression of charged monolayers.

Experimental

General. All measurements were made in a clean-air laboratory, whose specifications together with details of preparation of surface chemically pure water, sodium chloride, heptane, paraffin wax, and spreading solvents were given elsewhere (*9, 12, 13*). The water had a surface tension of 73.0 ± 0.1 dynes/cm at 20° ± 0.1°C and aged very little. Because of the open nature of the Langmuir trough, no special precautions were taken to exclude carbon dioxide; hence, all the experiments were done with water at pH 5.0–6.0. Salt solutions were made up with water distilled the same day. Prepared heptane/water interfaces gave interfacial tensions of 51.12 ± 0.02 dynes/cm at 20° ± 0.05°C.

Pure C_{18} sulfate and C_{18} TAB (trimethylammonium bromide) samples (*14*) were stored in the dark in clear glass bottles which were packed in silica gel.

Apparatus. A/W INTERFACE. Measurements on the C_{18} TAB at the A/W interface were made using a paraffin-coated silica trough, the floating boom method, and aqueous spreading solutions. Details of our procedures with this trough are given elsewhere (*9*).

Because paraffin wax contaminates monolayers by dissolving in the organic spreading solvents, the results on C_{18} sulfate at the A/W interface were obtained using a Teflon trough and a differential Wilhelmy plate method. Two matched, smooth glass plates were suspended from opposite pans of a Beckman LM600 electronic microbalance; the weight difference was measured with one plate dipping into the working interface and the other into the adjacent clean reference solution interface on the same trough. This procedure eliminates the effects of buoyancy changes caused by water evaporation during prolonged runs and lessens the effects of small drifts in surface temperature. Plates ($\sim 22 \times 11 \times 0.17$ mm), made from Chance microscope cover slips, were parallel to better than 1 part in 2000 and showed zero contact angle in the clean and C_{18} sulfate-covered interfaces. The output of the microbalance was fed into a potentiometric recorder and Π was read to 0.003 dynes/cm. Corner corrections at the plate were ignored in the small ranges of Π. Because conventional Teflon barriers leak, monolayers were compressed with siliconized glass slides that were identical to those described for work at the O/W interface (*12, 14*).

Both surface balances were housed in thermally insulated boxes. The temperature of the A/W interface was controlled by pumping thermostatted water through a clean glass serpent in the trough and through copper pipes on the inner walls and base of the box; the room temperature was held constant to $\pm 1°C$. Separate thermostatting units were used for the box and trough in the C_{18} sulfate work; in the older experiments on C_{18} TAB, the serpent and pipes were parallel on the same pump–thermostat unit, and the flow through the serpent was controlled by a screw clip. To facilitate thermostatting with the latter system, insulation was improved, and the time intervals between spreading and compression were increased. At 20°C the air and bulk solution temperatures were held constant to $\pm 0.1°C$ for both monolayers. At 5°C the solution temperatures were still held constant to $\pm 0.1°C$, and the air temperatures, at the same level as the A/W interface, varied less than 0.2°C from the mean substrate temperature.

O/W Interface. Surface pressures were measured by the dipping plate method which used a carbon-coated platinum plate suspended from an electronic microbalance for the C_{18} sulfate and C_{18} TAB runs and from a classical torsion balance for the mixtures. Monolayers were compressed by using the O/W trough described by Brooks and Pethica (*14*); an updated version is described by Taylor and Mingins (*12*). Values of Π were read to ± 0.001 dyne/cm for the charged films and ± 0.01 dyne/cm for the mixtures, again ignoring corner corrections. Compared with the A/W interface, thermostatting the O/W interface to $\pm 0.1°C$ was easy. This limit is quoted for all monolayers at all the temperatures studied.

Procedures. CLEANING. The Teflon trough was cleaned by successive washing and soaking in reagent grade carbon tetrachloride, redistilled ethanol, and pure water. Other items were cleaned as described elsewhere (*9, 12, 13*). We cleaned a working interface by sweeping it to a point or a line while at the same time sucking on the compressed interface with a clean glass capillary tube attached to a vacuum pump; the spreading and removal of two or three cleaning-up monolayers followed. Compression of cleaned interfaces to one-third or one-quarter of the area generated less than 0.005 dyne/cm in the experiments for C_{18} TAB (A/W), less than 0.04 dyne/cm for C_{18} SO_4 (A/W), less than 0.001 dyne/cm in the 0–1.0 dyne/cm range, and less than 0.005 dyne/cm in the 1.0–5.0 dyne/cm range for the charged films at the O/W interface.

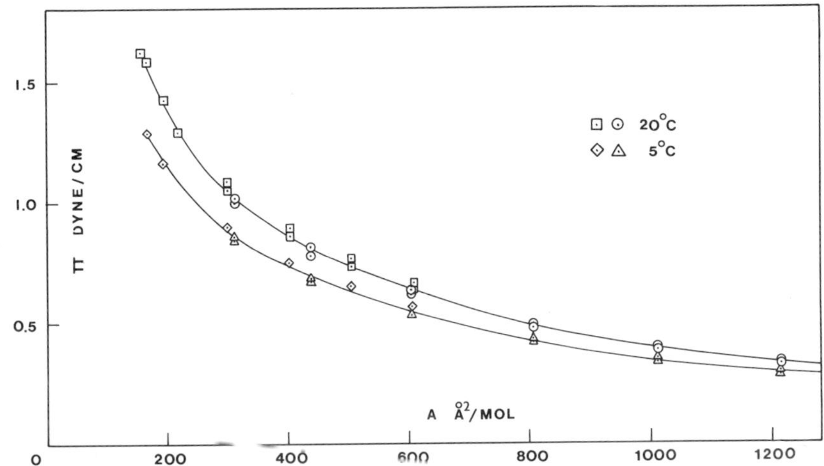

*Figure 1. Surface pressure–area isotherms for sodium octadecyl sulfate at the air–0.01*M *sodium chloride solution interface.* ◇ △ *5°C,* □ ○ *20°C.*

Spreading. Monolayers were spread from Agla micrometer syringes with fine bore glass or stainless steel needles. The zero reading of the syringe was taken after the tip was touched to another clean water surface. Area corrections from the menisci at the edges of the troughs and slides were neglected. Corrections were made for the area of interface on the Wilhelmy plates, and the calculated areas per molecule were accurate to better than 1%. Spreading was judged acceptable by the good overlap of Π–*A* curves from the different amounts spread, the lack of hysteresis on expansion of the Π–*A* curves, and their good reproducibility. The following spreading solutions were used.

(a) C_{18} TAB (A/W and O/W)—water (0.04–0.2 mg/ml, ultrasonicated and stored at 37°C to prevent precipitation).

(b) C_{18} TAB (O/W)–97/3 v/v, heptane/ethanol (0.05 mg/ml).

(c) C_{18} sulfate (A/W)–90/10/1 v/v, chloroform/methanol/ water (0.06 mg/ml).

(d) C_{18} sulfate (O/W)–70/30 v/v, water/ethanol (0.025 to 0.3 mg/ml).

(e) C_{18} sulfate + C_{18} TAB mixture (O/W)–75/25 v/v, ethanol/ water (0.2 to 0.37 mg total surfactant/ml).

Solutions were stored in special anti-evaporation bottles described elsewhere (*12*).

When various aliquots of solvents b–e were spread at a clean interface, small surface pressures were generated; these decayed to zero as the solvent evaporated or diffused into the heptane or water. Compres-

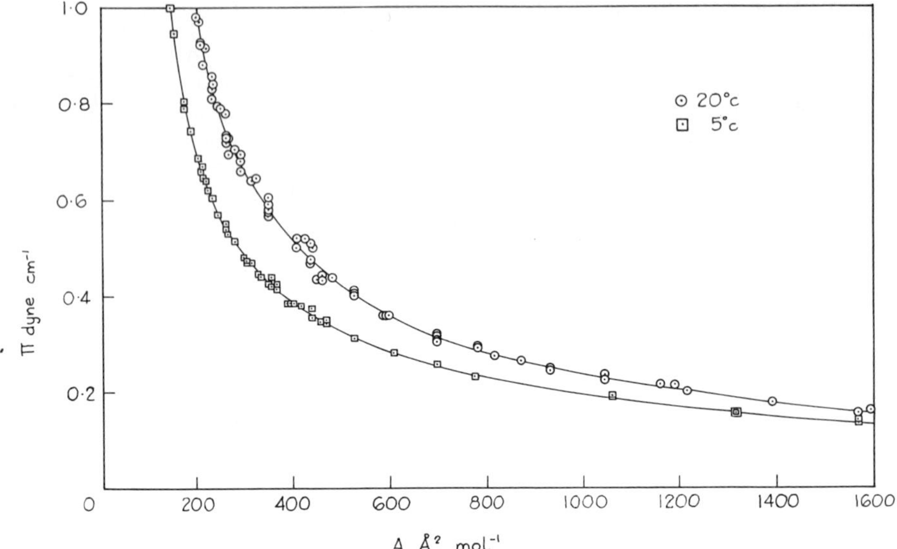

*Figure 2. Surface pressure–area isotherms for octadecyl trimethylammonium bromide at the air–0.1*M *sodium chloride solution interface.* □ *5°C,* ○ *20°C.*

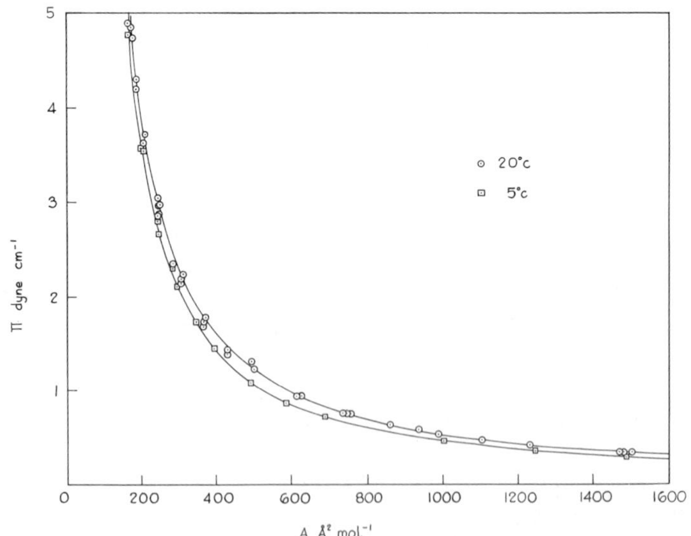

Figure 3. Surface pressure–area isotherms for sodium octadecyl sulfate at the n*-heptane–0.1*M *sodium chloride solution interface.* □ *5°C,* ○ *20°C.*

sion of the resulting aged interfaces gave changes in Π no different from those on the clean interfaces.

Π–*A* isotherms could only be obtained when sufficient times (5–60 min) were allowed after spreading the monolayer for thermal equilibrium to be re-established and spreading solvent to disappear from the interface. These times varied with monolayer density and concentration, the amount of spreading solution, the type of spreading solvent, the nature of the interface, and the temperature (*12*).

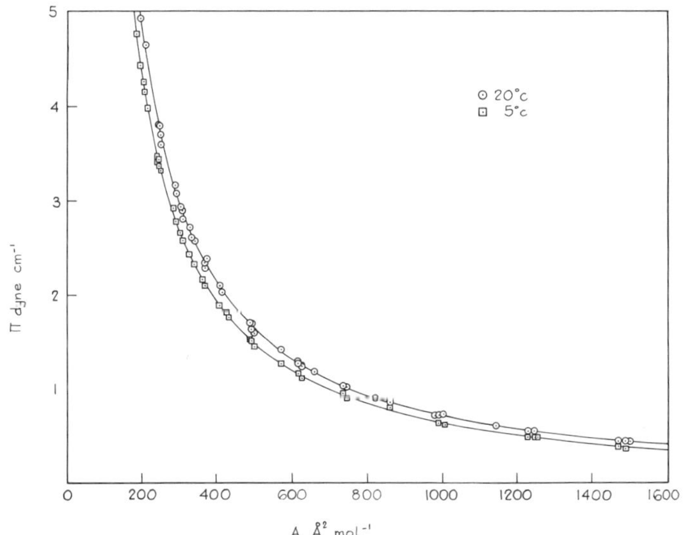

Figure 4. Surface pressure–area isotherms for sodium octadecyl sulfate at the n-*heptane–0.01*M *sodium chloride solution interface.* □ *5°C,* ○ *20°C.*

Results

Π–*A* isotherms for the monolayers at 5° and 20°C are shown in Figures 1–7. In the range of *A* depicted the monolayers are all gaseous-expanded. The temperature increase expands all the monolayers, and the increase in Π is well outside the scatter for each Π–*A* isotherm.

The results for C_{18} sulfate at the A/W interface are confined to 0.01*M* NaCl. At the O/W interface C_{18} sulfate desorbs at 250 A^2/molecule on 0.001*M* NaCl (Figure 5); consequently no data were collected at the A/W interface. The required accuracy (±0.01 dyne/cm) could not be attained on 0.1*M* NaCl by the Wilhelmy plate method because of the weight of "tears" that formed on the plate (smooth or roughened glass). Roughened glass plates were also unsuitable for this work owing to variability in the weight of "wicked" water. The results in Figure 1 compare with our results from the floating boom method (*8*, *15*). In the

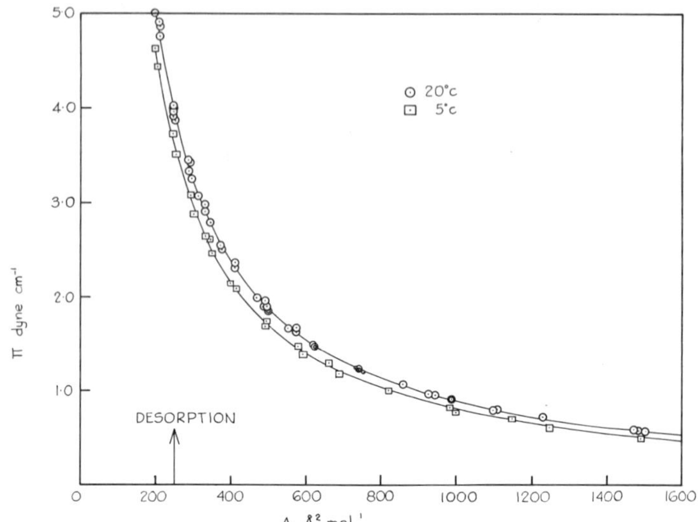

Figure 5. Surface pressure–area isotherms for sodium octadecyl sulfate at the n*-heptane–0.001*M *sodium chloride solution interface.* □ *5°C,* ○ *20°C.*

earlier work, only 3 min were allowed for the spreading solvents to disappear, with the box-door open and a forced draft across the surface. Thus air temperatures for the runs at 5°C were 4–5°C too high although thermal equilibrium was quickly established at 20°C. The new results at 20°C are within 0.02 dyne/cm of the best line through the earlier data, whereas at 5°C, agreement is only found at $A < 500$ A²/molecule. At high A the best line is about 0.06 dyne/cm higher than the present results although scatter in the earlier work would accommodate the new data. Apparently, time was insufficient in the earlier work at 5°C for the spreading solvents to disappear.

The results in Figure 2 for C_{18} TAB at 20°C are taken from an earlier paper (*9*) where good agreement was shown with results from crystal spreading experiments as well as with Saraga's results on monolayers spread to high A from organic solvents (*7*). The results at 5°C are from a thesis by one of us (*16*). There are no other data at this temperature available for comparison.

Isotherms for C_{18} sulfate at *n*-heptane/NaCl solution interfaces are given in Figures 3–5. The Π–A curves obtained by Brooks and Pethica (*14*) considerably exceed these isotherms; the discrepancies were ascribed (*12*) to the effect of ethanol on Π in the earlier work. The results of Phillips and Rideal (*17*) on C_{18} sulfate at the petroleum ether/water interface, obtained by the method of repeated addition, also exceed the isotherm in Figures 3 and 4. The retention of isopropyl alcohol in the interface is the likely cause.

Ethanol (*12*) caused the disagreement between the data of Brooks and Pethica for C_{18} TAB at the O/W interface (*14*) and the isotherms in Figure 6. Surprisingly, the one point quoted by Phillips and Rideal (*17*) at 200 A^2/molecule falls on the curve at 20°C.

The isotherms for the 1:1 mixture in Figure 7 are unpublished results associated with an earlier publication (*14*). Unfortunately, there are no accurate data available for $A > 700$ A^2/molecule. In view of the discrepancies in the earlier work on charged films (*12, 14*) it is possible that the spreading solvent remains in the mixtures. Therefore, although problems of desorption on 0.001*M* NaCl are eliminated with the mixtures, the isotherms must be treated with some circumspection. The only other published results on this kind of system at O/W are those of Phillips and Rideal (*18*), but unfortunately, the chain lengths do not match, the isotherms are limited to $A < 200$ A^2/molecule, and there is no temperature coefficient for Π. Since equimolar mixtures give solid-condensed monolayers at the A/W interface, no data can be obtained in the A range of interest here.

Monolayers are more expanded at the O/W interface than at the A/W interface; this has been generally confirmed since the early work of Davies (*10*). The condensing effect on fully ionized monolayers from increasing the electrolyte concentration (*19*) is shown in Figures 3–5. The equimolar C_{18} sulfate–C_{18} TAB monolayers are more condensed than the single-component charged monolayers shown by Phillips and Rideal (*17, 18*) and Brooks and Pethica (*14*) although in the present

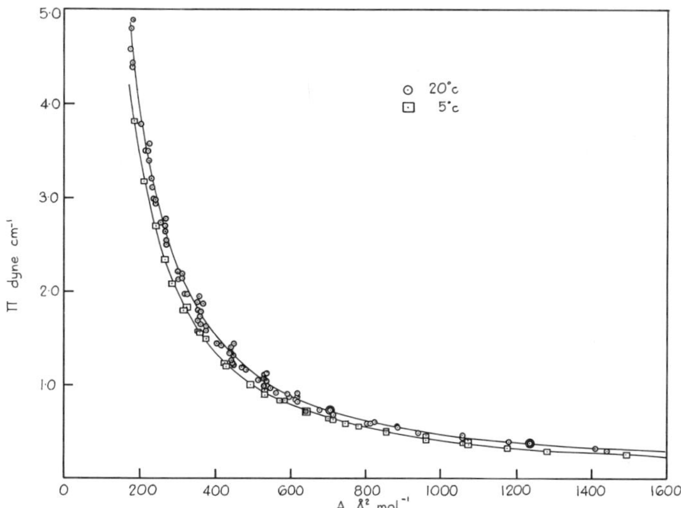

Figure 6. Surface pressure–area isotherms for octadecyl trimethylammonium bromide at the n-*heptane–0.1*M *sodium chloride solution interface.* □ *5°C,* ○ *20°C.*

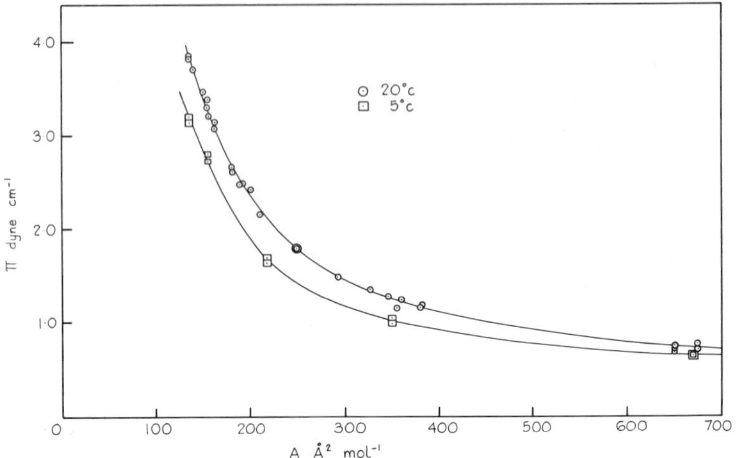

Figure 7. Surface pressure–area isotherms for an equimolar mixture of sodium octadecyl sulfate and octadecyltrimethylammonium bromide at the n*-heptane–0.001*M *sodium chloride interface.* □ *5°C,* ○ *20°C.*

work, Π is only marginally higher for the C_{18} TAB on 0.1*M* NaCl than for the mixture at $A > 600$ A^2/molecule. This shows the equivalent screening power of this electrolyte concentration at high separations of the monolayer ions. A comparison of Figures 3 and 6 shows that over most of the range of *A* depicted, the C_{18} sulfate monolayers are slightly more expanded than the C_{18} TAB monolayers for both temperatures. A common curve at each temperature would fit the data to ±0.03 dyne/cm over a large portion of the Π–*A* curves with a small region around 500–650 A^2/molecule on the curves at 20°C where ± 0.05 dyne/cm should be quoted.

Discussion

If the monolayer and its associated double layer occupy a separate phase which lies between two homogeneous bulk phases, then for constant temperature, volume, and total number densities of species in this phase, the total differential Helmholtz free energy for a planar interfacial region is given by (*20*):

$$dF^{\sigma} = \gamma dA, \tag{1}$$

where γ is the interfacial tension. The compression of a monolayer results in the creation of a clean interface behind the barrier with interfacial tension γ_0. The total Helmholtz free energy change (ΔF) on going from area A_1 to area A_2 can therefore be written

$$\Delta F = \int_{A_1}^{A_2} dF^{\sigma} = \int_{A_1}^{A_2} \gamma dA + \int_{0}^{A_1 - A_2} \gamma_0 dA =$$

$$\int_{A_1}^{A_2} \gamma dA - \int_{A_1}^{A_2} \gamma_0 dA = \int_{A_1}^{A_2} (\gamma - \gamma_0) dA \qquad (2)$$

The surface pressure of a monolayer is defined by:

$$\Pi = \gamma_0 - \gamma \qquad (3)$$

which, on substitution in Equation 2 gives

$$\Delta F = - \int_{A_1}^{A_2} \Pi dA \qquad (4)$$

Entropies of compression (ΔS_c) can now be calculated from the usual thermodynamic relation

$$\Delta S_c = \frac{-d(\Delta F)}{dT} \qquad (5)$$

By integrating the isotherms in Figures 1–7 from various chosen values of A down to a common reference area of 200 A^2/molecule and assuming that the resultant ΔF are linear with temperature, entropies of compression are calculated for the mean temperature 12.5°C. Any model must explain the following main features of the calculations. The entropies are negative, and scatter about two lines when plotted against the logarithm of the selected areas (as in Figure 8); one line is for the O/W interface, and a lower one is for the A/W although there is a slight tendency for the points to follow a curve at low surface densities. There is no distinction between C_{18} TAB and C_{18} sulfate; at the O/W interface, the entropies for the charged films agree with the limited data from the mixture. No ionic strength dependence can be distinguished within the scatter for the ΔS_c of C_{18} sulfate at O/W over the salt range 0.1–0.001M. This is consistent with the agreement at the A/W interface between C_{18} TAB on 0.1M NaCl and C_{18} sulfate on 0.01M NaCl.

If these expanded monolayers are considered as sets of particles that have two-dimensional translations, a compression from A_1 to A_2 gives a change in translational entropy:

$$\Delta S_{\text{trans}} = R \ln \frac{A_2}{A_1} \qquad (6)$$

where R is the gas constant. This entropy, plotted in Figure 9, is too small to account for the experimental entropies. The introduction of a finite area A_0 to allow for the hard sphere forces that limit the area available to the monolayer molecules gives an increased change in translational entropy described by:

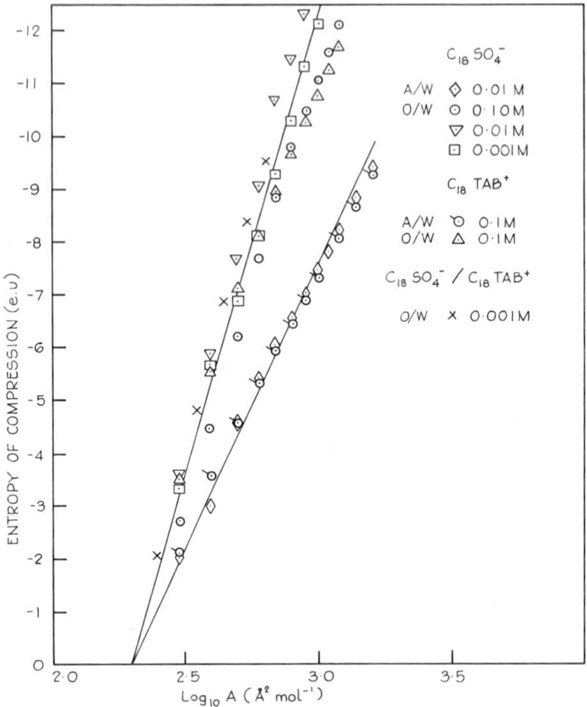

Figure 8. Entropies of compression from experiment.
A/W: C_{18} sulfate ◇ 0.01M NaCl; C_{18} TAB ⍥ 0.1M NaCl. O/W: C_{18} sulfate ○ 0.1M, ▽ 0.01M, □ 0.001M NaCl; C_{18} TAB △ 0.1M NaCl. (C_{18} sulfate + C_{18} TAB) mixture × 0.001M NaCl.

$$\Delta S^{\mathrm{HS}}_{\mathrm{trans}} = R \ln \frac{(A_2 - A_0)}{(A_1 - A_0)} \tag{7}$$

There is some uncertainty in the value of A_0 because we do not know the orientation of the monolayer molecules at different coverages. The usual practice is to take the limiting area from the high surface pressure end of the isotherm: for the TAB or sulfate head groups a reasonable estimate is 30 A^2/molecule. As shown in Figure 9, this A_0 would lessen the divergence between theory and experiment only marginally; it would not explain the difference between O/W and A/W. Other values of A_0 were chosen (Figure 9) to determine whether a fit can be obtained. Although higher values of A_0 give agreement over limited area ranges, the prime prerequisite of linearity throughout most of the $\log_{18} A$ plot is lost, and other contributions to the entropy should be sought.

By means of the Gouy-Chapman assumptions (*21, 22*) the entropy of compression of a "smeared-out" head group charge, and its associated diffuse layer counterion charge (ΔS_{DL}), can be calculated from established expressions for the Helmholtz free energy of the double layer (*19, 23, 24*). This entropy, which is plotted (G and H in Figure 9) for two salt concentrations, is comparable with the translational terms. It is nonlinear with log *A* and levels off at high *A*. In contrast to the experimental data, a distinct salt effect is evident, particularly at high *A*. Related work on the surface potentials of C_{18} sulfate monolayers (*8, 25*) suggests that the primary charge is effectively reduced by counterion adsorption over a large range of *A;* such localization may decrease the divergence between the curves for ΔS_{DL} on the two salt concentrations at the same time as it decreases the entropy contribution from this source. Adding the diffuse layer contributions for the two salts to the translational

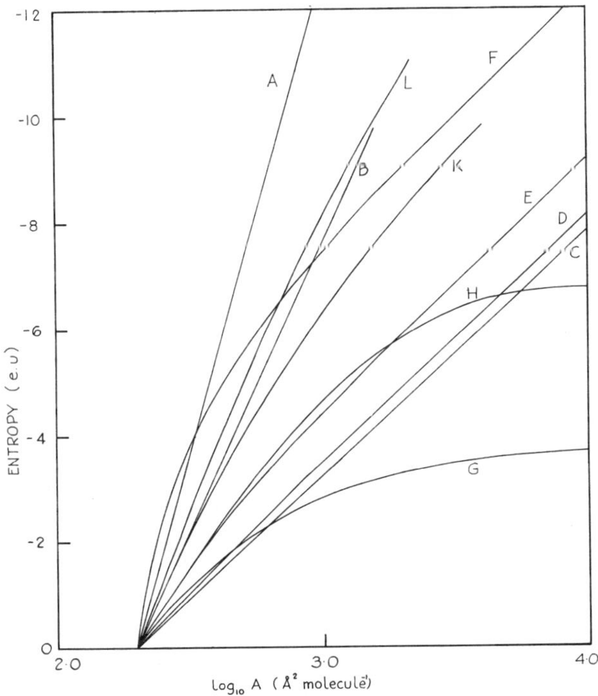

Figure 9. Contributions to the entropy of compression

(*A*) *Best line through the O/W data*
(*B*) *Best line through the A/W data*
(*C*) ΔS_{trans}
(*D*) $\Delta S_{trans}{}^{HS}$, $A_o = 30$
(*E*) $\Delta S_{trans}{}^{HS}$, $A_o = 100$
(*F*) $\Delta S_{trans}{}^{HS}$, $A_o = 180$
(*G*) *Double layer terms* ΔS_{DL} *for 0.1M NaCl*
(*H*) ΔS_{DL} *for 0.01M NaCl*
(*K*) $\Delta S_{trans}{}^{HS}$, $A_o = 30$ *plus* ΔS_{DL} *for 0.1M NaCl*
(*L*) $\Delta S_{trans}{}^{HS}$, $A_o = 30$ *plus* ΔS_{DL} *for 0.01M NaCl*

entropy, with A_0 equal to 30 A^2/molecule, gives two curves (K and L in Figure 9) which enclose the experimental entropies for the A/W interface. This suggests that the A/W data could, in principle, be explained by considerations of translation and charge alone. The slight difference between the dielectric constants of air and *n*-heptane means that little difference in the ion distributions at the two interfaces would be expected; hence, the divergence of ΔS_C for A/W and O/W is still unaccountable.

Because ΔS_c is larger at the O/W than at the A/W interface and considerably larger for O/W than for the combined simple translational and ionic double layer contributions, chain conformations could be changed as monolayers are compressed at the O/W interface. Over the area range studied, changes in chain conformation could be great enough to give the larger entropies required. The experimental data require merely that the chains at the A/W interface be more restricted than at the O/W interface so that the configurational entropy change can be potentially larger at the O/W interface, but the reasonable agreement between ΔS_c at the A/W and the sum of ΔS_{DL} and $\Delta_{trans}{}^{HS}$ requires that the chains are almost completely restricted at the A/W interface for all the *A* studied. This is feasible if the chains lie flat on the A/W interface. Calculations by Richmond (*26*) show that if only dispersion energies are considered, an isolated rigid molecule, lying flat on the surface, is about 50 orders of magnitude more probable than a perpendicular configuration. Evidence for the horizontal model was presented by Langmuir (*27, 28*) and Adam (*1*). The model of rolled-up chains, proposed by Ward (*29*) in his discussion of the Traube rule, would also explain the entropies calculated here, provided that the chains are almost completely restricted in this configuration; however, a Monte Carlo approach (*30*) has recently shown that a rolled-up 18-carbon chain is improbable. At O/W there is little distinction between the methylene groups of the *n*-heptane and those of the monolayer chains in their interaction with the water surface. Therefore, part of the chains should be free to kink in the bulk oil phase and suffer restrictions as the monolayer surface density is increased.

The consistency provided by the above models depends on a substantial contribution from the double layer; it is, therefore, difficult to see why the ΔS_c for the charged films coincides with the value for the equimolar mixture with its zero net charge. The onset of *n*-mer formation found at higher surface densities (*14, 18*) suggests that the attractive coulombic forces, aided by chain interactions, could engender some ion pairing at lower densities with attendant increased entropy losses on compression. However, further calculations are needed to confirm these speculations. In view of the importance of these findings the old, less reliable data on the equimolar mixture (Figure 7) must be repeated and extended to higher *A* values.

At very low monolayer densities, the plot of ΔS_c *vs.* log A must have the same slope as the line from the simple kinetic hard sphere contribution shown in Figure 9, but it will be extremely difficult to explore this ultradilute region experimentally.

Acknowledgments

The authors thank B. E. Houghton for constructing the surface balances and J. R. Brown for the O/W frames. We are also grateful to P. Richmond for showing us his calculations on the chain–surface interactions.

Literature Cited

1. Adam, N. K., "The Physics and Chemistry of Surfaces," 3rd ed., Oxford University Press, London, 1941.
2. Guastalla, J., *C.R. Acad. Sci., Paris* (1929) **189,** 241.
3. Allan, A. J. G., Alexander, A. E., *Trans. Faraday Soc.* (1954) **50,** 863.
4. Ter-Minassian-Saraga, L., *J. Colloid Sci.* (1956) **11,** 398.
5. Adam, N. K., Jessop, G., *Proc. Roy. Soc. Ser. A* (1926) **110,** 423.
6. Robb, I. D., Alexander, A. E., *J. Colloid and Interface Sci.* (1968) **28,** 1.
7. Ter-Minassian-Saraga, L., *Proc. Intern. Congr. Surface Activity, 2nd, London, 1957,* **1,** 36.
8. Mingins, J., Pethica, B. A., *Trans. Faraday Soc.* (1963) **59,** 1892.
9. Mingins, J., Owens, N. F., Iles, D. H., *J. Phys. Chem.* (1969) **73,** 2118.
10. Goddard, E. D., Kung, H. C., *J. Colloid and Interface Sci.* (1969) **30,** 145.
11. Mingins, J., Taylor, J. A. G., Owens, N. F., Brooks, J. H., ADVAN. CHEM. SER. (1975) **144,** 28.
12. Taylor, J. A. G., Mingins, J., unpublished data.
13. Yue, B. Y., Jackson, C. M., Taylor, J. A. G., Mingins, J., Pethica, B. A., unpublished data.
14. Brooks, J. H., Pethica, B. A., *Trans. Faraday Soc.* (1964) **60,** 208.
15. Mingins, J., Ph.D. thesis, University of Manchester (1962).
16. Owens, N. F., M.Sc. thesis, University of Salford (1971).
17. Phillips, J. N., Rideal, E. K., *Proc. Roy. Soc. Ser. A London* (1955) **232,** 159.
18. Phillips, J. N., Rideal, E. K., *Proc. Roy. Soc. Ser. A London* (1955) **232,** 149.
19. Davies, J. T., *Proc. Roy. Soc. Ser. A London* (1951) **208,** 224.
20. Guggenheim, E. A., "Thermodynamics," North-Holland, Amsterdam, 1949.
21. Gouy, G., *J. Phys.* (1910) **9,** 457.
22. Chapman, D. L., *Phil. Mag.* (1913) **25,** 475.
23. Derjaguin, B. V., Landau, L., *Acta Physicochim.* (1941) **14,** 633.
24. Verwey, E. J. W., Overbeek, J. Th. G., "Theory of the Stability of Lyophobic Colloids," Elsevier, Amsterdam, 1948.
25. Mingins, J., Pethica, B. A., *J. Chem. Soc., Faraday Trans., I* (1973) **69,** 500.
26. Richmond, P., unpublished data.
27. Langmuir, I., *J. Amer. Chem. Soc.* (1917) **39,** 1883.
28. Langmuir, I., "Colloid Symposium," Monograph 3, p. 75, 1925.
29. Ward, A. F. H., *Trans. Faraday Soc.* (1946) **42,** 399.
30. Lal, M., Spencer, D., *Mol. Phys.* (1971) **22,** 649.

RECEIVED October 3, 1974.

4

Surface Equation of State for Very Dilute Charged Monolayers at Aqueous Interfaces

J. MINGINS, J. A. G. TAYLOR, N. F. OWENS, and J. H. BROOKS[1]

Unilever Research Port Sunlight Laboratory, Unilever Ltd., Port Sunlight, Wirral Merseyside, England

Using surface pressure (Π)–area (A) isotherms for sodium octadecyl sulfate (C_{18} sulfate) and octadecyl trimethylammonium bromide (C_{18} TAB) monolayers spread at the air/water (A/W) and n*-heptane/water (O/W) interfaces, we show that the Davies–Guastalla and Davies surface equations of state are not followed over a wide range of temperature and salt concentration. We attribute the discrepancies to Davies' incorrect estimate of the chain-cohesion term at the A/W interface, the inadequacy of the electrical contribution to* Π *at all but the highest* A *studied, and a possible chain-cohesion term at the O/W interface.*

In his classic book (*1*) N. K. Adam discussed the behavior of very dilute monolayers at the air/water (A/W) interface; and using measurements published earlier by Jessop and himself (*2, 3, 4*), he showed that surface pressure (Π)–area (*A*) isotherms for insoluble uncharged species, when plotted on a Π*A vs.* Π basis, suggested a limit of $1kT$ at zero Π. The same limit was also suggested by Schofield and Rideal's plot (*5*) of Frumkin's surface tension data (*6*) using the Gibbs adsorption isotherm to calculate *A*. Adam (*1*) stressed that Π should be measured to the second decimal place to establish this limit unequivocally; Adam and Jessop (*4*) provide one of the few sound extrapolations to this limit with their data on the esters of some dicarboxylic acids.

The close resemblance between the Π*A*/Π plots at very low surface densities and the classical *PV*/*P* plots for gases led to the postulate of gaseous- and vapor-expanded states in monolayers. The kinetic theory

[1] Present address: Division of Textile Physics, C.S.I.R.O., Ryde, Sydney, N.S.W., Australia.

of gases, when applied in two dimensions, gave the ideal surface equation of state

$$\Pi = kT/A \tag{1}$$

which provided the limit of $1kT$ for the $\Pi A/\Pi$ plot (*1, 2, 7*). Implicit in this equation is the assumption that there are no interactions among the monolayer molecules; the $\Pi A/\Pi$ plots for uncharged monolayers at the A/W interface with their positive and negative deviations from Equation 1 show the untenability of this assumption over most of the range of A accessible to experiment.

The contribution to Π from the short range repulsive forces which dictate the distance of closest approach of the monolayer molecules can be allowed for by an excluded area term, A_o, as in the empirical equation introduced by Schofield and Rideal (*5*), following a suggestion by Volmer (*8*):

$$\Pi - kT/(A - A_o) \tag{2}$$

The $\Pi A/\Pi$ plot of Equation 2 gives a line of slope A_o with a limit of $1kT$ again at zero Π. For A/W data the range of A where the $\Pi A/\Pi$ plot is reasonably linear, commensurate with an A_o term, the intercept at zero Π is not kT; in the range of A where an intercept of kT is feasible, the slope is not A_o. These features allied with the increasing depth of the minimum in the $\Pi A/\Pi$ plot as the chain length increases suggested that a substantial attractive force operates between the monolayer molecules at the A/W interface; a cohesive term Π_s proportional to $1/A^2$ was introduced by analogy with the imperfect gases (*9, 10 11, 12, 13*). A statistical mechanical approach by Hill (*14*) also gave the $1/A^2$ term. However, using very accurate high A data on myristic acid at the A/W interface and subtracting the value of Π predicted from Equation 2 from the experimental Π, Guastalla (*15*) derived an empirical relation for this cohesive contribution with a different area dependence,

$$\Pi_s = -K/A^{3/2} \tag{3}$$

Guastalla attributed Π_s to dipole interactions, but Davies (*16*) showed that the data of Saraga (*17*), Phillips (*18*), and Guastalla (*15*) supported a chain length dependence of the Guastalla coefficient and more reasonably explained the cohesion in terms of hydrocarbon chain interaction. Davies claimed a linear dependence of Π_s on n, the number of methylene groups in the monolayer chain, and proposed the equation of state

$$\Pi = kT/(A - A_o) - 400\, n/A^{3/2} \tag{4}$$

for uncharged monolayers at the A/W interface. Hill's equation (*14*) also gave a linear term in n. By comparing Π–A isotherms for the same monolayer spread at A/W and hydrocarbon oil/water (O/W) interfaces under the same conditions of temperature and ionic strength, Davies (*19*) provided another approach for obtaining Π_s. Here, the kinetic and charge contributions to Π were assumed invariant with the type of interface, and Π_s was taken as zero at O/W, leaving Π_s at A/W as the difference in the two isotherms. Calculations on data for a single chain length cationic monolayer seemed to confirm the cohesive term in Equation 4 over the narrow area range 100–225 A^2/molecule (*16*). The $\Pi A/\Pi$ plots of Equation 4 again extrapolate to $1kT$ at zero Π; the same limit is obtained if the Guastalla–Davies cohesive term is replaced by Hill's term.

The introduction of a charge (or charges) into a monolayer head group expands the Π–A isotherm; consequently, many solid- or liquid-condensed monolayers become gaseous or vapor expanded when ionized. Although the charge confers increased solubility in the aqueous phase, it does preclude desorption into nonpolar oils so that monolayers can be safely spread at the O/W interface. The increase in Π caused by ionization (Π_{el}) depends on the primary surface charge density (σ) and the nature and amount of screening electrolyte in the aqueous phase. Early investigations were mainly empirical (*20, 21*); later, Davies, in an important paper (*22*), derived from the established theory for the free energy of the double layer (*23, 24*) the relation

$$\Pi_{el} = \int_0^{\psi_0} \sigma d\psi \tag{5}$$

where ψ_0 is the average potential in the plane of the primary charges. The Gouy–Chapman equation relating ψ and σ in terms of A, the area per primary ion, is for a uni-univalent electrolyte

$$\psi_0 = (2kT/e_0) \sinh^{-1} M/A\sqrt{C} \tag{6}$$

where $M = (10^{35}\pi e_0^2/2\epsilon kTN)^{1/2}$, C is the salt concentration, e_0–the proton charge, ϵ–the dielectric constant, and N–Avogadro's number. Substituting this equation into Equation 5, assuming that Equation 4 holds for uncharged monolayers at the A/W interface and Π_s is zero at O/W, gives the Davies equation of state for a fully ionized monolayer at the O/W interface:

$$\Pi = kT/(A-A_0) + (2kT/M)\sqrt{C}\ \{\cosh \sinh^{-1} (M/A\sqrt{C}) - 1\} \tag{7}$$

and the Davies–Guastalla equation for the A/W interface:

$$\Pi = kT/(A - A_o) - 400\, n/A^{3/2} + (2kT/M)\sqrt{C}\ \{\cosh \sinh^{-1}(M/A\sqrt{C}) - 1\} \qquad (8)$$

Davies (*16*) applied both equations to a wide collection of data at fairly low *A* for both soluble and insoluble ionized species and achieved only limited agreement between theory and experiment for some monolayers. Nevertheless, the Davies term is the basis of nearly every subsequent discussion on the isotherms of ionized monolayers. We discuss elsewhere (*25, 26*) the validity of Equations 7 and 8 for intermediate and high surface charge densities as well as other proposed equations of state (*14, 27, 28, 29, 30, 31, 32*). In this paper we establish whether these two equations are suitable limiting forms at high *A* where many of the assumptions used in their derivation should be more valid. In particular we are interested in the limit of ΠA at zero Π for both interfaces.

Experiments and Results

We describe in a companion paper (*33*) the salient features of our procedures for measuring low surface pressures, and we provide Π–A isotherms for octadecyl trimethylammonium bromide (C_{18} TAB) and sodium octadecyl sulfate (C_{18} sulfate) monolayers spread at A/W and O/W interfaces. Here we extend some of these data to higher *A* using, in some instances, some of our earlier reported measurements. We have also taken from literature the few reported isotherms for dilute charged monolayers and discuss them in the light of our results and the predictions of Equations 7 and 8.

Cationics at the Air/Water Interface. In Figure 1 we show on a Π–$1/A$ basis the various results obtained at the air/0.1*M* electrolyte solution interface. The results of Saraga (*34*) and Mingins, Owens, and Iles (M-O-I) (*35*) for C_{18} TAB on potassium bromide and sodium chloride are in good agreement; hence no specific counterion effects are evident. The results of M-O-I were obtained in the absence of organic spreading solvent whereas Saraga used a mixture of ethanol or methanol with chloroform. In the range of *A* up to 1200 A^2/molecule, alcohol takes up to half an hour to leave an A/W surface supporting a C_{18} sulfate monolayer, and considerably longer times would be needed at $A > 4000$ A^2/molecule. Saraga noted a time dependence of Π for salt concentrations less than $2 \times 10^{-1}M$ which she attributed to desorption, but this may only signify the slow departure of solvent on the higher salt concentrations. The results of Owens (*43*) at 5°C in Figure 1 have a larger scatter, and the effect of temperature cannot be discerned at $A > 2500$ A^2/molecule.

In Figure 1 we also give the results of Robb and Alexander (R-A) (*36*) for docosyl triethylammonium bromide (C_{22} ETAB) spread on 0.1*M* KBr together with a single result relevant to this range of *A* from Goddard and Kung's (G-K) isotherm (*37*) for C_{22} TAB on 0.1*M* NaCl. The legends on Figures 1 and 2 in R-A's work should be transposed. Again counterion effects would be difficult to distinguish within the scatter of

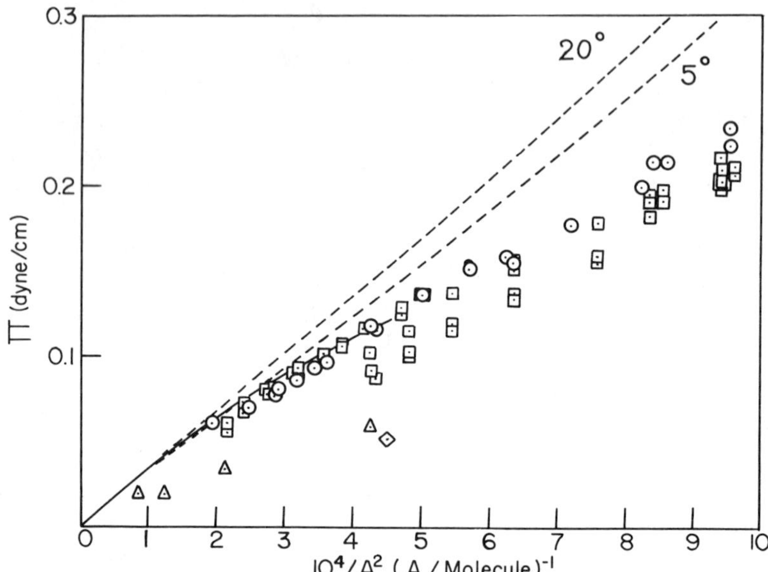

Figure 1. Π–1/A isotherm for cationics at the A/W interface—0.1M electrolyte. Continuous line describes the data of Saraga (34) *for C_{18} TAB on KBr at 20° ± 3°C.*

△ *C_{22} ETAB, KBr, 20°C [Robb and Alexander* (36)]
◇ *C_{22} TAB, NaCl, 25°C [Goddard and Kung* (37)]
○ *C_{18} TAB, NaCl, 20°C ± 0.1°C [Mingins, Owens, and Iles* (35)]
□ *C_{18} TAB, NaCl, 5° ± 0.1°C [Owens* (43)]

Dashed lines are the Davies–Guastalla equation for C_{18} TAB with A_o = 20 A^2/molecule.

these data; the same holds true for effects from differences in the size of head group. However, the condensing effect of increasing the chain length can clearly be seen with all four sets of data. We give a more extensive range of data on C_{18} TAB for the two temperatures 5° and 20°C with the ΠA/Π plots shown in Figure 7.

Results for cationic monolayers spread on aqueous substrates that contain no electrolyte are limited to those of R-A (*36*) and Saraga (*34*). Quaternary ammonium salts are notoriously soluble in water (much more so than sulfate films of the same chain length). M-O-I used pure water as a spreading solvent for C_{18} TAB, and although electrolyte in the substrate diminishes the solution of a charged monolayer, M-O-I found a significant rate of desorption of C_{18} TAB at 80 A^2/molecule on 0.1*M* NaCl—0.02 dyne/cm/min. Higher rates were found for $C_{18}SO_4$ (*38*) on 0.0001*M* NaCl—0.08 dyne/cm/min at 300 $A/^2$molecule. C_{22} ETAB was chosen by R-A to avoid this problem, but the results of G-K for C_{22} TAB showing crossovers on dilute salts solutions for several of the ΠA/Π plots as well as the surface potential curves almost certainly arise from monolayer desorption (as indicated by G-K). Saraga's isotherm on water crosses over those she obtained on 10^{-5} and $10^{-3}M$ KBr at ~ 12,000 A^2/

molecule (*34*). Alcoholic spreading solvents were used by R-A and G-K; the dilemma facing investigators of gaseous, ionic monolayers on pure water or very dilute electrolyte solutions is how to correct for monolayer desorption over the long periods which are often needed for the alcohol to disappear from the surface.

Anionic at the Air/Water Interface. In Figure 2 we show Π–$1/A$ isotherms obtained by one of us (*38*) for C_{18} sulfate on several salt solutions at 20°C. We also show a few points where the *A* values overlap with the results on 0.01*M* NaCl reported in our paper on entropies of compression (*33*). The favorable agreement between both sets of data at lower *A* is discussed in Ref. *33*. The most likely explanation of the slight discrepancies is trace residual alcohol in the older experiments. No other data are available for comparison at these large *A* values.

Cationics at the Oil/Water Interface. The results reported in Ref. *33* for C_{18} TAB on 0.1*M* NaCl at 20°C are extended to 5000 A^2/molecule; we compare them in Figure 3 with those of R-A on C_{22} ETAB and the one

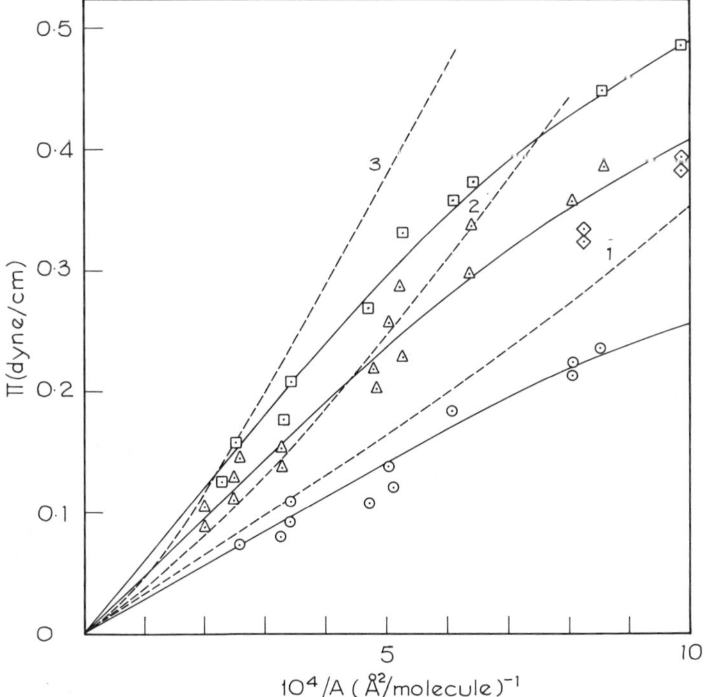

Figure 2. Π–1/A isotherms for C_{18} sulfate at the A/W interface at 20°C

○ *0.1*M *NaCl;* △ *0.01*M *NaCl;* □ *0.001*M *NaCl;* ◇ *results from Ref.* 33 *on 0.01*M *NaCl.*

*Dashed lines 1, 2, and 3 represent the Davies–Guastalla equation for 0.1, 0.01, and 0.001*M *NaCl.* $A_0 = 20\ A^2$*/molecule.*

point from Brooks' and Pethica's work (*39*) that lies in this range of high *A* for C_{18} TAB. The result of Brooks and Pethica is considerably higher than ours; we show elsewhere (*40*) that this arises from the reluctance of the alcohol to leave the interface in the time allowed by these workers. At high *A* our results almost merge with those of R-A, but at $A < 5000$ A^2/molecule the two sets of data diverge, and the results of R-A approach those of Brooks and Pethica. Likely differences in A_0 would only give a slight change in Π at these high *A*. We feel that solvent retention causes these discrepancies despite the checks made by R-A on their spreading solutions. R-A waited up to 20 min after spreading before compressing the monolayer (*41*); in this time a good proportion of the alcohol would diffuse from the surface. However, our results on many monolayer systems show that considerably longer times are needed before stable Π are reached when the monolayers are so dilute (*40*). These findings are supported by measurements of surface potentials (*42*).

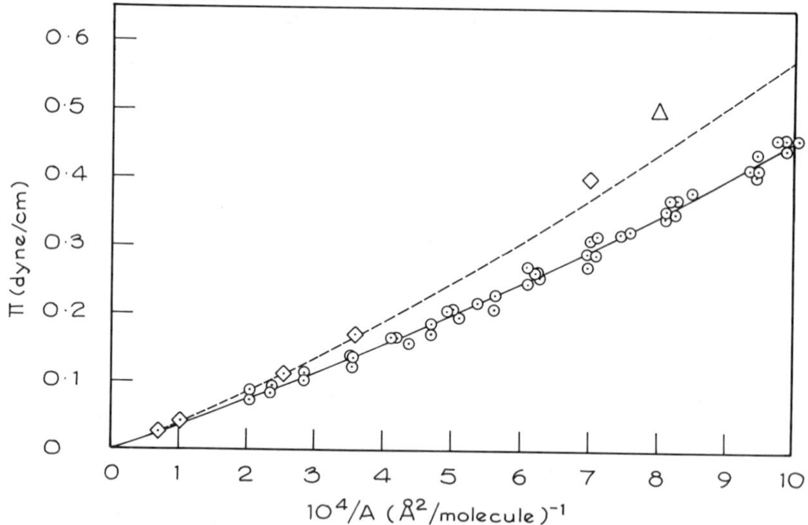

Figure 3. Π–*1/A isotherm for* C_{18} *TAB at the* n-heptane/water *interface at 20°C for 0.1*M *NaCl*

◇ *data of Robb and Alexander* (36) *on* C_{22} *ETAB;* △ *result of Brooks and Pethica* (39) *on* C_{18} *TAB.*

Dashed line is the Davies equation with $A_o = 20$ *A^2/molecule.*

Anionic at the Oil/Water Interface. Figures 4 and 5 show isotherms up to 10,000 A^2/molecule for C_{18} sulfate on 0.1*M* NaCl at 20°C and up to 3,500 A^2/molecule on 0.1, 0.01, and 0.001*M* NaCl at 5°C. These isotherms overlap those in Ref. 33. There are no data on C_{18} sulfate at high *A* at 5°C for comparison with Figure 5, and we discussed (*40*) the divergences from the Brooks and Pethica (*39*) data at 20°C.

Discussion

The usual choice of A_0 for the charged monolayers investigated here lies in the range 20–35 A^2/molecule. These values approximate either the area of the solid close-packed film or the limiting area. In either case the choice assumes that in the gaseous state an equivalent orientation of the monolayer dictates the free area available for its translation. A value of A_0 can only be calculated from the slope of a $\Pi A/\Pi$ plot in the absence of all but hard sphere forces. For highly interacting species such as long chain, charged molecules, it is impossible to assess A_0 by this method. One approach which avoided the long range, repulsive coulombic interactions utilized an equimolar mixture of anionic and cationic monolayers (*19, 28, 39*) where the coulombic forces were quickly screened. At the O/W interface, where the monolayer is expanded because of diminished chain–chain interaction, the $\Pi A/\Pi$ plot had an initial slope which yielded a mean A_0. Such calculated values fall in the range quoted above. It is questionable whether these values can be used in isotherms at the A/W interface where this approach cannot be taken because of the condensed nature of the equimolar mixture. An orientation of A_0 equal to 30 A^2/molecule, for example, which is constant over a large range of A, is perfectly feasible at the O/W interface; it is difficult to see how this value can hold at the A/W interface where the chains, in the absence of mixing with the oil, would either roll up or flop on to the water surface to give higher A_0. To test the limits of Equations 7 and 8 against the collected

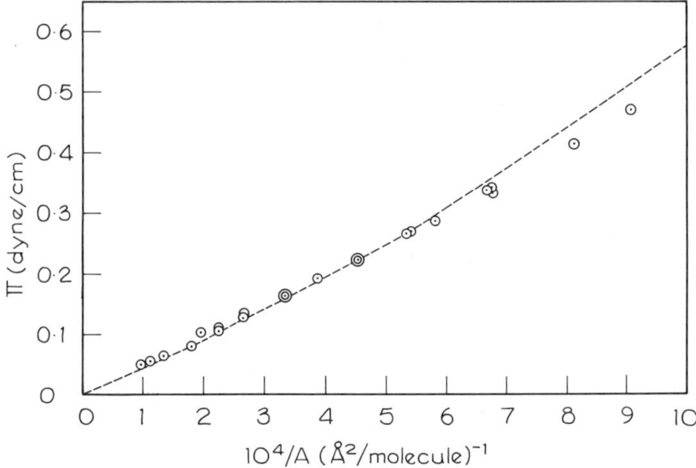

Figure 4. Π–*1/A isotherm for* C_{18} *sulfate at the* n-*heptane/water interface at 20°C for 0.1*M *NaCl. Dashed line is the Davies equation with* $A_0 = 20$ *A^2/molecule.*

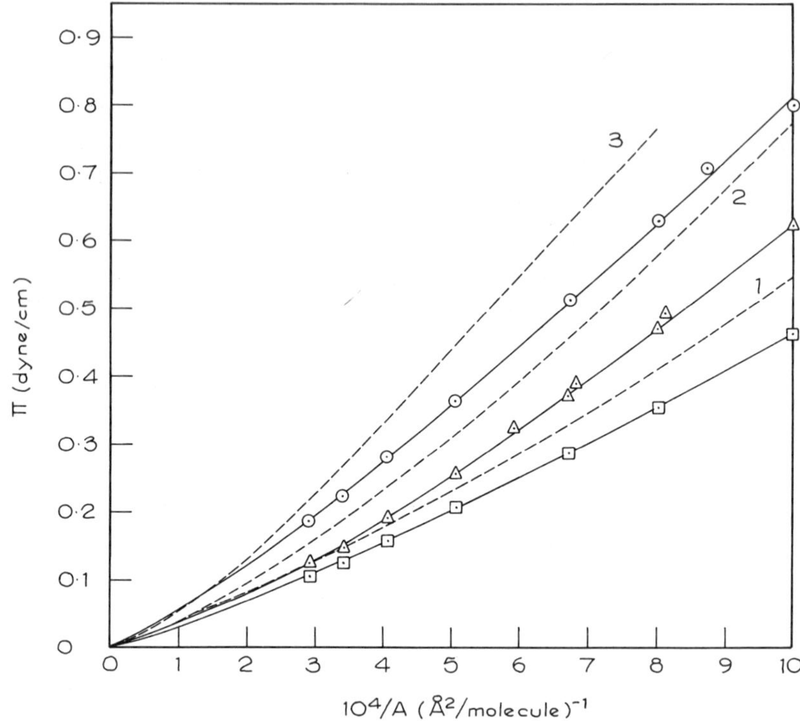

Figure 5. Π–1/A isotherms for C_{18} sulfate at the n-heptane/water interface at 5°C.

□ 0.1M NaCl; △ 0.01M NaCl; ○ 0.001M NaCl. Dashed lines 1, 2, and 3 represent the Davies equation for 0.1, 0.01, and 0.001M NaCl; A_o = 20 A^2/molecule.

data in Figures 1–5, the value of A_0 (except possibly the upper limit required by horizontal chains at the A/W interface) is not important because of the high *A* depicted. We chose an arbitrary value of 20 A^2/molecule for both interfaces.

We plotted the Davies–Guastalla equation (Equation 8) in Figures 1 and 2 for C_{18} TAB and C_{18} sulfate taking *n* as 18. The continuous line depicting Saraga's results was taken from a small scale graph in Ref. *34* which does not show data points. If Saraga's extrapolation to the origin means that there are results for $A > 10{,}000$ A^2/molecule, then such results fit Equation 8. At lower *A* Saraga's results lie below Equation 8 as do our own. Saraga's results on lower ionic strengths do not fit as well at high *A*. Increasing the chain length to 22 makes only a small difference to the position of the "theoretical" line at these high *A*; hence the results of Robb and Alexander (*36*) and Goddard and Kung (*37*) on the C_{22} cationics cannot be fitted by Equation 8. The Guastalla term is clearly in error;

it is possible that the increasing divergence between Equation 8 and both sets of data on C_{18} TAB as A decreases also arises from Davies' incorrect estimate of Π_s. The validity of Π_{el} cannot be assessed with these data. The experimental isotherm for C_{18} TAB on 0.1*M* NaCl at 20°C falls within the scatter of the results on C_{18} sulfate at the same salt concentration for all but the lowest A depicted; consequently, Equation 8 also fails with this monolayer as shown in Figure 2. The results on 0.01 and 0.001*M* NaCl are also lower than "theory" over most of the range shown in Figure 2 although the results on 0.01*M* scatter around the predictions of Equation 8 when $A > 2{,}000$ A²/molecule. Since Π varies with salt concentration, electrical forces are still operating at the highest A studied.

We compared the Davies equation (Equation 7) with the data on both monolayers at the O/W interface in Figures 3–5: the effect of head group is shown in Figures 3 and 4, the effect of temperature in Figures 4 and 5, and the effect of salt concentration in Figure 5. Except for C_{18} sulfate at 20°C on 0.1*M* NaCl at $A > 1{,}500$ A²/molecule, the data on all the monolayers are considerably lower than the Davies equation. The wide divergence of C_{18} TAB on 0.1*M* NaCl contradicts the good agreement shown by Robb and Alexander (*36*) but is supported by other measurements by one of us on C_{18} TAB at 5°C (*44*). The agreement between Davies and the data on C_{18} sulfate in Figure 4 is not seen at any salt concentration at 5°C (Figure 5). Results of similar precision for C_{18} sulfate on 0.01 and 0.001*M* NaCl at 20°C [not shown here (*45*)] confirm the poor agreement in general with the Davies equation, but the divergences are smaller, and the results on 0.01*M* NaCl agree with Davies when $A > 2{,}000$ A²/molecule. The isotherm in Figure 5 for C_{18} sulfate on 0.1*M* NaCl at 5°C agrees with that found for C_{18} TAB at the same temperature over nearly all the range of A investigated; this mirrors the agreement found between the two monolayers at the A/W interface at 20°C on this salt concentration. [The results in Figure 3 are from two different spreading solvents (pure water or 97% heptane and 3% ethanol v/v); hence the possibility of non-quantitative spreading from water is not the cause of the divergence.]

The dependence of Π on salt concentration at constant chosen A provides a test of Π_{el} provided that Π_s is either zero or has no salt dependence. A plot of Π *vs.* salt concentration by Saraga (*46*) for C_{18} TAB at 10,000 A²/ molecule approached the predictions of Equation 8 at 0.1*M* NaCl. On the more dilute salts the experimental points were significantly *higher* than calculated from Equation 8, but all these results may be affected by allyl alcohol in the spreading solution. The isotherms in Figure 5 partially agree with Davies' over the salt dependence at the highest A reached in these data; in unpublished results on C_{18} sulfate at

20°C (*45*) taken in conjunction with Figure 4, the Davies term *can* explain the difference in Π on 0.1 and 0.01*M* NaCl for $A > 2{,}000$ A²/molecule but fails on the change from 0.01 to 0.001*M* NaCl until an area of 5000 A²/molecule is reached. A similar picture at 20°C emerges from a comparison of the less precise results in Figure 2 for the A/W interface obtained before our spreading solvent problems were resolved.

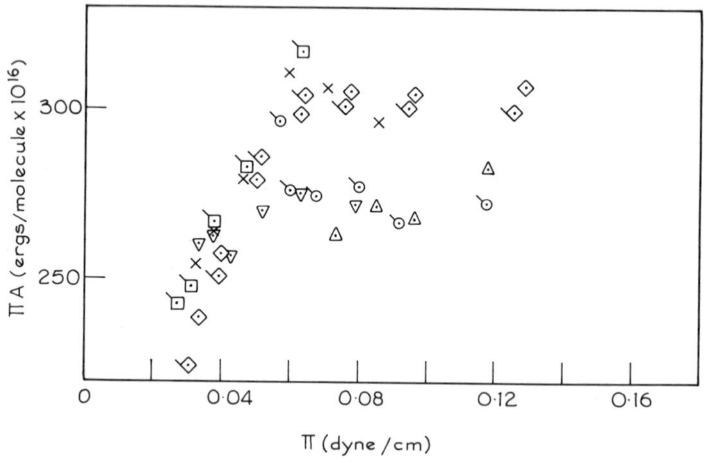

Figure 6. Effect of spreading errors on the ΠA vs. Π *plot for* C_{18} *TAB at the A/W interface at high* A *at 20°C for 0.1*M *NaCl*

We have already mentioned the limit of 1*kT* for Π*A* as Π tends to zero in Equations 1, 2, and 4. The Davies and Davies–Guastalla equations also extrapolate unambiguously to this limit for finite salt concentrations. However, extrapolation of experimental results is not easy; at low Π we show in Figure 6 some dubious results for C_{18} TAB spread to 9000 A²/molecule. Because of some undetermined error, the Π–1/*A* plot does not extrapolate to the origin; the deviation from the best line through the origin and the points in the range 3000–5000 A²/molecule, where the data are more reliable, amount to about 0.007 dynes/cm whereas the scatter around the curve through all the points in the range 3000–9000 A²/molecule is less than 0.005 dynes/cm. Both figures are less than the specification suggested by Adam (*1*), and yet the Π*A*/Π plot in Figure 6 is unintelligible. If an extrapolation to zero Π is attempted, the deviation at high *A* forces a limit of 0.5 *kT* whereas a Π*A*/Π plot of the more established data from Figures 1 and 2 of Ref. *33* as shown in Figure 7 suggests the normal limit of 1 *kT*. A precision of 0.001 dyne/cm with an accuracy to match is needed for measurements out to 10,000 A²/molecule for sensible $\Pi A/\Pi$ plots with which to establish limits for the charged systems.

The plots in Figure 7 for C_{18} TAB dip below the kT line; hence the cohesive forces here overwhelm the double-layer ones on $0.1M$ NaCl. The results for C_{18} sulfate on the more dilute salt with their larger repulsive term are nearer to the kT line. The C_{18} sulfate plots at both temperatures tend to the limit of 1 kT, and in fact the monolayer at 20°C would appear to behave like an ideal gas when $A > 1{,}000$ A^2/molecule. The results at high A in Figure 2 scatter too much to plot in Figure 7.

The $\Pi A/\Pi$ plots in Figures 8 and 9 for the O/W interface are all on the repulsive side of the 1 kT line when $\Pi > 0.3$ dyne/cm. For $\Pi < 0.3$ dyne/cm, the plots on C_{18} TAB at 20°C and the C_{18} sulfate on the two higher salt concentrations at 5°C extrapolate to below 1 kT; this indicates that some cohesive term is operating at the O/W interface. In view of our earlier comments that slight errors alter trends in $\Pi A/\Pi$ plots, this conclusion must be treated with some caution, particularly since the $\Pi A/\Pi$ plot for C_{18} sulfate in Figure 8 fails to swing down towards 400

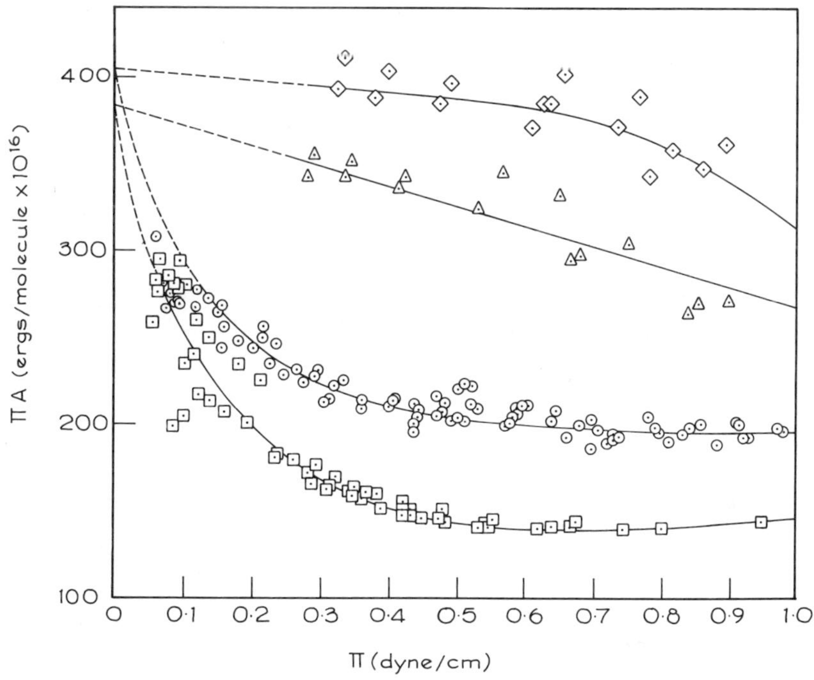

Figure 7. ΠA vs. Π *plot for* C_{18} *TAB and* C_{18} *sulfate at the* A/W *interface*

◇ C_{18} *sulfate, 0.01*M *NaCl at 20°C;* △ C_{18} *sulfate, 0.01*M *NaCl at 5°C;* ○ C_{18} *TAB, 0.1*M *NaCl at 20°C;* □ C_{18} *TAB, 0.1*M *NaCl at 5°C.*

at low Π. A likely explanation is that even longer times are needed for ethanol to disappear from the surface at 20,000 A^2/molecule. Because of the importance of such a conclusion, it is essential to investigate this limiting region again more thoroughly, since such a cohesive term could account for the divergence between the Davies equation and nearly all the isotherms we have reported at high *A*. It would also mean that Davies' approach to calculate Π_s at A/W using charged monolayers would

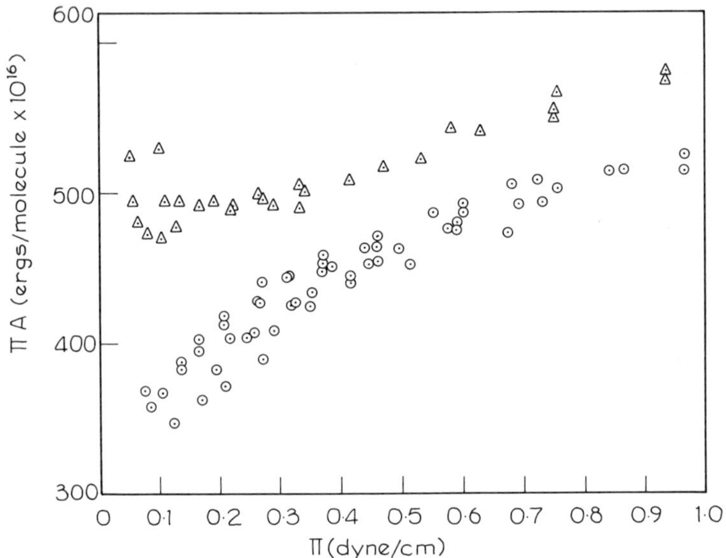

Figure 8. ΠA vs. Π *plot for* C_{18} *sulfate and* C_{18} *TAB at the* n-*heptane/water interface at 20°C for 0.1*M *NaCl*

△ C_{18} *sulfate;* ○ C_{18} *TAB.*

underestimate Π_s; hence the gap between the Davies–Guastalla equation and our A/W data would be narrowed. Interaction between the monolayer chains at O/W at lower *A* is required to explain the entropies of compression calculated in Ref. *33*. The fact that monolayer chains can, in principle, interact in heptane is shown by the chain length-dependent phase transitions in phospholipid monolayers at the *n*-heptane/water interface (*47, 48, 49*). It seems that chain interactions were also involved in the phase transitions found for equimolar cationic/anionic mixtures. We suggest that this is the system to use at high *A* to check whether the known interactions of chains at high surface densities are still operating in very dilute gaseous monolayers at the O/W interface.

The above discussion by no means implies that other contributions to Π_{el}, such as counterion adsorption or discrete ion effects, are not oper-

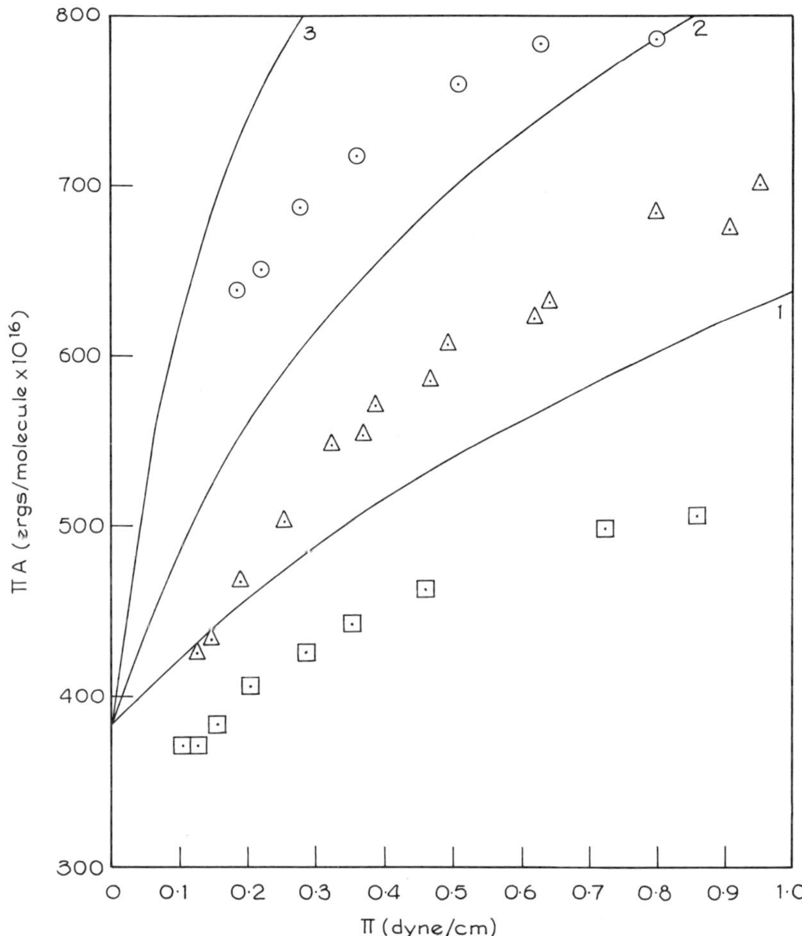

Figure 9. ΠA vs. Π *plot for C_{18} sulfate at the* n-*heptane/water interface at 5°C*

□ *0.1M NaCl,* △ *0.01M NaCl,* ○ *0.001M NaCl*
Continuous lines 1,2,3 represent the Davies equation for 0.1, 0.01, and 0.001M NaCl; A_o = *20 A^2/molecule.*

ating and that these could not in principle explain some of the divergence between Davies–Guastalla and our experiments. These effects will certainly assume more importance at high monolayer densities, but it is difficult to judge at what *A* they become significant. We have already shown that if Π_s is zero or constant with salt concentration at the O/W interface, Π_{el} inadequately explains the salt dependence of Π as *A* is decreased; it may be that only this approach with very accurate data will resolve the electrical terms. If for some reason Π_s varies with salt, then that leaves a nice problem for some student who after reading Adam's book (*1*) is as captivated by monolayers as we were.

Conclusions

Considerable charge interaction of the head groups is shown by Π–$1/A$ isotherms for very dilute charged monolayers spread at both A/W and O/W interfaces; substantial monolayer chain–chain cohesive interactions are also evident at the A/W interface. There is also some indication of a small chain–chain cohesive term at the O/W interface, particularly at low temperatures and high salt concentrations; consequently, none of the isotherms obey the ideal two-dimensional equation of state. The salt dependence of Π at both interfaces is less than that predicted by Davies' electrical term, and except for one isotherm at $A > 2000$ A^2/molecule, the divergence of the isotherms at the O/W interface from those calculated from the Davies equation of state can be attributed to an overestimate of this electrical term and possibly to neglect of the small chain-cohesive term under some experimental conditions. There are large discrepancies between the isotherms at the A/W interface and the Davies–Guastalla equation of state; these arise from Davies' underestimate of the chain–chain cohesive terms at the A/W interface as well as inadequacies in the electrical term.

Acknowledgments

We thank I. D. Robb for useful discussions of his results.

Literature Cited

1. Adam, N. K., "The Physics and Chemistry of Surfaces," 3rd ed., Oxford University Press, London, 1941.
2. Adam, N. K., Jessop, G., *Proc. Roy. Soc. Ser. A* (1926) **110,** 423.
3. Adam, N. K., Jessop, G., *Proc. Roy. Soc. Ser. A* (1926) **112,** 362.
4. Adam, N. K., Jessop, G., *Proc. Roy. Soc. Ser. A* (1926) **112,** 376.
5. Schofield, R. K., Rideal, E. K., *Proc. Roy. Soc. Ser. A* (1925) **109,** 57.
6. Frumkin, A. N., *Z. Phys. Chem.* (1925) **116,** 476, 485.
7. Adam, N. K., *Proc. Roy. Soc. Ser. A* (1928) **119,** 628.
8. Volmer, M., *Z. Phys. Chem.* (1925) **A115,** 253.
9. Volmer, M., Mahnert, P., *Z. Phys. Chem.* (1925) **115,** 239.
10. Magnus, A., *Trans. Faraday Soc.* (1932) **28,** 386.
11. Magnus, A., *Z. Phys. Chem.* (1929) **A142,** 401.
12. De Boer, J. H., "The Dynamical Character of Adsorption," Oxford University Press, London, 1953.
13. Hedge, D. G., *J. Colloid Sci.* (1957) **12,** 417.
14. Hill, T. L., "An Introduction to Statistical Thermodynamics," Addison-Wesley Publishing Co., Reading, Mass., 1960.
15. Guastalla, J., *J. Chim. Phys.* (1946) **43,** 184.
16. Davies, J. T., *J. Colloid Sci.* (1956) **11,** 377.
17. Saraga, L. Ter Minassian, *J. Colloid Sci.* (1956) **11,** 406.

18. Phillips, J. N., Ph.D. thesis, University of London (1954).
19. Davies, J. T., *Trans. Faraday Soc.* (1952) **48,** 1052.
20. Lyons, C. G., Rideal, E. K., *Proc. Roy. Soc. Ser. A* (1929) **124,** 322.
21. Alexander, A. E., Teorell, T., *Trans. Faraday Soc.* (1939) **35,** 727.
22. Davies, J. T., *Proc. Roy. Soc. Ser. A* (1951) **208,** 224.
23. Derjaguin, B. V., Landau, L., *Acta Physicochim.* (1941) **14,** 633.
24. Verwey, E. J. W., Overbeek, J. Th. G., "Theory of the Stability of Lyophobic Colloids," Elsevier, Amsterdam, 1948.
25. Mingins, J., Taylor, J. A. G., Owens, N. F., Pethica, B. A., unpublished data.
26. Mingins, J., Taylor, J. A. G., Brooks, J. H., Owens, N. F., Pethica, B. A., unpublished data.
27. Bell, G. M., Levine, S., Pethica, B. A., *Trans. Faraday Soc.* (1962) **58,** 904.
28. Phillips, J. N., Rideal, E. K., *Proc. Roy. Soc. Ser. A* (1955) **232,** 149.
29. Phillips, J. N., Rideal, E. K., *Proc. Roy. Soc. Ser. A* (1955) **232,** 159.
30. Vrij, A., "Colloquim over Grenslaagverschijnselen. Vloeistoffilmenschuimen-emulsies," p. 13, Koninklijke Vlaamse Academie voor Wetenschappen, Brussels, 1956.
31. Smith, T., *J. Colloid Interface Sci.* (1967) **23,** 27.
32. Bell, G. M., Levine, S., Stephens, D., *J. Colloid Interface Sci.* (1972) **38,** 609.
33. Mingins, J., Owens, N. F., Taylor, J. A. G., Brooks, J. H., Pethica, B. A., ADVAN. CHEM. SER. (1975) **144,** 14.
34. Saraga, L. Ter Minassian, *Proc. Intern. Congr. Surface Activity, 2nd, London,* 1957, p. 36.
35. Mingins, J., Owens, N. F., Iles, D. H., *J. Phys. Chem.* (1969) **73,** 2118.
36. Robb, I. D., Alexander, A. E., *J. Colloid Interface Sci.* (1968) **28,** 1.
37. Goddard, E. D., Kung, H. C., *J. Colloid Interface Sci.* (1969) **30,** 145.
38. Mingins, J., Ph.D. thesis, University of Manchester (1962).
39. Brooks, J. H., Pethica, B. A., *Trans. Faraday Soc.* (1964) **60,** 208.
40. Taylor, J. A. G., Mingins, J., submitted to *J. Chem. Soc. (London) Faraday Trans. I.*
41. Robb, I. D., private communication.
42. Mingins, J., Taylor, J. A. G., unpublished data.
43. Owens, N. F., M.Sc. thesis, University of Salford (1971).
44. Owens, N. F., unpublished data.
45. Taylor, J. A. G., Mingins, J., unpublished data.
46. Saraga, L. Ter Minassian, *Proc. Intern. Congr. Surface Activity, 2nd, London,* 1957, p. 138.
47. Taylor, J. A. G., Mingins, J., Pethica, B. A., Yue, B. Y., Jackson, C. M., *Biochim. Biophys. Acta* (1973) **323,** 157.
48. Yue, B. Y., Jackson, C. M., Taylor, J. A. G., Mingins, B. A., unpublished data.
49. Mingins, J., Taylor, J. A. G., Pethica, B. A., Yue, B. Y., Jackson, C. M., unpublished data.

RECEIVED October 3, 1974.

5

Estimation of Ionization in Unstable Fatty Acid Monolayers from Desorption Kinetics

Relationships between Ionization, Field Strength, and Cation Selectivity

GAJANAN S. PATIL, RICHARD H. MATTHEWS, and DAVID G. CORNWELL

Department of Physiological Chemistry, The Ohio State University, Columbus, Ohio 43210

The surface area, A, *of a fatty acid monolayer that desorbed after ionization was measured as a function of time,* t, *at constant surface pressure,* π. *The initial desorption coefficient,* K_i, *was calculated from these data. The apparent surface* pK_a *of the carboxylic acid was estimated from the reciprocal plot of* K_i *and* $[H^+]$ *data. The apparent* pK_a *was an index of carboxylate field strength. Field strength, a function of charge and surface area, was altered experimentally by varying pH, ionic strength,* π, *and fatty acid structure. Selectivity sequences for binding alkali metal and alkaline earth cations to the ionizing film were explained by the field strength theory. Some exceptions to this theory suggested that steric fit or polarizability could contribute to the selection process.*

In 1922 Adam (*1*) published the third paper in his extraordinary series on surface film structure. He observed that fatty acid monolayers greatly expanded on alkaline subphases. He also suggested that fatty acid anions desorbed or dissolved from the monolayer into the alkaline subphase. In 1933 he and Miller (*2*) showed that the composition of the subphase buffer significantly affected the monolayer; thus palmitic and stearic acid monolayers were more condensed on 2*N* sodium hydroxide than on 2*N* potassium hydroxide. The expansion, desorption, and cation selectivity of ionizing monolayers are the subjects of this investigation.

Adam's and Miller's observation (*2*) that monolayers were more condensed on Na^+ than on K^+ subphases is now explained by the concept that the counterion which binds more strongly to the monolayer than other ions reduces the surface area, *A*, of the monolayer at any given surface pressure (π), pH, ionic strength, and temperature, *i.e.*, Na^+ was more strongly bound than K^+ with completely ionized monolayers. Sears and Schulman (*3*) and Goddard, Kao, and Kung (*4*) expanded the selectivity sequence to $Li^+ > Na^+ > K^+$. Christodoulou and Rosano (*5*) completed the sequence—$Li^+ > Na^+ > K^+ > Rb^+ > Cs^+$. Alkaline earth cations received fewer investigations, but Goddard and Ackilli (*6*) showed that Ca^{2+} was more strongly bound to a stearic acid monolayer at pH 10 than was Mg^{2+}.

Selectivity was affected by several variables. The $Na^+ > K^+$ selectivity (*4*), evident at higher values of π (smaller *A*), lessened at lower values of π (larger *A*). The $Ca^{2+} > Mg^{2+}$ selectivity (*6*), evident at lower values of π (larger *A*), lessened at higher values of π (smaller *A*). The $Na^+ > K^+$ selectivity on a strong alkaline subphase reversed on a weakly alkaline subphase (*3*). The selectivity for a completely ionized weak acid was $Li^+ > Na^+ > K^+ > Rb^+ > Cs^+$ (*5*); it reversed for a completely ionized strong acid—$K^+ > Na^+ > Li^+$ (*7*). These data and our laboratory studies (*8*) suggested that ionizing monolayers behaved like strong-field and weak-field surfaces and were influenced by changing surface area and subphase pH.

Surface area depends on π, pH, ionic strength, and temperature, but wide variations in surface area are most readily attained with shorter-chain and unsaturated fatty acids (*9*). Since these fatty acids do not form stable monolayers, they are difficult to study using conventional procedures. So we took advantage of their instability and used desorption kinetics to estimate the apparent surface pK_a, an indicator of charge density and field strength (*10*).

The apparent pK_a for ionizing groups on stable surfaces is estimated by many techniques. In early studies, Peters (*11*) and Danielli (*12*) measured the changes in interfacial tension at the oil–water interface as a function of pH. They found that the apparent pK_a for carboxylic acids adsorbed to a surface was higher than that for carboxlic acids dissolved in solution. Later Schmidt-Nielson (*13*) and Mattson and Volpenhein (*14*) titrated oleate soaps and found apparent pK_a values of 7.8 and 8.0.

Parameters such as surface area are easier to control at the air–water interface than at the oil–water interface. Ionization at the air–water interface was first investigated by Schulman and Hughes (*15*), who introduced a surface-potential measurement. Recent investigations include the infrared spectra of skimmed films (*16, 17*), the collapse pressure of films (*18*), and the expansion of films (*8*). Hartley (*19*) used indicator

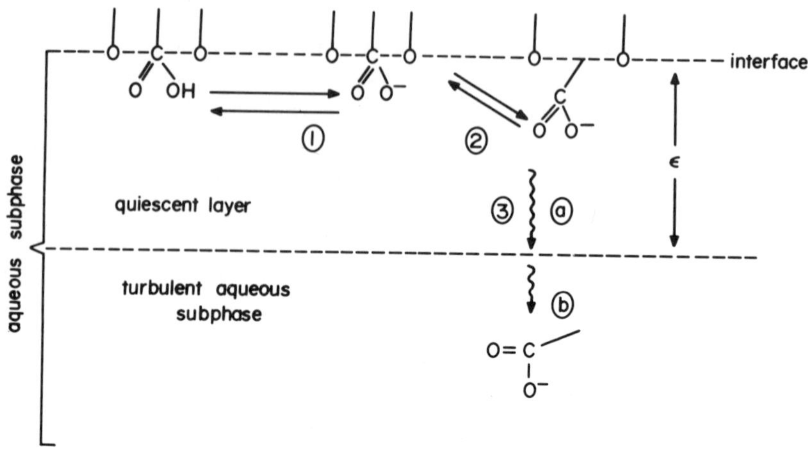

Figure 1. Schematic (26) of the processes which result in monolayer contraction. 1, ionization; 2, desorption from the surface into an unstirred subphase layer of thickness ε; 3a, diffusion through the unstirred layer; 3b, escape from the unstirred layer.

dyes 40 years ago to study ionization at the oil–water interface. In 1973 Montal and Gitler (*20*) measured ionization at the water–liposome interface with the fluorescent dye umbelliferone, and Fromherz (*21*) introduced an insoluble derivative, 4-heptadecylumbelliferone, for ionization studies at the air–water interface. In 1974 Goddard (*22*) presented a detailed analysis of ionizing monolayers. The newer techniques confirmed the early ionization studies at the oil–water interface.

Unstable contracting monolayers form when either the fatty acid or its anion desorbs from the surface into the subphase. The desorption rate is controlled by π, ionic strength, and temperature. These parameters can be varied so that the unionized fatty acid desorbs slowly, if at all, while the fatty acid anion desorbs rapidly. In a previous study, we (*23*) suggested that the rate of contraction, R, of an unstable fatty acid monolayer maintained at constant π provided a means to estimate ionization in the monolayer. R was defined by Equation 1:

$$R = -\frac{1}{A}\frac{dA}{dt} \tag{1}$$

A is the surface area, and t is time. A sigmoidal curve, reminiscent of a titration curve, was obtained when the initial rate of monolayer contraction, R_i, was plotted as a function of pH (*23*). The apparent pK_a, estimated from the midpoint of the sigmoidal curve, depended on the chain length and the degree of unsaturation of the fatty acid. The apparent pK_a always exceeded the pK_a for soluble carboxylic acids.

R varies with time, and there are experimental difficulties in defining $t = 0$ and $A = 0$ and in measuring R_i when the rate of monolayer contraction is high (*23*). Ter Minassion-Saraga (*24*) and later Gershfeld and Patlak (*25*) found that log A of a contracting monolayers was linear with $\sqrt{t}$ in the initial temporal phase of contraction. The initial desorption coefficient, K_i, was expressed by Equation 2:

$$K_i = -\frac{d \log A}{d\sqrt{t}} \quad (2)$$

When the anion is the desorbing species, the desorption process can be represented by the diagram in Figure 1. In a previous study (*26*) we suggested that K_i could be used to estimate apparent surface pK_a for two reasons: K_i was directly proportional to the concentration of the diffusing species immediately adjacent to the monolayer, and ionization and initial desorption were fast, quasi–equilibrium processes. This suggestion is examined in the present study.

Materials and Methods

Palmitic and myristic acids (Applied Science Laboratories, State College, Pa.) and oleic acid (Hormel Institute, Austin, Minn.) were applied to the Langmuir trough in a hexane solution. Constant pressure–variable area measurements were obtained at 25°C with a floating barrier and piston oils as previously described (*23*). Castor oil, tri-*m*-tolylphosphate, and linoleyl alcohol (Hormel Institute, Austin, Minn.) were the piston oils; they exerted surface pressures of 17 ± 0.7, 9.5, and 33.5 dynes/cm. Variable pressure–variable area measurements were obtained at 24°–26°C with a movable barrier propelled by a high-torque motor (*27*). π was measured by the Wilhelmy plate technique (*27*).

Subphase buffers at pH 9–11 and 0.1 to 0.2 ionic strength were prepared with sodium or potassium bicarbonate–carbonate mixtures (*8*, *26*). Subphase buffers at pH 8–11 were prepared by adding 0.01*M* tris to 0.1*M* or 1.0*M* sodium chloride and adjusting the pH with concentrated hydrochloric acid or concentrated sodium hydroxide. Another alkaline subphase contained 0.1*N* sodium hydroxide (pH 12.7). The acid subphase contained 0.01*N* hydrochloric acid and 0.1*M* sodium chloride (pH 2.1). All subphases contained 0.1 m*M* EDTA.

Results and Discussion

π-A Isotherms for Desorbing Monolayers. In a previous study (*8*) we showed that the stable stearic acid monolayers which were spread on alkaline subphases underwent phase transitions during compression and that the surface pressures of the phase transitions varied directly with pH. The fatty acid seemed completely ionized in the expanded

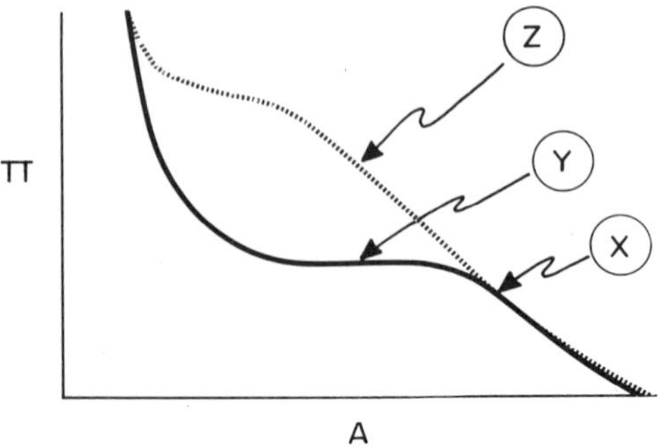

Figure 2. Schematic of the relationships between ionization and surface area, A, in a monolayer: solid line, moderately alkaline subphase; dashed line, strongly alkaline subphase; X and Z, complete ionization in expanded region of π–A isotherm; Y, partial ionization in plateau region of π–A isotherm

region of the π–A isotherm (*see* X in Figure 2). The charge density and field strength of the monolayer increased with compression, and the fatty acid was only partially ionized in the plateau region of the π–A isotherm (*see* Y in Figure 2). The fatty acid was completely ionized at this lower

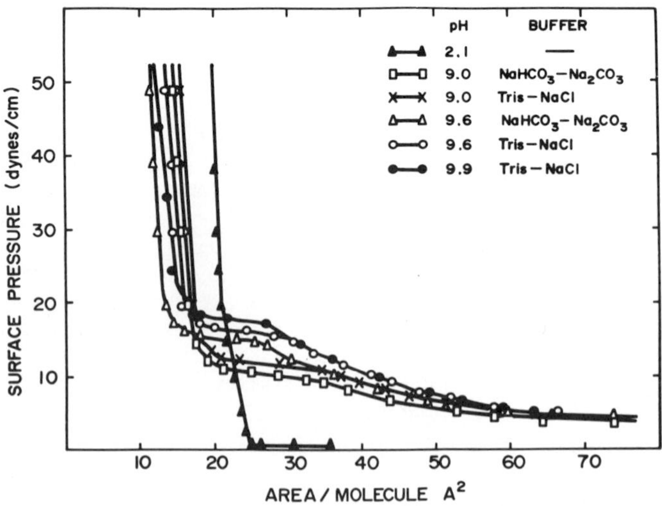

Figure 3. π–A isotherms for palmitic acid spread on the designated subphase buffers. Films were compressed at a rate of 20 A^2/molecule/min. Data represent mean values from at least three isotherms.

area when the monolayer was spread on a more alkaline subphase (*see* Z in Figure 2).

Palmitic acid films contracted rapidly on strongly alkaline subphases (pH > 10). The palmitate anion also desorbed in moderately alkaline subphases (pH < 10) giving unusually low limiting areas at high surface pressures (Figure 3). π–A isotherms had typical phase transitions. The actual area/molecule at a given pH, and consequently the pH of the phase transition could not be ascertained for the desorbing film. The phase transition seemed to depend on the composition of the subphase buffer. The limiting areas were lower for films spread on bicarbonate buffers than for those spread on tris buffers (Figure 3). Tris buffers gave irregular expanded films (Figure 3). Subsequent experiments showed that film expansion on a tris buffer resulted from a decrease in the desorption rate.

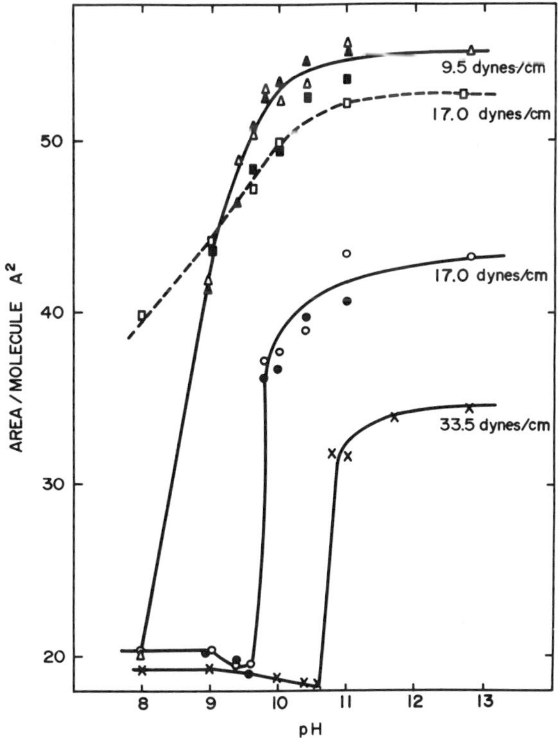

Figure 4. Surface areas for palmitic acid (solid line) and oleic acid (dashed line) extrapolated from desorption data. Open symbols are tris buffers. Closed symbols are bicarbonate buffers. Only tris buffers were used at 33.5 dynes/cm.

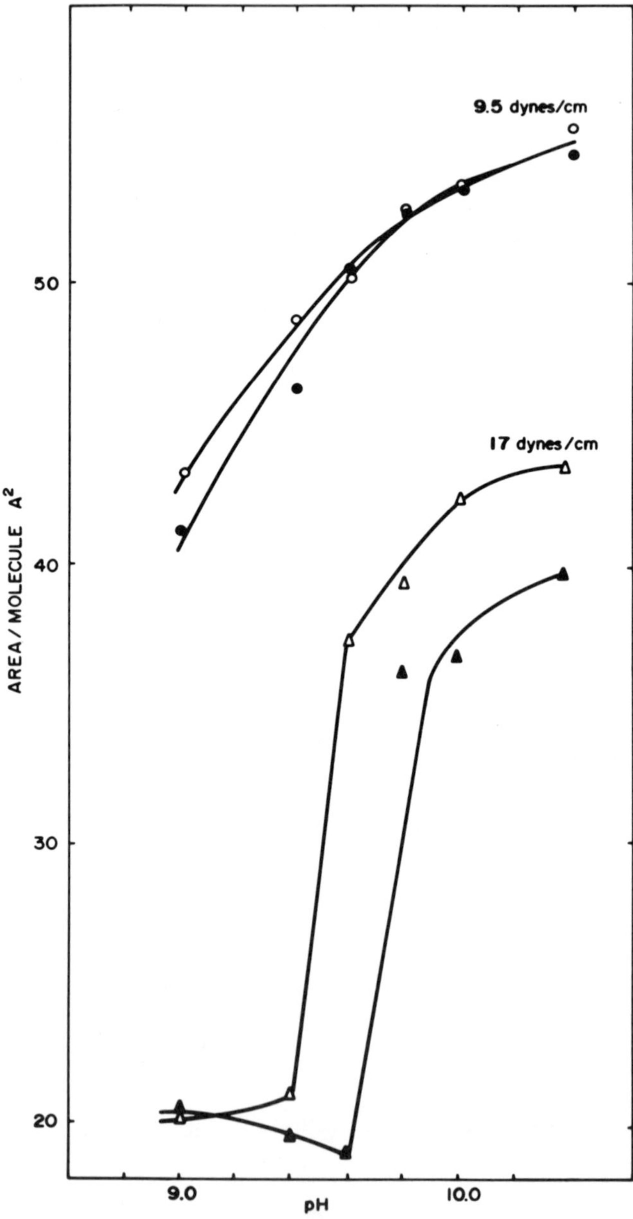

Figure 5. Surface areas extrapolated from desorption data for palmitic acid spread on bicarbonate buffers. Open symbols are 0.2 ionic strength. Closed symbols are 0.1 ionic strength.

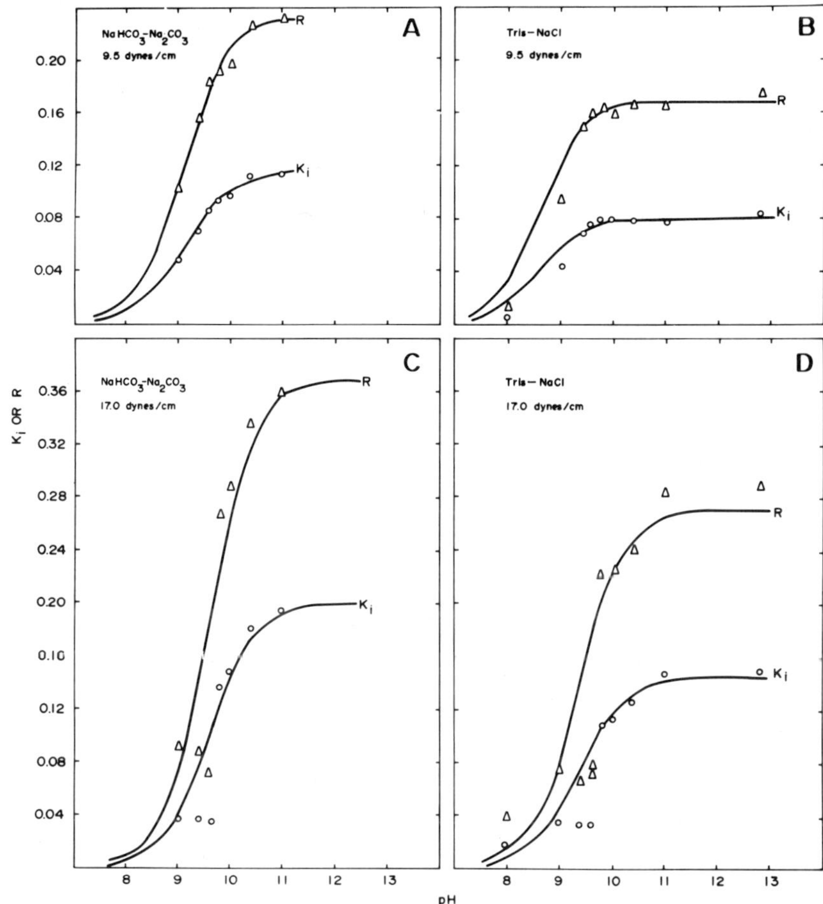

Figure 6. K_i and R for palmitic acid spread on the designated buffers (see text) and maintained at the designate π. K_i and R were mean values from at least five desorption experiments. R was estimated by measuring film contraction over the period t = 0 to 1 min.

***A* as a Function of pH and π.** Since the area/molecule could not be ascertained from the π–A isotherm of a desorbing film, the A values were calculated by extrapolating kinetic data (Equation 2) to zero t (*8*). Abrupt phase transitions occurred with palmitic acid films, and the π of the phase transition varied directly with pH (Figure 4). The anomalous decrease in A that was found immediately before film expansion (Figure 4) suggested partial ionization and the formation of condensed acid soaps (*6, 8, 28*). Since fatty acid monolayers are partially ionized in the condensed state (*8*), the apparent pK_a of the fatty acid could not be estimated from these pH–A isotherms. However, it was evident that field-strength increased as π increased.

Extrapolated surface areas were identical for palmitic acid films desorbing into bicarbonate and tris buffers (Figure 4). These data suggested that these buffers had similar effects on the ionization and expansion of the monolayers.

***A* as a Function of Fatty Acid Structure.** Unsaturated fatty acids have larger surface areas than saturated fatty acids (*9*); consequently, the unsaturated fatty acid anions form weaker fields than the saturated ones. Thus the expansion of an oleic acid film, maintained at the same π as a palmitic acid film, was less abrupt and occurred at a lower pH than the expansion of a palmitic acid film (Figure 4).

***A* as a Function of Ionic Strength.** The degree of ionization in a monolayer varies with ionic strength (*8, 29, 30*); consequently, partially ionized films may expand as the ionic strength of the buffer increases. This proposition was confirmed when stable stearic acid monolayers were spread on 0.1 and 1.0 ionic strength buffers (*8*). Desorbing palmitic acid films maintained at 17 dynes/cm also expanded when the ionic strength of the buffer was increased from 0.1 to 0.2 (Figure 5). In films with increased ionization—those were maintained at a lower π—the ionic strength effect disappeared (Figure 5).

***R* and K_i as Functions of π, pH, and Buffer Composition.** R (Equation 1) and K_i (Equation 2) for palmitic acid varied directly with π and pH (Figure 6). Desorption occurred, yielding measurable R and K_i values even at pH 9 and π at 17 dynes/cm where the film was condensed (Figure 4). Furthermore, both R and K_i were lower for the fatty acid spread on tris buffers than for the fatty acid spread on bicarbonate buffers (Figure 6), showing that apparent film expansion on tris buffers (Figure 3) was, indeed, a desorption rate effect.

R and K_i for oleic acid also varied directly with pH (Figure 7). Again R and K_i were lower for the fatty acid spread on tris buffers than for the fatty acid spread on bicarbonate buffers (Figure 7).

Since both R and K_i increased markedly when the monolayer was spread on a high pH buffer, they seemed potentially useful for describing ionization in unstable monolayers. R and K_i are related by:

$$\frac{R}{K_i} = \frac{1.15}{\sqrt{t}} \tag{3}$$

It was impossible to measure R_i for rapidly desorbing monolayers; consequently we estimated it by measuring the film contraction over the period $t = 0$ to 1 min. However, this method, as seen in the following calculation, underestimated R. The ratio R/K_i possesses an average value over the interval $t =$ 0–1 min.

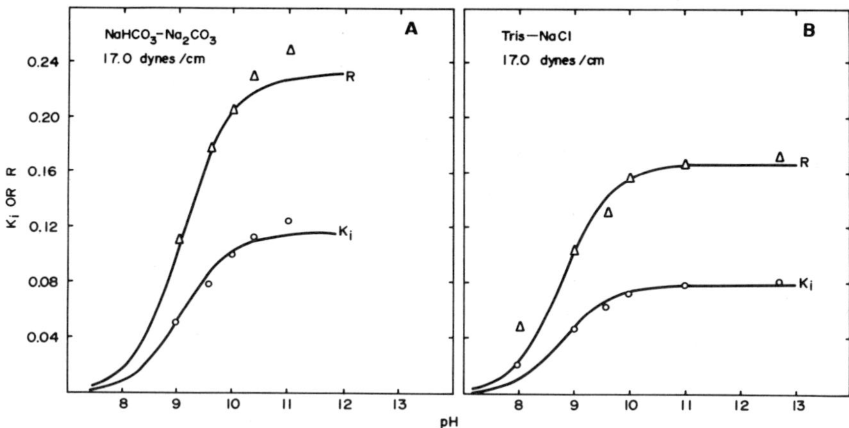

Figure 7. K_i and R for oleic acid spread on the designated buffers (see *text) and maintained at 17 dynes/cm. K_i and R were mean values from at least five desorption experiments.* R *was estimated by measuring film contraction over the period* t = *0 to 1 min.*

$$\frac{R}{K_i}\ \text{av.} = \int_0^1 1.15\ t^{-1/2}\ dt = 2.3 \tag{4}$$

The experimental data yielded R/K_i ratios near 2 and only approached the limiting value of 2.3 in a few experiments. Since K_i was constant over a short time interval and since experimental data tended to underestimate R, we adopted the parameter K_i as an index of ionization.

K_i and Apparent pK_a. Since palmitic and oleic acids formed stable monolayers when they were spread on acid subphases (*see* Ref. *23* and limiting surface areas in Figures 3 and 4), the fatty acid anion, X^-, was the only significant desorbing species from an unstable monolayer. Previous studies (*24, 25*) indicated that K_i varied directly with the surface concentration, $[X^-]$, of a fatty acid anion when ionization and desorption were fast, quasi–equilibrium processes. Gershfeld and Patlak (*25*) described a test for quasi–equilibrium. They noted that the activity coefficient of the monolayer, γ^*, may be estimated from π–A isotherms,

$$\frac{d \ln \gamma^*}{d \pi} = \frac{A}{RT} - C_s \tag{5}$$

where the surface compressibility, C_s, is given by,

$$C_s = -\frac{1}{A}\left[\frac{\partial A}{\partial \pi}\right]_T \tag{6}$$

The activity coefficient may also be estimated from desorption kinetics,

$$\frac{d \ln \gamma^*}{d\pi} = \frac{d \ln K_s}{d\pi} \quad (7)$$

where the rate constant, K_s, for desorption at constant π in the steady state is given by,

$$K_s = -\frac{d \ln A}{dt} \quad (8)$$

Gershfeld and Patlak (*25*) found that the activity coefficient obtained from equilibrium data (Equation 5) was similar to that from desorption data (Equation 7) when desorption was an equilibrium process.

We calculated activity coefficients for palmitic acid with the A values extrapolated from kinetic data (*see* Figure 4). These coefficients are 0.063 cm/dyne at pH 10 and 0.088 cm/dyne at pH 11. Activity coefficients calculated from K_s data in a previous study (*26*) are 0.048 cm/dyne at pH 10 and 0.078 cm/dyne at pH 11. The similarities in the coefficients calculated by the two methods suggest that the desorbing palmitic acid monolayer was in a quasi–equilibrium and therefore,

$$K_i = k\,[X^-] \quad (9)$$

We recast K_i and pH data into a linear form to estimate the apparent surface pK_a. Let Y be the total surface concentration of the fatty acid, HX, and its anion,

$$Y = [X^-] + [HX] \quad (10)$$

and

$$Y = [X^-] + [X^-]\frac{[H^+]}{K_a} \quad (11)$$

or

$$[X^-] = \frac{Y}{1 + \frac{[H^+]}{K_a}} \quad (12)$$

where K_a in Equations 11 and 12 is the fatty acid dissociation constant. Now K_i/k (Equation 9) may be substituted for $[X^-]$ in Equation 12:

$$K_i = \frac{kY}{1 + \frac{[H^+]}{K_a}} \quad (13)$$

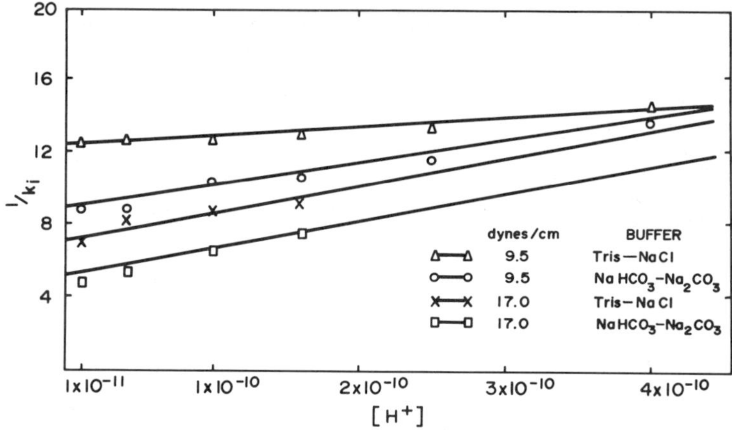

Figure 8. Reciprocal plots (Equation 16) of K_i *and* $[H^+]$ *data for palmitic acid monolayers spread on the designated buffers (see text) and maintained at the designated* π*. Linear regression equations (solid lines) were calculated by least squares analyses of the data.*

K_i approaches its maximum when the fatty acid is completely ionized (Figures 3 and 4), and

$$K_{imax} = kY \tag{14}$$

When Equations 13 and 14 are combined,

$$K_i = \frac{K_{imax}}{1 + \frac{[H^+]}{K_a}} \tag{15}$$

Equation 15 may be rearranged in a linear form as the reciprocal equation,

$$\frac{1}{K_i} = \frac{1}{K_{imax}} + \frac{[H^+]}{K_{imax}K_a} \tag{16}$$

where $1/K_{imax}$ is the intercept and $1/K_{imax}K_a$ is the slope of a $1/K_i$ *vs.* $[H^+]$ plot.

Reciprocal plots (Equation 16) of the experimental data for monolayers in the expanded state yielded linear regression equations. Representative plots are in Figure 8. Slope and intercept data from regression equations were used to calculate the apparent surface pK_a of the fatty acid in the monolayer. These pK_a values are tabulated in Table I. The apparent surface pK_a was several pH units above the pK_a of carboxylic acids dissolved in solution. Thus desorption kinetics generated data for

Table I. Apparent Dissociation Constants[a] for Palmitic Acid and Oleic Acid in Monolayers: Effect of Buffer[b] Composition and Surface Pressure

Buffer	π *(dynes/cm)*	pK_a
Palmitic Acid		
Tris–NaCl	9.5	8.6
$NaHCO_3$–Na_2CO_3	9.5	9.1
Tris–NaCl	17.0	9.3
$NaHCO_3$–Na_2CO_3	17.0	9.6
Tris–NaCl	33.5	9.9
Oleic Acid		
Tris–NaCl	17.0	8.8
$NaHCO_3$–Na_2CO_3	17.0	9.1

[a] K_a was calculated from slope and intercept values obtained by a least squares analysis of K_i and $[H^+]$ data in a reciprocal plot (Equation 16).
[b] Buffers were 0.1 ionic strength.

unstable monolayers that agreed with many studies involving stable monolayers (*8, 11–18*).

The apparent pK_a can be used to evaluate how well experimental R and K_i data fit a sigmoidal curve. The percent of ionization at a specified pH is obtained when the surface pK_a is substituted in the Henderson-Hasselbalch equation. R_{max} and K_{imax} were obtained from the reciprocal plots of R and K_i data. When R_{max} or K_{imax} was multiplied by percent ionization, a sigmoid curve relating R or K_i and pH resulted. These sigmoid curves appear as the solid lines in Figures 6 and 7. Deviations from the sigmoid curve appeared in the lower pH region where films contracted slowly and in the upper pH region where films contracted rapidly. Experimental measurements in both regions of the sigmoid curve were difficult to obtain.

Desorption kinetics provide additional information that contributes to our understanding of surface phenomena such as the specific effects of tris and bicarbonate buffers on monolayers. Stable condensed monolayers were expanded on tris buffers, and ionization appeared enhanced (*8*). Unstable, expanded monolayers did not expand further on tris buffers, but K_i data (Table I) showed a consistent decrease in the apparent pK_a when fatty acids were spread on tris buffers.

Surface potential (*31*) and infrared (*16*) studies indicate that monolayers spread on a subphase buffer containing the bicarbonate anion behave differently than do monolayers spread on buffers containing other anions. Goddard, Smith, and Kao (*31*) suggested that the bicarbonate anion interacts with the monolayer forming acid soaps. Desorption kinetics show that the desorption rate increases—that is, R and K_i are higher (Figures 6 and 7) when monolayers are spread on bicarbonate subphases. These studies suggest that solubilization is enhanced by the bicarbonate anion which may act as a chaotropic ion (*32*).

Apparent pK_a and Field Strength. The character of anionic sites in a surface determines the nature of the charge field (*10, 33*). Strong-field surfaces bind protons more firmly than weak-field surfaces; consequently, the apparent surface pK_a is an indicator of field strength. In studies of desorbing monolayers (Table I), the field strength increased when the charge density of the surface was increased by decreasing the area/molecule of anions in the surface. In expanded monolayers, *A* was readily controlled by π, ionic strength, and the structure of the fatty acid (Figures 4 and 5).

Area/pH effects are shown in Figure 9. Myristic acid spread on tris–1.0*M* sodium chloride buffers expanded to the same area/molecule

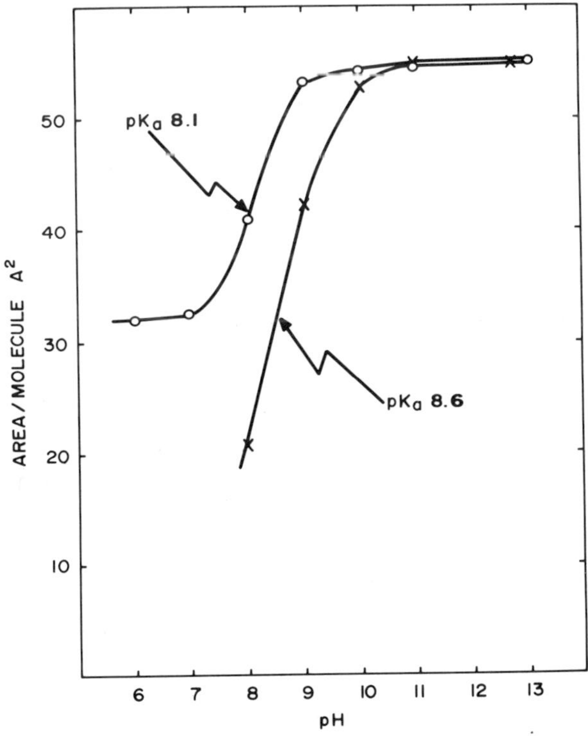

*Figure 9. Surface areas for myristic acid (–O–O–) and palmitic acid (–X–X–) extrapolated from desorption data. Myristic acid was spread on tris–1.0*M *sodium chloride, and palmitic acid was spread on tris–0.1*M *sodium chloride.* π *was maintained at 9.5 dynes/cm.*

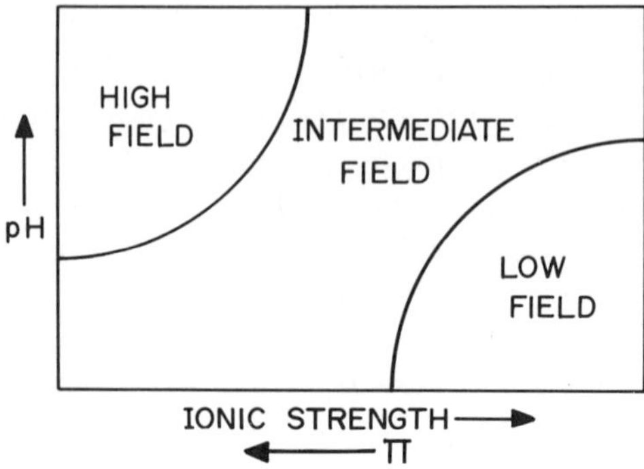

Figure 10. Schematic of the relationships between the experimental variables: pH, ionic strength, π, and the field strength of the monolayer surface

Table II. Selectivity Sequences Found Experimentally for Alkali Metal Cations (33)

I	$Cs^+>$	$Rb^+>$	$K^+>$	$Na^+>$	Li^+	Weak field
II	$Rb^+>$	$Cs^+>$	$K^+>$	$Na^+>$	Li^+	
III	$Rb^+>$	$K^+>$	$Cs^+>$	$Na^+>$	Li^+	
IV	$K^+>$	$Rb^+>$	$Cs^+>$	$Na^+>$	Li^+	
V	$K^+>$	$Rb^+>$	$Na^+>$	$Cs^+>$	Li^+	
VI	$K^+>$	$Na^+>$	$Rb^+>$	$Cs^+>$	Li^+	
VII	$Na^+>$	$K^+>$	$Rb^+>$	$Cs^+>$	Li^+	
VIII	$Na^+>$	$K^+>$	$Rb^+>$	$Li^+>$	Cs^+	
IX	$Na^+>$	$K^+>$	$Li^+>$	$Rb^+>$	Cs^+	
X	$Na^+>$	$Li^+>$	$K^+>$	$Rb^+>$	Cs^+	↓
XI	$Li^+>$	$Na^+>$	$K^+>$	$Rb^+>$	Cs^+	Strong field

at a high pH as palmitic acid spread on tris–0.1M sodium chloride buffers. However, the myristic acid film expanded more on weakly alkaline subphases, and the apparent pK_a (Figure 9), calculated from K_i data, was significantly lower for myristic acid.

Table III. Selectivity Sequences Found Experimentally for Alkaline Earth Cations (10)

I	$Ba^{2+}>$	$Sr^{2+}>$	$Ca^{2+}>$	Mg^{2+}	Weak field
II	$Ba^{2+}>$	$Ca^{2+}>$	$Sr^{2+}>$	Mg^{2+}	
III	$Ca^{2+}>$	$Ba^{2+}>$	$Sr^{2+}>$	Mg^{2+}	
IV	$Ca^{2+}>$	$Ba^{2+}>$	$Mg^{2+}>$	Sr^{2+}	
V	$Ca^{2+}>$	$Mg^{2+}>$	$Ba^{2+}>$	Sr^{2+}	
VI	$Ca^{2+}>$	$Mg^{2+}>$	$Sr^{2+}>$	Ba^{2+}	↓
VII	$Mg^{2+}>$	$Ca^{2+}>$	$Sr^{2+}>$	Ba^{2+}	Strong field

The character of anionic sites in a surface may be altered by screening, as in the case of glass electrodes (*10, 33*). Screening, which alters charge density, was done by adding neutral molecules and oppositely charged ions to monolayers and liposomes (*8, 20, 21, 22*). A decrease in pH has the same effect by interposing unionized fatty acid molecules between the fatty acid anions (*8*). A number of screening techniques are available, and it may be possible to test Seufert's suggestion (*34*) that cation selectivity for a lipid bilayer model system depends on asymmetries of carrier location, a form of screening, in the membrane.

The apparent pK_a data demonstrate that a strong-field surface has a high charge density. A strong-field surface is obtained experimentally by increasing the pH and decreasing the surface area through a decrease in

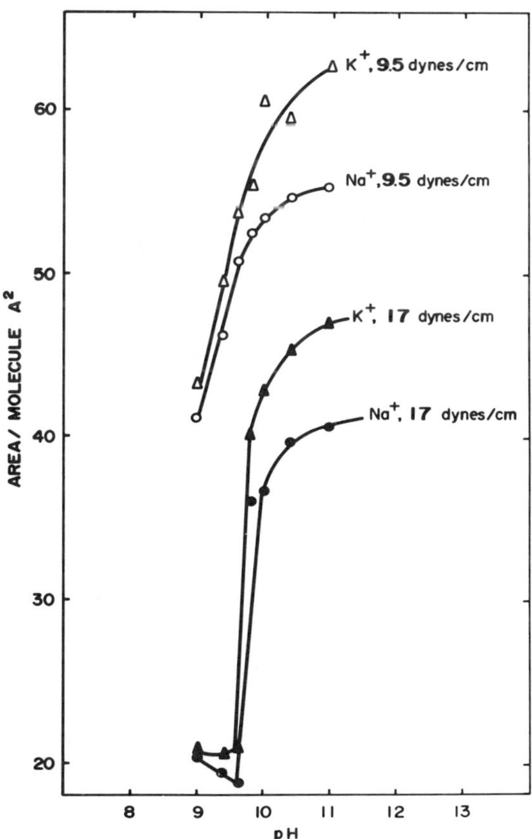

Figure 11. Surface areas for palmitic acid spread on 0.1 ionic strength bicarbonate buffers containing either Na^+ or K^+

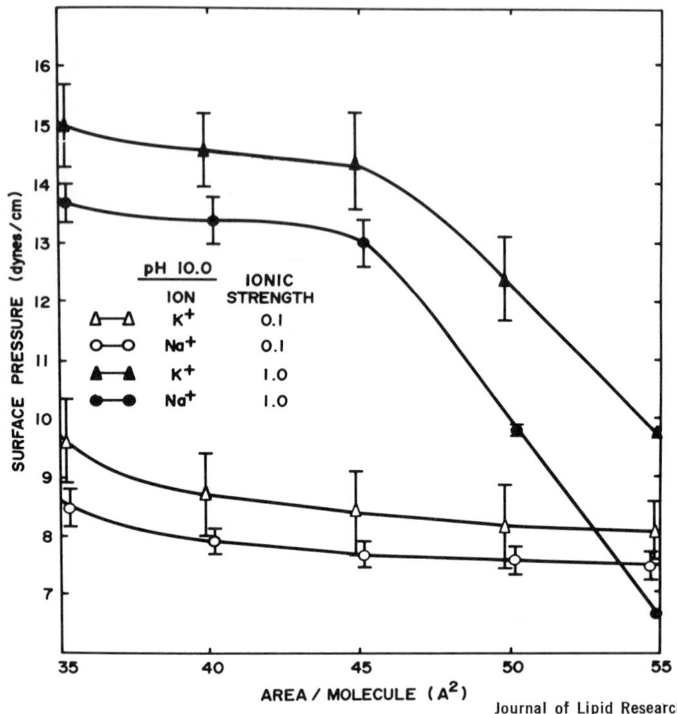

Figure 12. Plateau regions of π–A isotherms for stearic acid at pH 10 on an expanded scale. Data represent means ± SD for seven isotherms (8).

ionic strength or an increase in π (Figure 10). Conversely, a weak-field surface is obtained experimentally by decreasing pH, decreasing π, and increasing ionic strength (Figure 10).

Cation Selectivity and Field Strength. Cation selectivity sequences for anionic binding sites have been discussed for years by investigators. In 1962 Eisenman (*33*) published a general theory based on field strength that explained the selectivity sequences that were observed with glass electrodes, collodion membranes, various minerals, and several biological membranes. Specific sequences were predicted from the free energies involved in the transfer of ions from aqueous solution to the anionic site. There were evidently only two significant terms: the coulombic interaction of the cation with the anionic site and the removal of water of hydration from the cation. Weak fields selected larger cations because the hydration term predominated and water removal from the larger cations was easier. Strong fields selected smaller cations because the coulombic interaction term predominated. Eisenman selectivity sequences (*33*) for alkali metal cations are presented in Table II. Selectivity sequences for alkaline earth cations, attributed to H. S. Sherry by Diamond and Wright (*10*), are presented in Table III. Since the field strengths of fatty acid monolayers can be varied systematically (Figure 10), the validity of the field strength theory can be tested by using these monolayers.

Cation selectivity data for monolayers (summarized in the introductory section) coincide with the predictions of the field strength theory. Fatty acids have high apparent pK_a values, and completely ionized monolayers (*2, 3, 4, 5, 6*) select a strong field sequence (XI in Table II). Alkyl sulfates have low apparent pK_a values, and completely ionized mono-

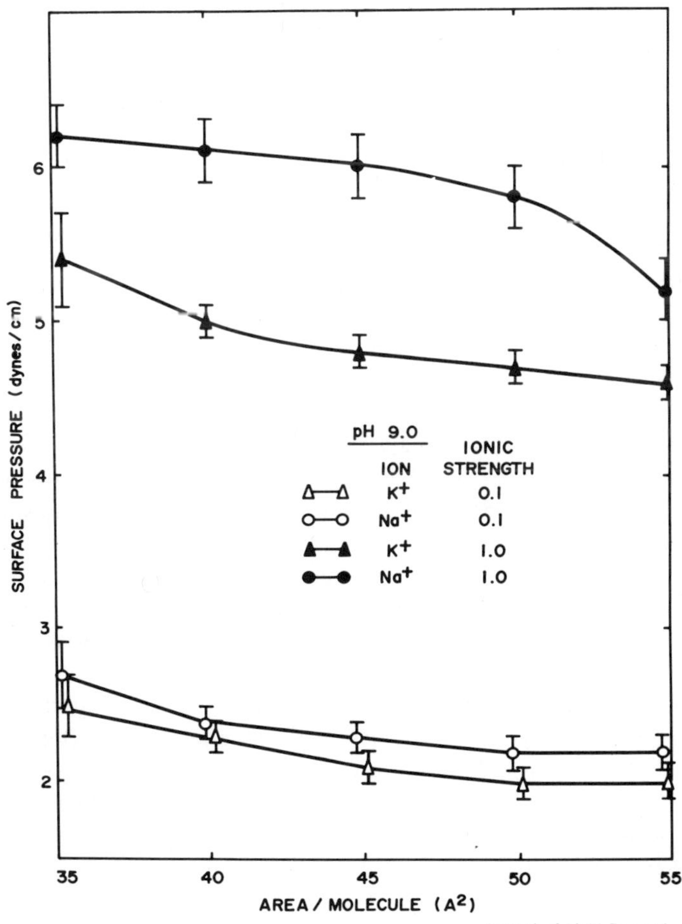

Figure 13. Plateau regions of π–A isotherms for stearic acid at pH 9 on an expanded scale. Data represent means ± SD for seven isotherms at 0.1 ionic strength and 16 isotherms at 1.0 ionic strength (8).

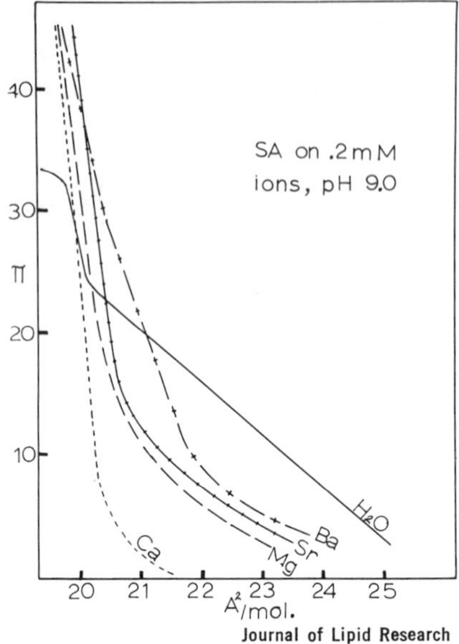

Figure 14. π–A isotherms for stearic acid (SA) spread on subphases containing different alkaline earth ions (0.2mM) at pH 9. "H_2O" isotherm is for pure water (17).

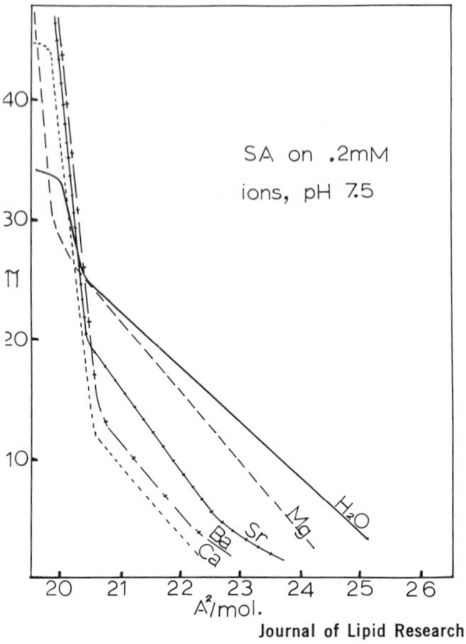

Journal of Lipid Research

Figure 15. π–A isotherms for stearic acid (SA) spread on subphases containing different alkaline earth ions (0.2mM) at pH 7.5 (17)

layers select weak field sequences (I through VI in Table II).

The initial observation of Sears and Schulman (*3*) that $Na^+ > K^+$ selectivity on strongly alkaline subphases reversed on a weakly alkaline subphase was confirmed in several studies. Desorption experiments (Figure 11) showed that palmitic acid monolayers with essentially complete ionization—strong fields—were condensed more by Na^+ than by K^+. At lower pH, the partially ionized monolayers—weaker fields—were condensed the same amount by Na^+ and K^+.

π–A isotherms provided similar data for stable monolayers (*8*). Thus Na^+ was preferred to K^+ at pH 10 with monolayers spread on both 0.1 and 1.0 ionic strength buffers (Figure 12). By lowering the pH to 9 (Figure 13), the selectivity pattern reversed when the field was weakened. The reversal was more apparent on the 1.0 ionic strength buffer where a higher ionic strength further weakened the field.

In previous studies (*17, 35*), we explained selectivity sequences for alkaline earth cations by specific binding sites, rather than by strong and weak fields. A re-examination of these data shows that selectivity data coincides with the field strength theory. The strong field sequence (VI in Table III) which was found at 10 dynes/cm and pH 9 (Figure 14) was replaced by an intermediate field sequence (III in Table III) at pH 7.5 (Figure 15) and a weak field sequence (I in Table III) at pH 6 (Figure 16). When field strength was increased by increasing π from 10 dynes/cm to 30 dynes/cm at pH 7.5 (Figure 15), the intermediate field sequence (III in Table III) was replaced by a strong field sequence (VII in Table III).

Other properties of fatty acid monolayers such as the phase transition temperature are consistent with the field strength theory; stearic acid monolayers formed rigid films on alkaline earth subphases (*17, 35*). The temperature of the phase transition from rigid to fluid monolayers, estimated by the Devaux talc test, was a function of pH and buffer composition (Figure 17). Thus transition temperature at pH 6 decreased in a weak field sequence (I in Table III) while transition temperatures at pH 8 decreased in an intermediate field sequence (III or IV in Table III). Since variables such as π are not controlled in these experiments, it is surprising that transition temperature data followed these sequences.

Even though both π–A (Figures 14–16) and transition temperature (Figure 17) data followed the field strength theory, the infrared spectra of skimmed films presented a different pattern (Figure 18). IR data indicated that stearic acid monolayers at pH 6 interacted with alkaline earth cations in a strong field sequence (VI in Table III). These data may have reflected the compression generated by the skimming process. For example, Sr^{2+} had an anomalous condensing effect at pH 6 and 30 dynes/cm (Figure 17). The anomalous sequences found with skimmed

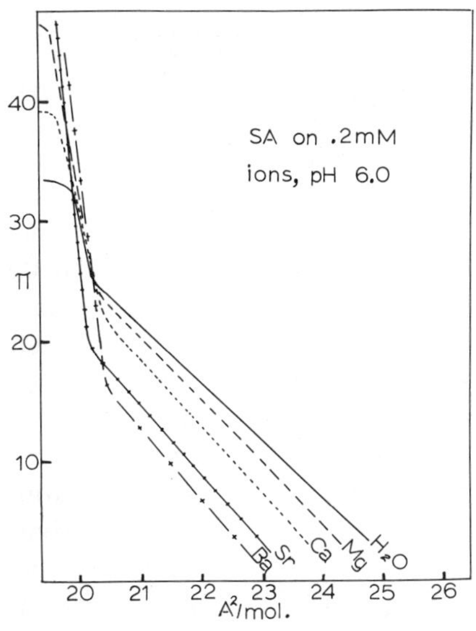

Journal of Lipid Research

Figure 16. π–A isotherms for stearic acid (SA) spread on subphases containing different alkaline earth ions (0.2mM) at pH 6.0 (17)

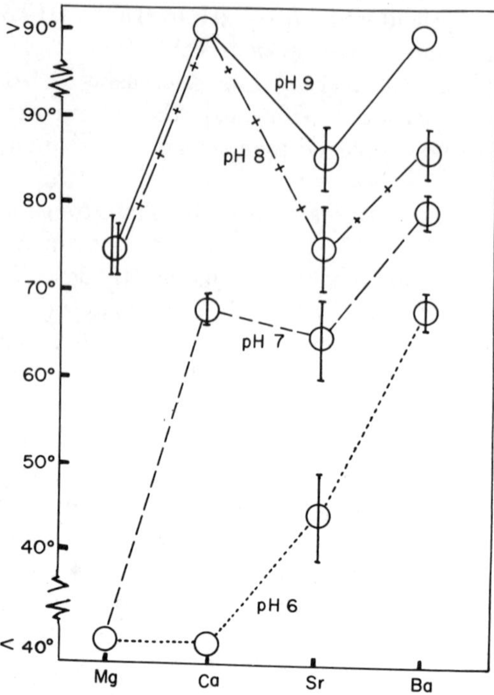

Journal of Lipid Research

Figure 17. Phase transition temperatures for stearic acid monolayers spread on subphases containing different alkaline earth ions (0.2 mM) at the specified pH (17)

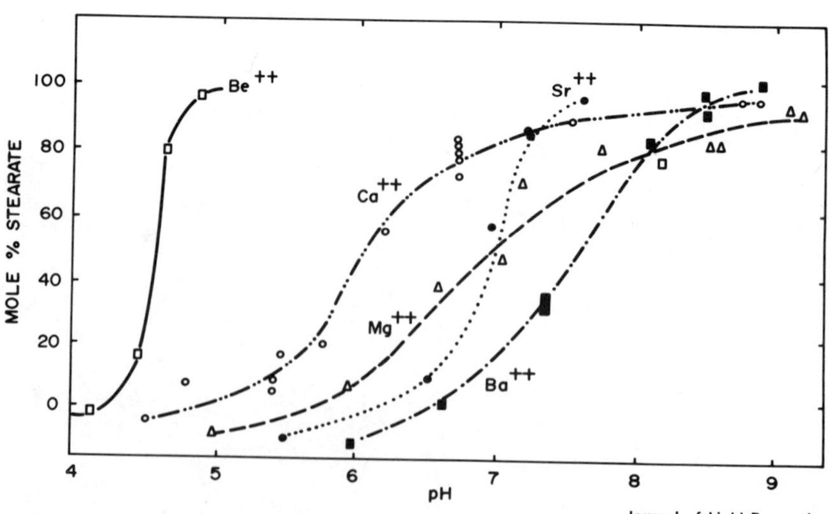

Journal of Lipid Research

Figure 18. Stearate content, determined by infrared spectroscopy of monolayers spread on alkaline earth (0.2 mM) subphases plotted as a function of pH (17)

films were expected by Diamond and Wright (*10*) and explained by the polarizability of larger cations.

Anomalous sequences suggest that steric fit may be required as an additional term in the general field strength theory. The most compelling evidence to support a steric fit term occurred from studies using detergents and antibiotics as ionophores. For example, polyoxyethylene ether detergents followed an alkali metal cation selectivity sequence in lowering the dc resistance of synthetic bilayers (*36*). However, $K^+ > Na^+$ selectivity was greater than the single order of magnitude predicted by the field strength theory (*33*). Haynes *et al.* (*37, 38*) found that antibiotics had a similar high K^+ selectivity. They concluded that it was based partly on the conformation of the complex. Steric fit is undoubtedly important in K^+–nonaction complexes (*39*) and alkali metal–valinomycin complexes (*40, 41*). Some indication of the magnitude of the steric fit term will be supplied by techniques such as internal reflection spectroscopy, where the monolayer is undisturbed. These studies are in progress.

Acknowledgment

We thank F. J. Kezdy for helpful discussions concerning methods for plotting ionization as a function of hydrogen ion concentrations.

Literature Cited

1. Adam, N. K., *Proc. Roy. Soc. Ser. A* (1922) **101,** 516.
2. Adam, N. K., Miller, J. G. F., *Proc. Roy. Soc. Ser. A* (1933) **142,** 401.
3. Sears, D. F., Schulman, J. H., *J. Phys. Chem.* (1964) **68,** 3529.
4. Goddard, E. D., Kao, O., Kung, H. C., *J. Colloid Interface Sci.* (1967) **24,** 297.
5. Christodoulou, A. P., Rosano, H. L., ADVAN. CHEM. SER. (1968) **84,** 210.
6. Goddard, E. D., Ackilli, J. A., *J. Colloid Sci.* (1963) **18,** 585.
7. Rogers, J., Schulman, J. H., "Proceedings 2nd International Congress on Surface Activity," Vol. 3, p. 243, Butterworth, London, 1957.
8. Patil, G. S., Matthews, R. H., Cornwell, D. G., *J. Lipid Res.* (1972) **13,** 574.
9. Adam, N. K., "The Physics and Chemistry of Surfaces," Dover Publications, New York, 1968.
10. Diamond, J. M., Wright, E. M., *Ann. Rev. Physiol.* (1969) **31,** 581.
11. Peters, R. A., *Proc. Roy. Soc. Ser. A* (1931) **133,** 140.
12. Danielli, J. F., *Proc. Roy. Soc. Ser. B* (1937) **122,** 155.
13. Schmidt-Nielson, K., *Acta Physiol. Scand.* (1946) **12,** *Suppl.* **37,** 1.
14. Mattson, F. H., Volpenhein, R. A., *J. Amer. Oil Chemists' Soc.* (1966) **43,** 286.
15. Schulman, J. H., Hughes, A. H., *Proc. Roy. Soc. Ser. A* (1932) **138,** 430.
16. Bagg, J., Haber, H. D., Gregor, H. P., *J. Colloid Interface Sci.* (1966) **22,** 138.
17. Deamer, D. W., Meek, D. W., Cornwell, D. G., *J. Lipid Res.* (1967) **8,** 255.
18. Joos, P., *Bull. Soc. Chim. Belges* (1971) **80,** 277.
19. Hartley, G. S., *Trans. Faraday Soc.* (1934) **30,** 444.
20. Montal, M., Gitler, C., *Bioenergetics* (1973) **4,** 363.

21. Fromherz, P., *Biochim. Biophys. Acta* (1973) **323**, 326.
22. Goddard, E. D., *Advan. Colloid Interface Sci.* (1974) **4**, 45.
23. Heikkila, R. E., Deamer, D. W., Cornwell, D. G., *J. Lipid Res.* (1970) **11**, 195.
24. Ter Minassian-Saraga, L., *J. Chim. Phys.* (1955) **52**, 181.
25. Gershfeld, N. L., Patlak, C. S., *J. Phys. Chem.* (1966) **70**, 286.
26. Patil, G. S., Matthews, R. H., Cornwell, D. G., *J. Lipid Res.* (1973) **14**, 26.
27. Heikkila, R. E., Kwong, C. N., Cornwell, D. G., *J. Lipid Res.* (1970) **11**, 190.
28. Goddard, E. D., Goldwasser, S., Goliken, G., Kung, H. C., ADVAN. CHEM. SER. (1968) **84**, 67.
29. Davies, J. T., Rideal, E. K., "Interfacial Phenomena," Academic Press, New York, 1963.
30. Payens, T. A. J., *Philips Res. Rep.* (1955) **10**, 425.
31. Goddard, E. D., Smith, S. R., Kao, O., *J. Colloid Interface Sci.* (1966) **21**, 320.
32. Dandliker, W. B., de Saussare, V. A., *in* "The Chemistry of Biosurfaces," Hair, M. L., Ed., pp. 1-43, Marcel Dekker, New York, 1971.
33. Eisenman, G., *Biophys. J.* (1962) **2**, 259.
34. Seufert, W. D., *Biophysik* (1973) **10**, 281.
35. Deamer, D. W., Cornwell, D. G., *Biochim. Biophys. Acta* (1966) **116**, 555.
36. VanZutphen, H., Merola, A. J., Brierley, G. P., Cornwell, D. G., *Arch. Biochem. Biophys.* (1972) **152**, 755.
37. Haynes, D. H., Pressman, B. C., *J. Membrane Biol.* (1974) **18**, 1.
38. Haynes, D. H., Wiens, T., Pressman, B. C., *J. Membrane Biol.* (1974) **18**, 23.
39. Kilbourn, B. T., Dunitz, J. D., Pioda, L. A. R., Simon, W., *J. Mol. Biol.* (1967) **30**, 559.
40. Krigbaum, W. R., Kuegler, F. R., Oelschlaeger, A., *Biochemistry* (1972) **11**, 4548.
41. Grell, E., Funck, Th., *J. Supramolecular Struct.* (1973) **1**, 307.

RECEIVED September 24, 1974. Work supported in part by research grant GM-09506 from the National Institutes of Health.

6

The Influence of Calcium and Magnesium Ions on Dodecyl Sulfate and Other Surfactant Anions

E. D. GODDARD[1]

Unilever Research Laboratory, Port Sunlight, England

Various properties of aqueous solutions of calcium and magnesium dodecyl sulfate (DS) are compared. These include surface tension, solubility, and electrical conductivity as well as force area characteristics of DS monolayers spread on high ionic strength solutions of calcium and magnesium chloride. The latter data and those for absorbed monolayers derived from surface tension measurements point to a stronger interaction of DS with Ca than with Mg in view of the more expanded nature of the monolayers on the Mg subsolutions. This is supported by a higher Krafft point of the Ca salt and the higher binding of Ca to DS micelles as revealed by conductance measurements. The results are compared with published data on the differences in the effect of alkaline earth metal ions on sulfonate, phosphate, and carboxylate monolayers.

In recent years, the influence of counterions on the properties of ionized monolayers has received much attention. Even though Davies' (*1*) application of the Gouy–Chapman double layer theory to ionized monolayers represented a major advance in the understanding of the properties of these systems, it has been increasingly recognized that we must account for the different effects (*i.e.*, specific counterion effects) that counterions of the same net charge may have on the charged monolayer. Because of counterion sequence inversions which have been ob-

[1] Present address: Union Carbide Corp., Tarrytown Technical Center, Tarrytown, N. Y. 10591.

served when the nature of the charged monolayer is changed, no facile explanation (*2, 3*), based, for example, on counterion size, adequately explains the observations. The results strongly parallel those observed in the interaction of ions with exchange resins and polyelectrolytes (*2, 3*), and the measurements of the activity coefficient of simple electrolytes in aqueous solution (*4*).

Although charged, spread monolayers have received chief emphasis, many other properties of long chain electrolytes also manifest a counterion sequence dependence. These properties include solubility, lowering of critical micelle concentration (*5, 6, 7*), surface tension depression (*8*), and counterion binding (*9*) by micelles. Most reports on specific ion effects are restricted to monovalent counterions. This report provides data on two representative bivalent cations—*viz.*, calcium and magnesium; their dodecyl sulfate salts were chosen for a detailed study of solubility, electrical conductivity, surface tension, and spread films. The results are compared with literature data on the effect of bivalent cations on other long chain anions. The solubility and surface tension data augment published information (*10*).

Experimental

Materials. High purity reagents were used. Dodecyl sulfuric acid was prepared from carefully fractionated dodecyl alcohol by the method of Dreger *et al.* (*11*), and neutralizations were performed with magnesium oxide and calcium hydroxide. The products were recrystallized several times: the magnesium salt (MgDS) from wet methyl ethyl ketone and the calcium salt (CaDS) from 90% ethanol; thereafter, both specimens were extracted with light petroleum ether. It was confirmed (*10*) that the MgDS existed as the hexahydrate. The $MgCl_2$, $CaCl_2$, and NaCl used in the subsolutions for the spread monolayer work were foam purified in concentrated aqueous solution and then recrystallized. In addition, the calcium and sodium chlorides, after drying at 105°C, were heat treated at 300°C for 6 hr before use. The water in these studies had a conductivity of 2×10^{-6} ohm^{-1} cm^{-1} at 25°C.

Surface Tension. Measurements were carried out by the duNouy ring method, and the corrections of Harkins and Jordan (*12*) were applied. Aging effects in the measured surface tensions were negligible. Each surface tension point corresponded to a solution individually prepared by weighing.

Monolayers. The measurements used a standard Langmuir–Adam film balance. As a precaution against contamination, the strong (4*M*) $MgCl_2$, $CaCl_2$, and NaCl subsolutions were swept repeatedly before the dodecyl sulfate (DS) was spread. We also checked that a blank compression did not develop surface pressure. The DS was spread as its sodium salt from a 1:4 water:ethanol solution, and compression time was approximately 20 min. Reproducibility was very good although at the highest surface pressure values there was slight tendency for the pressure

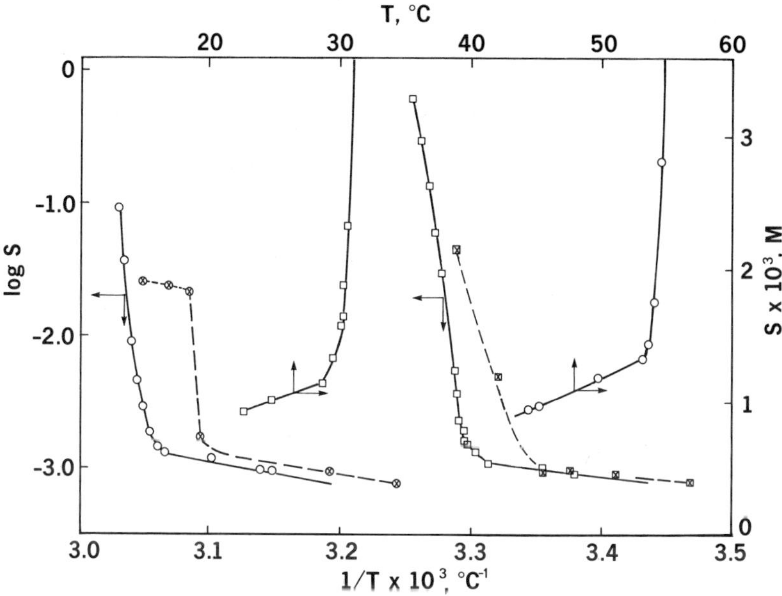

Figure 1. Solubility vs. *temperature and log solubility* vs. *reciprocal temperature plots for MgDS* (□) *and CaDS* (○). *Data from Ref.* 10 *included* (×).

to decrease on standing. Since comparative data for the spread monolayers were required, this was not a serious limitation (*see* Results and Discussion).

Solubility. The rotating tube method of Pankhurst and Adam (*13*) was used. Solid and 10-ml quantities of water were introduced into borosilicate glass tubes which were then drawn off to yield closed containers of *ca.* 20-ml capacity. Four to six tubes were attached to a slowly rotating, inclined stirrer shaft, and the temperature was increased at *ca.* 0.1°C/24 hr on approaching complete solution of a particular tube's content. For the very low solubilities, the final rate was even lower—*viz., ca.* 0.1°C/96 hr. No hydrolysis was detected in any of the tubes after the solubility point had been established.

Conductivity. A 900-ml dilution cell was used. The platinum electrodes had a light platinization, and the cell design ensured wide separation of the electrical leads to minimize capacitance effects.

Results and Discussion

The solubility values are plotted against temperature and are also presented as a logarithm of reciprocal temperature plot in Figure 1. The Krafft points, derived from this figure, are 29.2° and 53.4°C for MgDS and CaDS and are somewhat higher than those in Ref. *10—viz.,* 25° and

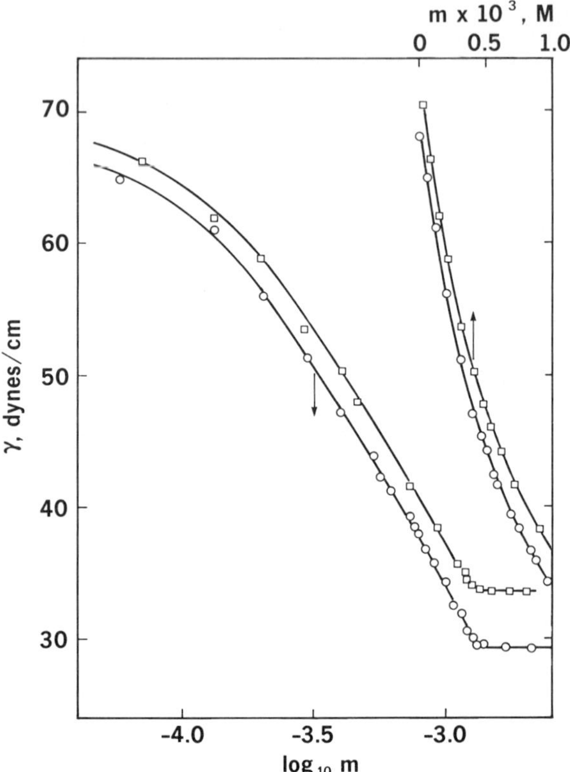

Figure 2. Surface tension vs. *molarity and log molarity for aqueous solutions of MgDS at 40°C* (□) *and CaDS at 55°C* (○)

50°C. If, as is generally assumed, the solubility at the Krafft point represents the critical micelle concentration (CMC) at that temperature, the derived values are 1.16 and $1.33 \times 10^{-3}M$. For comparison, Miyamoto's solubility data (*10*) are included in Figure 1. Agreement is moderately good in the high and very low solubility region, especially for MgDS, but at intermediate values around the Krafft point, serious discrepancies exist. No obvious explanation accounts for these differences.

By applying the Clausius–Clapeyron equation to the observed solubilities, S, of these di-univalent salts in the form:

$$\frac{d\ln S}{d(1/T)} = -\frac{2}{3}\frac{\Delta H_s}{R} \quad (1)$$

one can estimate the enthalpy of solution, ΔH_s. The values obtained per long chain DS mole are +12.45 and +10.0 kcal for the Ca and the Mg

Table I. CMC Values of MgDS and CaDS ($M \times 10^{-3}$)

	MgDS					*CaDS*		
	29°C	*35°C*	*40°C*	*45°C*	*55°C*	*50°C*	*53°C*	*55°C*
$k - N$		1.262	1.302	1.357	1.500			1.342
$\Lambda - \sqrt{N}$		1.230	1.276	1.322	1.470			1.315
$\gamma - \ln m$			1.280					1.318
Solubility	1.16						1.33	
[a]$\gamma - \ln M$			1.100		1.100*			1.300*
[a]Dye			1.100					1.300*
[a]Solubility	1.0**					1.3		

[a] Miyamoto (*10*): *54°C, **25°C

salt. No calorimetric data are available for comparison, but these values are of the same order of magnitude as the enthalpy of solution of NaDS —*viz.*, +8 kcal/mole (*14, 15*). Although it is well known that a heat of solution essentially represents the difference between two large quantities (the energy of hydration and the energy of ion separation), it is reasonable to conclude that the higher value of ΔH_s for CaDS than for $MgDS \cdot 6H_2O$ is consistent with a more stable arrangement of ions in the lattice of the former material.

The temperatures chosen for the surface tension (γ) measurements were 40° (MgDS) and 55°C (CaDS); the surface tension *vs.* concentration and surface tension *vs.* logarithm concentration plots are given in Figure 2. The derived CMC values, 1.28 and $1.32 \times 10^{-3}M$, compared with $1.1 \times 10^{-3}M$ (40°C) and $1.3 \times 10^{-3}M$ (54°C), obtained by Miyamoto (*10*). Like the latter author, we observed that the surface tension of the CaDS solution in the plateau region was appreciably less than that of the MgDS. This effect is generally associated with a greater density of hydrocarbon chains in the interface brought about, in this case, by a greater degree of binding of the calcium ion. Note that Miyamoto found no temperature effect on the CMC of MgDS in the range 40°–54°C (*see* conductivity results below and Table I) or upon the surface tension in the plateau region.

To calculate surface pressure–area isotherms, following the Gibbs convention, we used the adsorption equation relating surface tension to surface excess, Γ, and chemical potential, μ, in the form

$$-d\gamma = d\pi = \Sigma\, \Gamma_i d\mu_i \qquad (2)$$

This becomes, for a single solute of activity (a):

$$\begin{aligned} d\pi &= \Gamma\, d\mu \\ &= RT\, \Gamma\, d\ln a \end{aligned} \qquad (3)$$

When the solute is a di-univalent electrolyte, the equation is

$$d\pi = 3RT\ \Gamma\ d\ln\ (Mf_{\pm}) \tag{4}$$

where M is the molarity of the electrolyte, and $f_{\pm}$ is the mean activity coefficient.

In deriving Γ values, Equation 4 was used to estimate $f_{\pm}$ values for solutions of ionic strength μ':

$$-\ln f_{\pm} = \frac{2.3\ A'\ z_{+}\ z_{-}\sqrt{\mu'}}{1 + d\ \beta\ \sqrt{\mu'}} \tag{5}$$

where z_{+}, z_{-} are the valence of the positive and negative ions; A' and β are constants and d is an estimated value of the distance of closest approach of the oppositely charged ions, which in this case is assumed to be 5A (*i.e.*, 5 Angstroms). This equation predicts $f_{\pm}$ more accurately than the simple Debye–Hückel limiting law expression when comparing such values with experimental data (*16*) for $CaCl_2$ in the concentration range of interest. Differentiating Equation 5 and making the substitution in Equation 4 lead to the final expression

$$2A = \frac{1}{\Gamma} = -3\ RT\ \frac{d\ln M}{d\gamma}$$

$$\left[1 - 2.3\left(\frac{(1 + \sqrt{3M}\ d\ \beta)\sqrt{3M}\ A' - 3M\ A'd\ \beta)}{(1 + \sqrt{3M}\ d\ \beta)^2}\right)\right] \tag{6}$$

where A is the area per DS chain. The bracketed factor in Equation 6 introduced correction factors of 0.94 to 0.98 to the computed A. In deriving A, both γ–$d\ln M$ and γ–M plots were used, the latter facilitating determination of tangential slopes at the lower concentrations.

Since we wished to compare the monolayer properties of CaDS and MgDS directly, it was necessary to allow for the temperature difference between the two derived π *vs.* A isotherms. It was found empirically that the CaDS curve obeyed the following equation of state quite accurately:

$$\pi\ (A - A_o) = nkT \tag{7}$$

with $A_o = 25\ A^2$ and $n = 1$. Accordingly, on the assumption that the co-area term is independent of temperature over the range considered, the π values of CaDS were corrected to 40°C. The resulting π–A curve is compared with that of MgDS in Figure 3. Although differences between the two curves are small, note the more expanded condition of the MgDS

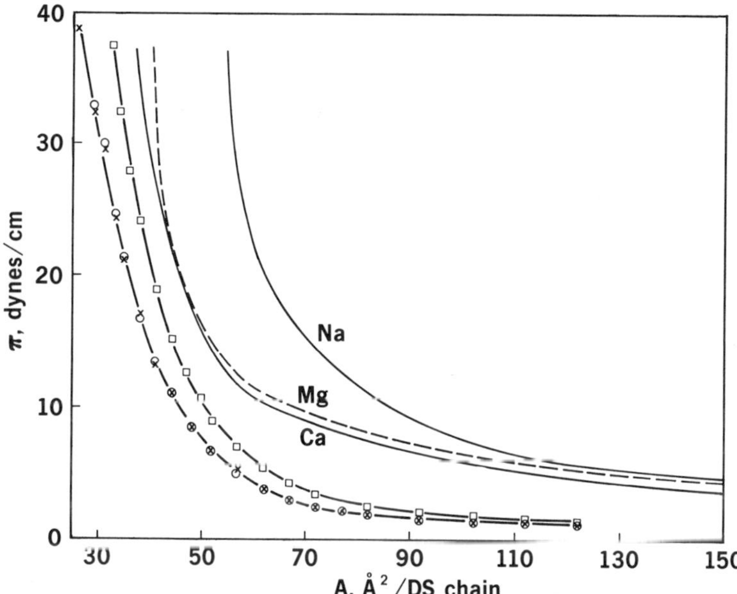

Figure 3. π–A curves for DS monolayers. Bottom three curves represent spread monolayers on 4M subsolutions of $MgCl_2$ (□), $CaCl_2$ (○), and NaCl (×) at 25°C. Top three curves represent adsorbed monolayers at 40°C; Ca and Na (17) curves are temperature corrected (see *text).*

film at high areas and its higher limiting area of ~40 A²/molecule *vs.* 37.5 A²/molecule for the CaDS. These trends are supported by the spread monolayer results below.

We also compare the above film characteristics with those of NaDS absorbed from a simple, salt-free, aqueous solution. For this purpose we use the "composite" π–A curve of Pethica (*17*) calculated from the "2" form of the Gibbs adsorption equation—*viz.*,

$$d\pi = 2\Gamma\ RT\ d\ln a \tag{8}$$

For direct comparison with the MgDS and CaDS, the π–A curve for NaDS, which followed an equation of the form of Equation 7, was recomputed for 40°C on the assumption that the co-area term is invariant over the appropriate temperature range; the curve is presented in Figure 3. Pressures for the NaDS are somewhat higher than those for MgDS and CaDS in the high area range and markedly higher in the low area range. By the criteria (*2, 3*) of pressure at a particular area and of limiting area of these charged monolayers, the interaction sequence is $Ca^{2+} \geq Mg^{2+} >> Na^{+}$.

Considerable discussions question which form of the Gibbs equation should be applied to uni-univalent surface-active electrolytes in simple aqueous solution. The essential point is that if hydrogen ions compete to any appreciable degree with the electrolyte cation (*e.g.*, Na^+) for positions in the double layer, then the coefficient n of Equation 8 would not be 2 but would move toward unity. Thus, all areas computed from the "2" form of Equation 8 would be too high, and the "correct" curve of NaDS in Figure 3, for example, would be shifted to the left. The correction would be progressively less as surface pressure increased, corresponding to more concentrated solutions of the surface-active electrolyte.

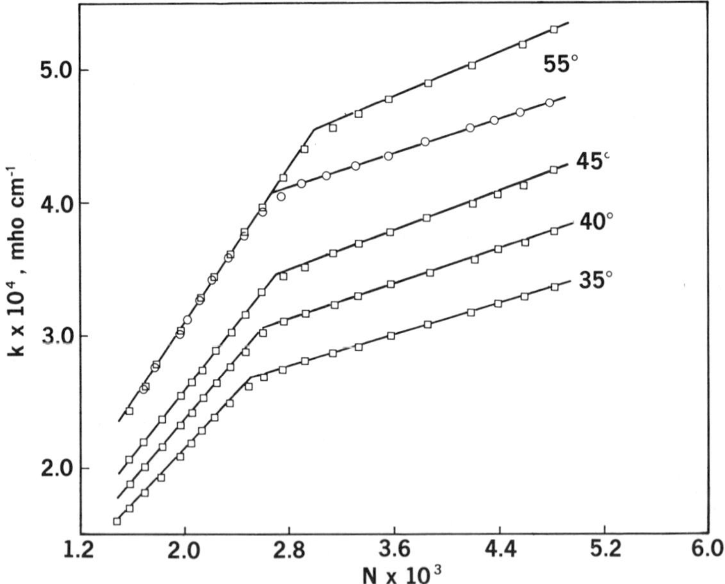

Figure 4. Specific conductivity–normality plots for MgDS (□) and CaDS (○)

Bujake and Goddard (*18*), however, established that "hydrolysis" is far less than was originally supposed and that it is probably not serious above a concentration of $\sim 10^{-4} M$. Thus most of the π–A curve for NaDS presented in Figure 3, especially the section above $\pi = \sim 10$ dynes/cm, is reliable. Because of the relative enrichment of bivalent counterions over monovalent counterions in the double layer, conditions are correspondingly less favorable for "hydrolysis" with CaDS and MgDS, implying that the "no hydrolysis" form of Equation 4 will apply to the concentration range in the present work.

Figure 3 also includes the isotherms of DS monolayers spread on strong salt solutions. This technique was used first by Brady (*19*) to study the properties of soluble surface-active electrolytes. An interesting feature is that, within experimental error, the π–A isotherms for DS on $CaCl_2$ and NaCl coincide. Under these conditions of swamping electrolyte concentration, the usual exponential expression, $c^+_s = c^+_b \exp(-z_+ e\psi_0/kT)$, employed by Cassie and Palmer (*20*) and others to relate the surface and bulk concentrations of counterion, c^+_s and c^+_b, in terms of the surface potential, ψ_0, no longer has validity. This emphasizes the fact that the observed differences between the MgDS and CaDS are not explicable on simple electrical grounds. Note that the curve for MgDS is again the more expanded, *i.e.*, in a direction opposite to that expected if solubility of the spread monolayers were involved, and the limiting area value is some 15% higher than the value for CaDS.

Unsuccessful attempts were made to fit a single equation of state to the spread monolayer π–A data. The methods used included Fowkes' plots (*21*) and plots of various functions of π and A. For example, a graph of π *vs.* πA revealed two linear segments that intersected around $\pi = 15$ dynes/cm. A similar "two-segment" pattern was found by Pethica (*17*) for $1/\pi$ *vs.* A for DS monolayers spread on strong $(NH_4)_2SO_4$ solutions. In the present case the equations fitted to the high (h) and low (l) pressure segments of the πA *vs.* π plots are for MgDS and CaDS:

(h)	$\pi\,(A - 21.5) \approx 400 = 1\,kT$	(l)	$\pi\,(A - 38) \approx 100 = 1/4\,kT$
	$\pi\,(A - 16.6) \approx 400 = 1\,kT$		$\pi\,(A - 35) \approx 100 = 1/4\,kT$

Both the low A_o terms in the h segments and low pressure, area products in the l segments suggest the formation of aggregates in these spread monolayers.

Electrical conductivity results for MgDS at 35°, 40°, 45°, and 55°C and for CaDS at 55°C are presented as specific conductance (k) *vs.* normality (N) plots in Figure 4. They exhibit behavior typical of long chain electrolytes—*viz.*, two linear segments exhibiting a break and a considerable reduction in slope as the concentration is increased through the CMC. Likewise, equivalent conductivity (Λ) *vs.* $\sqrt{N}$ plots, not shown, have a sharp drop in conductance in this region. Note the virtually identical conductance values of the CaDS and MgDS below the CMC at 55°C and the appreciably lower slope of the k–N plot (higher slope of the Λ–$\sqrt{N}$ plot) for CaDS above the CMC. Unfortunately, comparative data,especially the effect of temperature, on the conductivity of Ca^{2+} and Mg^{2+} are meager. However, data for $Ca(NO_3)_2$ and $Mg(NO_3)_2$ show that the difference of *ca.* 6% in Λ values at 25°C is reduced to *ca.*

2% at 50°C (*22*). The approximately 6% greater values of $\Lambda_{\frac{1}{2}Ca}$ over $\Lambda_{\frac{1}{2}Mg}$ at 25°C have been confirmed for the chloride salts (*23*).

Regarding the second point above, we used the slope differences above the CMC to estimate the difference in relative binding of Mg^{2+} and Ca^{2+} to the micelle by a method similar to that developed by Evans (*24*). If the micelle is composed of p long chain ions, X, and q univalent counterions, Me, where Me = Na^{+}, ½ Mg^{2+}, ½ Ca^{2+}, etc., the aggregation process can be represented by

$$p\ \mathrm{MeX} \rightleftharpoons p\ \mathrm{Me}^{+} + p\mathrm{X}^{-} \rightleftharpoons (\mathrm{X}_p\mathrm{Me}_q)^{-(p-q)} + (p-q)\ \mathrm{Me}^{+} \qquad (9)$$

The slope, S_2, of the k *vs.* N plot above the CMC can be expressed as

$$1000\ S_2 = \frac{1}{p}\ \Lambda_{\mathrm{mic.}} + \frac{p-q}{p}\ \Lambda\mathrm{Me} \qquad (10)$$

If the conductance of the ions and micelles is proportional to their charge and inversely proportional to their charge and inversely proportional to their radius, then

$$\frac{\Lambda_{\mathrm{mic}}}{\Lambda_{\mathrm{x}}} = \frac{p-q}{1} \times \frac{1}{p^{1/3}} \qquad (11)$$

Equation 10 becomes

$$1000\ S_2 = \frac{p-q}{p}\left(\frac{\Lambda_{\mathrm{x}}}{p^{1/3}} + \Lambda\mathrm{Me}\right) \qquad (12)$$

Since $\Lambda_{\frac{1}{2}Mg} \approx \Lambda_{\frac{1}{2}Ca}$, and assuming that $p_{\mathrm{MgDS}} = p_{\mathrm{CaDS}}$, we have

$$\frac{(p-q)_{\mathrm{CaDS}}}{(p-q)_{\mathrm{MgDS}}} = \frac{(S_2)_{\mathrm{CaDS}}}{(S_2)_{\mathrm{MgDS}}} = 0.77 \qquad (13)$$

Thus, the effective charge of the CaDS micelle emerges as three-quarters that of the MgDS micelle at 50°C; computation of relative rather than absolute values reduces the effect of possible errors in the theoretical analysis.

The results of the CMC determinations are given in Table I which also includes the data of Miyamoto (*10*). As is generally found, the CMC values obtained from the k *vs.* N plots are slightly, but definitely, higher than those from the $\Lambda - \sqrt{N}$ plots which show satisfactory agreement with the $\gamma - \ln M$ derived values at the two temperatures where comparison is possible. Furthermore, agreement with the values of Miyamoto obtained by surface tension and dye solubilization is satisfactory for CaDS but not so for MgDS.

Utilizing the specific conductivity–temperature data and the following expression (*25*)

$$\Delta H_{\mathrm{m}} = -\frac{3}{2} RT^2 \frac{d\ln(\mathrm{CMC})}{dT} \tag{14}$$

leads to −2.20 kcal (35°–45°C) and −3.18 kcal (45°–55°C) for the heat of micelle formation, ΔH_m, of MgDS, expressed per mole of long chain ion. Unfortunately, no comparable values are available or can be deduced from the present data for CaDS.

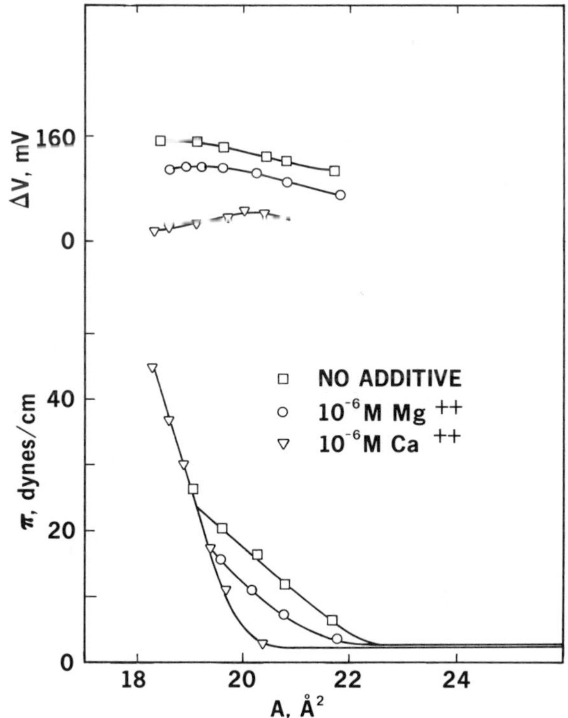

Figure 5. π–A and ΔV–A characteristics of stearic acid monolayers on 0.01M NaCl subsolutions at pH 10 (NaOH) containing small amounts $MgCl_2$ and $CaCl_2$ (32)

Summation and Comparison with Other Data

Results of this work consistently point to a stronger interaction between calcium than magnesium with the dodecyl sulfate ion. The evidence (referring to calcium) includes:

(a) A lower critical micelle concentration at the same temperature

(b) A higher degree of binding to the DS micelle

(c) A lower surface tension in the plateau region of the γ *vs* log M plot

(d) Less expanded monolayers, both spread and absorbed, at the same temperature, and lower A_0 values

(e) Lower solubility and higher Krafft point of the long chain salt.

Considering published information on the interaction of alkaline earth metals with long chain anions, we turn first to ionized fatty acids—the most widely investigated systems of this type. That calcium soaps are less soluble than magnesium soaps is well known; the strong interaction of calcium ions with carboxylate monolayers was recognized by Langmuir (*26*) and by many subsequent investigators who used direct chemical (*27*), radiochemical (*28*), and infrared analyses (*29, 30*). Webb and Danielli (*31*), for example, found that the amount of calcium associated with a palmitate film is far higher than can be accounted for by a Gibbs–Donnan equilibrium mechanism. Goddard and Ackilli (*32*) demonstrated the extreme sensitivity of stearate monolayers to trace quantities of Mg^{2+} and Ca^{2+}; both the ability to condense the monolayer and to alter its surface potential showed that calcium ions clearly had the greater tendency to associate with the monolayer [*see* Figure 5 (Ref. 32)]. Of special interest was that both Ca ions and Mg ions led to a lower (more negative) potential; this is the opposite of what is expected on electrostatic grounds and indicates a gross reorientation or change of electronic configuration of the carboxylate group on interaction with these alkaline earth metal ions. This subject has been recently investigated by Deamer, Meek, and Cornwell (*33*), who used monolayer and infrared data for stearic acid films spread on substrates containing Be, Mg, Ca, Sr, or Ba salts. Infrared spectroscopy titration data gave the following order of stability constants of the stearates: Be $>$ Ca $>$ Mg $>$ Sr $>$ Ba; data from condensation of the monolayers gave: Ca $>$ Mg $>$ Sr $>$ Ba although differences in sequence were observed if the subsolution pH was below 9. The departure of Ca from the series Be $>$ Mg $>$ Ca $>$ Sr $>$ Ba, predicted on the basis of electronegativity and ionic radius, was explained (*33*) by a "geometric factor"—the possession by calcium of a favorable ionic radius for binding to the carboxylate ion.

Abrahamson *et al.* (*34*) observed a similar departure from the latter sequence of calcium and magnesium in their reaction with a phosphatidic acid dispersion at pH 7. Shah and Schulman (*35*) also found a higher tendency of Ca^{2+} than Mg^{2+} to interact with a monolayer of dicetyl phosphate at pH 5.6 as judged by surface pressure measurements although surface potential measurements indicated little difference between the two ions as did potential measurements on phosphatidyl serine mem-

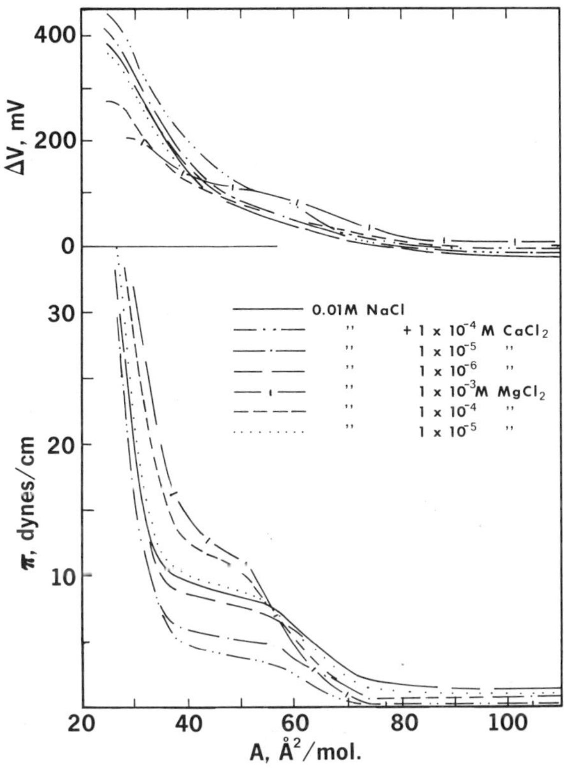

Figure 6. Effect of Ca and Mg ions on π–A *and* ΔV–A *isotherms of nonadecylbenzene-sulfonate monolayer isotherms; subsolutions contain 0.01*M *NaCl* (41)

branes (*36*). From the limited available data, a direct comparison of the influence of Ca^{2+} and Mg^{2+} on phosphatidylcholine monolayers indicates little difference in the effect of the two ions (*37*). The proximity of the charged nitrogen in the choline groups undoubtedly modifies the interaction of these cations with the phosphate groups in the monolayer in this case. In this respect results (*38, 39, 40*) obtained with ortho-, di-, tri-, and various adenosine phosphates in solution are of interest in that they generally indicate stronger binding of Mg than Ca. One infers that the presence of an alkyl group and the orientation it imposes in the various long alkyl chain phosphates (referred to above) may be responsible for modifying the inter-ionic interaction from that observed with phosphates in solution.

We turn finally to the effect of Ca^{2+} and Mg^{2+} on monolayers of the long chain sulfonate, nonadecylbenzenesulfonate (*41*). The influence of small amounts of $MgCl_2$ and $CaCl_2$ added to a subsolution of 0.01*M* NaCl is shown in Figure 6 (*41*). Nonadecylbenzenesulfonate, spread on 0.01*M* NaCl, gives an expanded monolayer which undergoes a transition to a condensed state at an area/molecule between 50 and 60 A^2. In the high area region, the presence of Mg^{2+} or Ca^{2+} leads to effects predictable on electrical grounds—*i.e.*, a lowering of the surface pressure and an increase in the surface potential. Below the transition points there is a marked difference in the counterion effect; Ca^{2+} leads to a decrease in surface pressure and an increase in surface potential; Mg^{2+} does just the opposite, and these effects increase with concentration of the added salts. The lower pressures observed with Ca^{2+} are again consistent with a higher degree of association between this ion and the sulfonate ion of the monolayer. Similar effects have been obtained by Dreher and Wilson (*42*).

It is tempting to explain these phenomena in terms of the size of Ca and Mg ions in solution. Nightingale (*43*), for example, lists their Stokes radius values as 3.10 and 3.47 A and their hydrated radii as 4.12 and 4.28 A. The influence of size would be seen chiefly in the low area region of the isotherm where physical restrictions on the inhabitation by the long chain ion and counterion of the surface layer are to be expected; it is particularly significant that the expansion effect caused by Mg ions can be seen even at very low concentrations of added ion. This expansion illustrates not only the ion size effect but also the substantial attraction the sulfonate group has for the Mg ion. The lower (more negative) surface potential observed with the Mg^{2+} is consistent with a structure in which more of these counterions are positioned, on the average, further below the sulfonate groups than are the calcium ions. The propensity of Mg ions to hydrate is also illustrated by the hydrated state of the MgDS salt used in these studies.

On the other hand, despite the information about long chain sulfates, sulfonates, phosphates, and carboxylates that indicates stronger interaction with Ca^{2+} than with Mg^{2+} (*i.e.*, in apparent harmony with the sequence of the Hofmeister (*44*) series), several difficulties remain. For example, while Miyamoto's data for DS (*10*) indicate the interaction sequence Mg $<$ Ca $<$ Sr $<$ Ba from solubility measurements (as well as from temperature/CMC measurements if one accepts the Mg–Ca sequence of the present paper), this sequence, with the exception of the position of Mg and Ca, is the opposite of that found by Deamer *et al.* (*33*) from condensation effects on the force/area curves of ionized fatty acids. At the same time, the ion sequence obtained by these authors from phase transition temperatures of spread fatty acids (*33*) differs from that deduced from the above-mentioned condensation effects, and the latter depended strongly on pH. Lastly, definite differences in ion sequence effects exist for the alkaline earth metals in their interaction with long chain phosphates from those found with phosphates in solution (*38, 39, 40*).

One must conclude as for the monovalent counterion sequences referred to in the Introduction that interaction between bivalent cations and anions is a complex phenomenon not determined solely by simple electrostatic considerations and ion size. Of the current hypotheses that explain ion sequences, that of Eisenman (*45*), based on anion field strength and recently expanded to include divalent cation effects (*46, 47*), seems the most comprehensive. However, we believe a fundamental understanding of these systems awaits better knowledge of the role of such factors as water structure, hydration, and geometric effects, including orientation associated with the presence of an alkyl group and of pH. Studies on a more extensive range of cations are required for all the cited systems, with the possible exception of those based on fatty acids.

Acknowledgments

The author thanks Patricia Hunt for her assistance with the measurements; T. D. Heyes for preparing the dodecyl sulfate salts, and H. C. Kung for confirmatory studies on the spread monolayers.

Literature Cited

1. Davies, J. T., *Proc. Roy. Soc., London, 1951,* **A208,** 224.
2. Goddard, E. D., Kao, O., Kung, H. C., *J. Colloid Interface Sci.* (1968) **27,** 616.
3. Goddard, E. D., Kao, O., Kung, H. C., *J. Colloid Interface Sci.* (1967) **24,** 297.
4. Harned, H. S., Owen, B. B., "The Physical Chemistry of Electrolytic Solutions," Reinhold, New York, 1950.
5. Goddard, E. D., Jones, T. G., Harva, O., *Trans. Faraday Soc.* (1953) **49,** 980.
6. Lange, H., *Kolloid Z.* (1951) **121,** 66.
7. Mukerjee, P., Mysels, K. J., Kapauan, P., *J. Phys. Chem.* (1967) **71,** 4166.
8. Weil, I., *J. Phys. Chem.* (1966) **70,** 133.
9. Feinstein, M. E., Rosano, H. L., *J. Colloid Interface Sci.* (1967) **24,** 73.
10. Miyamoto, S., *Bull. Chem. Soc. Japan* (1960) **33,** 371, 375.
11. Dreger, E. E., Keim, G. I., Miles, G. D., Shedlovsky, L., Ross, J., *Ind. Eng. Chem.* (1944) **36,** 610.
12. Harkins, W. D., Jordan, H. F., *J. Amer. Chem. Soc.* (1930) **52,** 1751.
13. Adam, N. K., Pankhurst, K. G. A., *Trans. Faraday Soc.* (1946) **42,** 523.
14. Hutchinson, E., Manchester, K. E., Winslow, L., *J. Phys. Chem.* (1954) **58,** 1127.
15. Goddard, E. D., Benson, G. C., *Trans. Faraday Soc.* (1956) **52,** 409.
16. Shedlovsky, T., McInnes, D., *J. Amer. Chem. Soc.* (1937) **59,** 503.
17. Pethica, B. A., *Trans. Faraday Soc.* (1954) **50,** 413.
18. Bujake, J. E., Goddard, E. D., *Trans. Faraday Soc.* (1965) **61,** 190.
19. Brady, A. P., *J. Colloid Sci.* (1949) **4,** 417.
20. Cassie, A. B. D., Palmer, R. C., *Trans. Faraday Soc.* (1941) **37,** 156.

21. Fowkes, F. M., *J. Phys. Chem.* (1962) **66,** 385.
22. "International Critical Tables," Vol. 6, p. 238, 1929.
23. Shedlovsky, T., Brown, A. S., *J. Amer. Chem. Soc.* (1934) **56,** 1066.
24. Evans, H. C., *J. Chem. Soc.* (1956) 579.
25. Goddard, E. D., Benson, G. C., *Can. J. Chem.* (1957) **35,** 986.
26. Langmuir, I., Schaefer, V. J., *J. Amer. Chem. Soc.* (1936) **58,** 284.
27. Havinga, E., *Rec. Trav. Chim.* (1952) **71,** 72.
28. Sobotka, H., Demeny, M., Chanley, J. D., *J. Colloid Sci.* (1958) **13,** 565.
29. Bagg, J., Abrahamson, M. B., Fichman, M., Haber, M. D., Gregor, H. P., *J. Amer. Chem. Soc.* (1964) **86,** 2759.
30. Ellis, J. W., Pauley, J. L., *J. Colloid Sci.* (1964) **19,** 755.
31. Webb, J., Danielli, J. F., *Nature* (1950) **146,** 197.
32. Goddard, E. D., Ackilli, J. A., *J. Colloid Sci.* (1963) **18,** 585.
33. Deamer, D. W., Meek, D. W., Cornwell, D. G., *J. Lipid Res.* (1967) **8,** 255.
34. Abrahamson, M. B., Katzman, R., Gregor, H. P., Curci, R., *Biochemistry* (1966) **5,** 2207.
35. Shah, D., Schulman, J. H., *J. Lipid Res.* (1965) **6,** 341.
36. McLaughlin, S. G. A., Szabo, G., Eisenman, G., *J. Gen. Physiol.* (1971) **58,** 667.
37. Anderson, P. J., Pethica, B. A., *Int. Conf. Biochem. Prob. Lipids Ghent.* (1955) 24.
38. Smith, R. M., Alberty, R. A., *J. Amer. Chem. Soc.* (1956) **78,** 2376.
39. Lambert, S. M., Watters, J. I., *J. Amer. Chem. Soc.* (1957) **79,** 5606.
40. Lambert, S. M., Watters, J. I., *J. Amer. Chem. Soc.* (1959) **81,** 3201.
41. Goddard, E. D., Kung, H. C., *J. Colloid Interface Sci.* (1971) **37,** 585.
42. Dreher, K. D., Wilson, J. E., *J. Colloid Interface Sci.* (1970) **32,** 248.
43. Nightingale, E. R., *J. Phys. Chem.* (1959) **63,** 1381.
44. Mysels, K. J., "Introduction to Colloid Chemistry," Interscience, New York, 1959.
45. Eisenman, G., *Biophys. J.* (1962) **2,** 259.
46. Eisenman, G., *Proc. Intern. Congr. Physiol. Sci., 23rd,* Excerpta Medica International Congress Series No. 87, p. 489, 1965.
47. Sherry, H. S., in "Ion Exchange," J. A. Mirensky, Ed., Vol. 2, p. 89, Marcel Dekker, New York, 1969.

Received October 18, 1974.

7

Salt Effects in Interfacial Films Formed from Nonionic Surfactants

R. BUSCALL[1] and R. H. OTTEWILL

School of Chemistry, University of Bristol, Bristol, BS8 1TS, England

The equilibrium properties of foam films formed from aqueous solutions of decylmethyl sulfoxide have been studied in the presence of sodium chloride and potassium thiocyanate. Stable films were formed whose thicknesses depended on the electrolyte concentration. As the electrolyte concentration was increased, a sudden increase in film thickness occurred but gradually decreased with further electrolyte addition. Examination of the electrophoretic mobility of dodecane droplets stabilized by decylmethyl sulfoxide showed an increase in mobility at about the same concentration. These data indicated that the thicker foam films were charge stabilized owing to the adsorption of the anions. The surface pressures and surface potentials of monolayers of octadecyl sulfoxide were also investigated.

Aqueous foam films stabilized by long chain, surface active agents are useful model systems for investigating surface forces. Such films consist of two monolayers of surface active agent separated by an aqueous core. Most studies have been carried out with ionic surface active agents, and the literature has been extensively reviewed (*1*, *2*). Only a few studies have been reported using films stabilized by nonionic surface active agents, although these materials can form quite stable foam films. Donaldson (*3*) studied in detail films prepared from dodecyldimethylamine oxide at low pH values where the surface active agent was positively charged and at high pH where the molecules were uncharged. Films stabilized by decylmethyl sulfoxide (DMS) have been investigated

[1] Present address: Department of Pharmacy, University of Aston in Birmingham, Gosta Green, Birmingham, B4 7ET, England.

by Clunie *et al.* (*4*) and by Ingram (*5*). It was found that inorganic electrolytes had a considerable influence on the equilibrium thickness of the film. At low concentrations of 1:1 electrolytes, the film thickness (*ca.* 50 A) was almost independent of electrolyte concentration; the films were considered to be in an equilibrium state in a potential energy primary minimum (second black film). However at a particular concentration of salt, the film thickness increased, and with further increases in salt concentration it decreased gradually in the manner expected for a charge stabilized film; the charge was considered to arise from the adsorption of anions to the film surface. These films were considered to be stabilized in a secondary minimum of potential energy and were classified as first black films (*5*).

A systematic study of the influence of salts on foam films formed from nonionic surface active agents was carried out in these laboratories (*3, 6*). This paper reports an investigation of the effects of sodium chloride and potassium thiocyanate on the thickness of foam films formed from DMS. In addition to measurements on films, the electrophoretic mobilities of dodecane droplets stabilized with DMS were determined as a function of salt concentration, and the properties of insoluble monolayers of octadecylmethyl sulfoxide (OMS) at the air–water interface have been examined using the classical methods largely developed by N. K. Adam (*7*).

Experimental

Materials. Fresh water doubly-distilled from a borosilicate glass apparatus was used. The sodium chloride (Analar material) was roasted to remove surface active contaminants. Analar potassium thiocyanate was recrystallized from water and was washed with diethyl ether.

Pure decylmethyl sulfoxide (DMS) was kindly supplied by T. Walker (*8*) of Procter and Gamble Ltd.

Octadecylmethyl sulfoxide (OMS) was prepared in the following manner: 15 g of octadecyl bromide and 3.5 g of thiourea were dissolved in 150 cm^3 of ethanol. The mixture was refluxed for 1 hr after which 100 cm^3 of ethanol were removed by distillation. The residue, after cooling, was made alkaline by an ethanol/sodium hydroxide solution (5 g sodium hydroxide in a minimum volume of ethanol). Methyl iodide, 3.7 g, was then added, and the mixture was left to stand overnight. The resulting thioether was recovered by filtration and washed with cold ethanol. The thioether was dissolved in glacial acetic acid; then small quantities of 20 vol % hydrogen peroxide, constituting the stoichiometric amount, were added over several hours. The product was precipitated with water, and the precipitate was collected by filtration. The crude product was washed successively with water, ice-cold ethanol, and a small quantity of diethyl ether; it was finally dried at 70°C. OMS was purified by successive recrystallization from benzene; purification was continued until the product gave UV and IR spectra which showed no bands characteristic of a sulfone.

Measurement of Film Thickness. The apparatus used to measure the thickness of foam films, formed in the horizontal direction was that developed by Donaldson (*3*). The films were supported by a stainless steel frame which was made from a rectangular block by boring two concentric countersunk holes, one from each direction, so that a thin ring *ca.* 4 mm in diameter was left in the center. The steel frame was mounted on a hollow Perspex rod; a thermistor was inserted in the hollow center to monitor the temperature of the frame. A cell was built to house the film holder; it consisted of a glass bulb, approximately 10 cm^3 capacity, with an optical glass window at the top. The Perspex rod on the foam film holder was inserted into a neck in the side of the glass cell by an inert plastic sleeve which formed a vapor-tight seal. It was possible with this device to rotate the film frame. The glass cell in turn was mounted in a Perspex box which was double-jacketed to allow thermostatic control. Films were formed by dipping the stainless steel frame into a surface active agent solution and allowing the excess liquid to drain off. The frame was then inserted into the cell. A small pool of surface active agent solution was kept in the bottom of the cell to maintain a saturated atmosphere.

The film thickness was determined from measurement of the intensity of light reflected from the film at near to normal incidence. A wavelength of 5095 A was selected using the appropriate filter. The procedure of measuring the ratio of the intensity of the reflected light from the first bright fringe and from the black film was adopted (*9*). Assuming that the film was a triple layer of hydrocarbon–water–hydrocarbon and applying the treatment of Cabellero (*10*) the overall thickness of the film was given by

$$d = 513.6 \left[\frac{I_R}{I_{max}}\right]^{1/2} + 1.9 \text{ A} \quad (1)$$

where I_R = the intensity of reflection from the black film and I_{max} = the reflected intensity from the first bright fringe both after correction for background intensity. The total film thickness is also given by: $[d = 2d_1 + d_2]$ where d_2 = the thickness of the aqueous core, and d_1 = the length of the hydrocarbon chain of the surface active agent. Measurement of the latter value from Catalin models gave 13 A, but taking the head group area as 30 A^2 (*see* later) and calculating d_1 from the molar volume gave 10.7 A. This value was used in Equation 1 to calculate d together with a value of 1.4119 for the refractive index of the hydrocarbon layers and 1.333 for the water layer in the film. At least three measurements were made on each solution and the mean value was taken. The reproducibility was within $\pm$ 5%.

Microelectrophoresis. Mobility measurements were made using an apparatus similar to that described by Alexander and Saggers (*11*) but modified to use slit ultramicroscope illumination. The cell was of the Mattson type (*12*) and was mounted in a water bath. Refraction effects in the cell were allowed for by applying the Henry correction (*13*). All mobility measurements were made at the stationary level in the cell at room temperature, 20° ± 0.5°C.

The emulsions for electrophoresis measurements were prepared by adding one drop of dodecane to about 20 cm^3 of a DMS solution and by subjecting the mixture to ultrasonic treatment for 30 sec. The emulsion was then aged overnight and remixed by shaking just prior to use. This procedure produced drops approximately 1–2 μm. Zeta potentials (ζ) were calculated from the Smoluchowski equation,

$$u = \frac{\varepsilon\zeta}{4\pi\eta} \tag{2}$$

where ϵ = dielectric constant of the medium, η = viscosity, and u = electrophoretic mobility.

Surface Pressure and Surface Potential Measurements. The surface balance was Langmuir–Adam type with a glass trough as previously described (*14*). The sensitivity of the balance in terms of surface pressure was 0.1 dyne/cm.

Surface potentials were measured using an Americium 241 electrode (Nuclear Radiation Developments Ltd., type A 1011) which was surrounded by an earthed copper mesh cage. A silver–silver chloride wire electrode was the reference electrode; the potential diffrence between the two electrodes was measured with a Vibron electrometer. The apparatus was housed in an earthed dust-free box.

The OMS solution was prepared in benzene and added to the surface by an Agla micrometer syringe. Two to 3 min were allowed for evaporation of the solvent before readings were taken.

Results

Foam Film Measurements. Figure 1 shows the results obtained for the thickness of foam films drawn from solutions of DMS containing sodium chloride at various concentrations. At low salt concentrations, below 10^{-2} mole/dm^3 (dm = decimeter), films 60 A thick were formed (second black films). This thickness is slightly greater than that of 50 A reported by Ingram (*5*). The difference could in part arise from the choice of optical model for the calculations although for the second black films of sodium dodecyl sulfate our values are in exact agreement with theirs. In addition our measurements were made on horizontal films, and Ingram's were made on vertical films (*5*) where a gravitational component would be experienced.

At a sodium chloride concentration of 10^{-2} mole/dm^3, two coexisting films were observed; the thicker, first black, was *ca.* 85 A thick. With a further increase in salt concentration the thickness of the first black film decreased until it reached 60 A at a salt concentration of *ca.* 5×10^{-1} mole/dm^3—*i.e.*, it became a second black film.

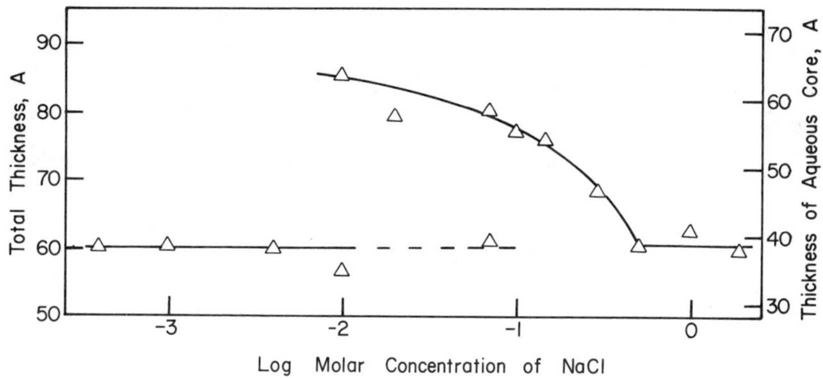

Figure 1. Thickness of DMS foam film formed from sodium chloride solutions at various concentrations

The equilibrium thickness of foam films drawn from solutions of DMS containing potassium thiocyanate at various concentrations are shown in Figure 2. In this case first black films were formed with a thickness of ~ 975 A at a salt concentration of 4×10^{-4} mole/dm³. The thickness decreased with increasing salt concentration and reached 59 A, those of a second black film, at salt concentrations of the order of 0.5 mole/dm³ potassium thiocyanate.

Effect of Temperature. The thickness of foam films drawn from a 10^{-3} mole/dm³ sodium chloride solution was also investigated as a func-

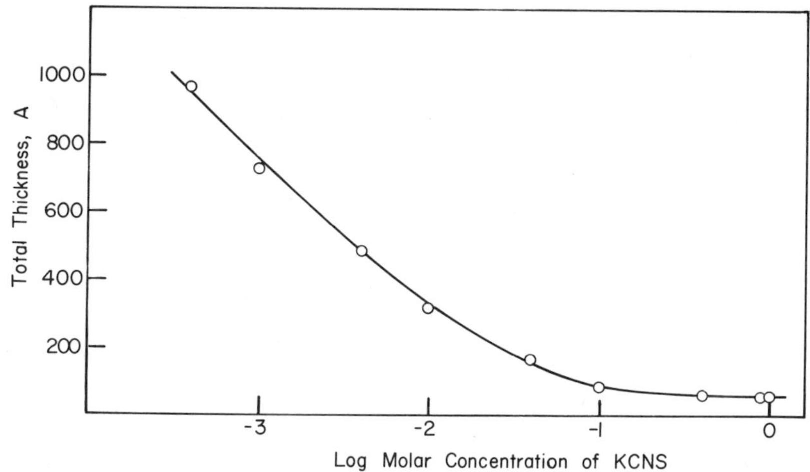

Figure 2. Thickness of DMS foam film formed from potassium thiocyanate solutions at various concentrations

tion of temperature. The results are given in Figure 3. At this salt concentration, second black films are formed above 24°C and thick, first black films, below 22°C. The thickness of the latter films ($\sim$ 600 A) did not change markedly over the temperature range 0°–22°C. A similar effect was observed with a 10^{-4} mole/dm^3 sodium chloride solution where the thickness transition occurred between 12°–13°C; the films below the transition temperature were *ca.* 950 A thick (*6*). The increase in thickness of the film with decreasing electrolyte to a thickness of 10^3 A suggests that at lower temperatures the behavior of the films with various sodium chloride concentrations would resemble that obtained with potassium thiocyanate at 25°C. This also implies that the transition temperature for the potassium thiocyanate systems was above 25°C even at low electrolyte concentrations, but this point has not been investigated.

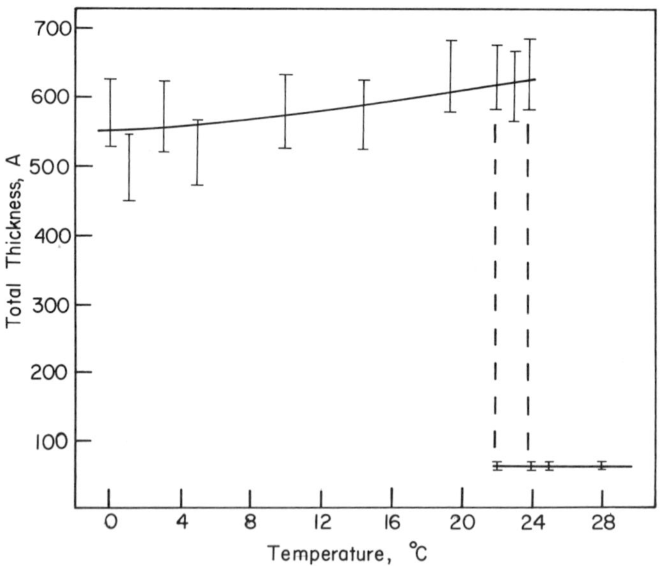

Figure 3. Thickness of DMS foam films formed from solutions containing 10^{-3} mole/dm^3 sodium chloride as a function of temperature

Electrophoretic Mobility. To estimate the possible charge on the sulfoxide film surface, we investigated the effects of sodium chloride and potassium thiocyanate on the electrophoretic mobility of dodecane droplets stabilized with DMS (*see* Figure 4). In the sodium chloride system the droplets were essentially uncharged at low concentrations, but between 3–5 $\times$ 10^{-3} mole/dm^3, the mobility increased to -1.2 mμ cm/V sec. In the potassium thiocyanate system, the droplets had a mobility of

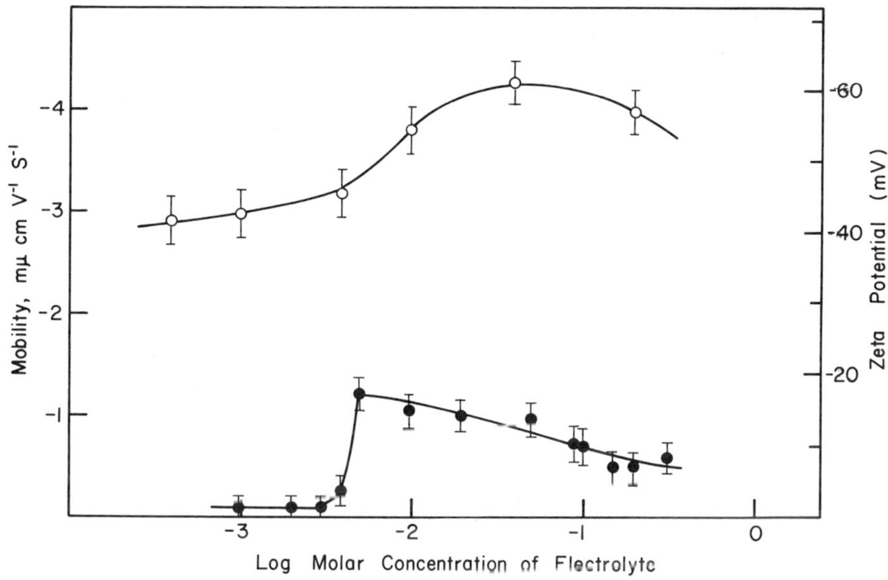

Figure 4. Electrophoretic mobility vs. *log. Molar concentration of electrolyte for DMS stabilized dodecane droplets (●) in sodium chloride solutions, (○) in potassium thiocyanate solutions.*

$-3m\mu$ cm/V sec at a salt concentration of 4×10^{-4} mole/dm³. The mobility increased to a maximum of $-4.30\ m\mu$ cm/V/sec at 4×10^{-2} mole/dm³ and then decreased.

The remarkable similarity between the sodium chloride concentration at which the mobility obtained a reasonable value and that at which a first black film was formed suggests that at this concentration chloride ions adsorb to the sulfoxide groups and thus produce a potential which can provide electrostatic repulsion between the monolayers in the soap film. In the case of potassium thiocyanate, adsorption appears to commence at a lower salt concentration, suggesting a stronger free energy of adsorption for the thiocyanate ion and consequently a greater extent of adsorption. Thus a higher potential would be obtained at the interface.

Monolayer Studies. Surface pressure *vs.* area curves for OMS spread on sodium chloride solutions are given in Figure 5. The range of ionic strengths corresponds with those used in the film measurements. However there is no apparent significant influence of ionic strength on the form of the curve. The films formed are condensed, indicating, the predominance of cohesion between the octadecyl chains. The curves of surface pressure (π) *vs.* area per molecule (A) resemble those obtained with stearic acid (7). The data at the lower surface pressures extrapolate to an area per

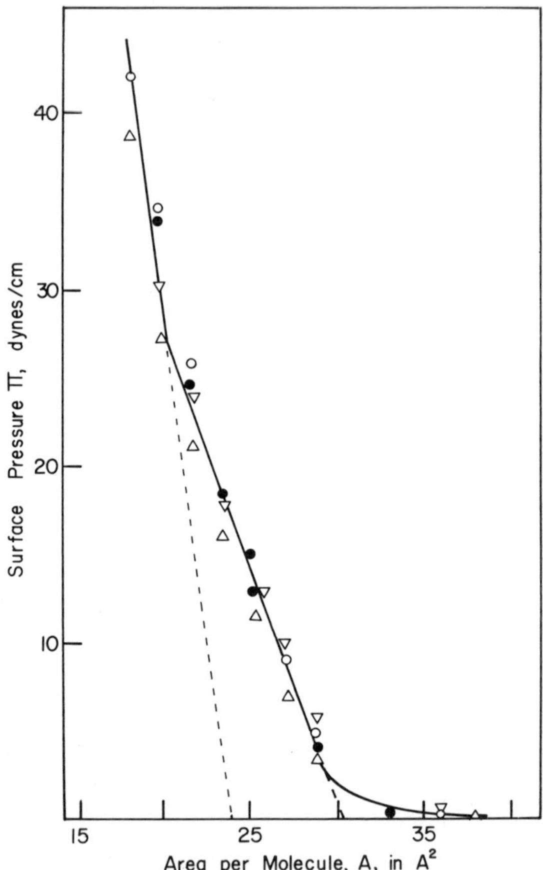

Figure 5. Curves of surface pressure vs. *area per molecule for films of OMS on sodium chloride solutions:* (○) 10^{-4} *mole/dm³;* (△) 10^{-3} *mole/dm³;* (●) 10^{-2} *mole/dm³;* (▽) 10^{-1} *mole/dm³*

molecule of 30.4 A^2 which corresponds to the head groups lying in the same plane. This value is in good agreement with that obtained from models and the values obtained by Ingram (5). At a surface pressure of *ca.* 27 dynes/cm, a change in $d\pi/dA$ occurs, and the head groups rearrange to give an area per head group of 24 A^2; this appears to correspond to a staggered packing of the head groups as first suggested by Adam to explain the π *vs.* A curves obtained for stearic acid monolayers (7).

The surface potential (ΔV) *vs.* area per molecule (A) curves obtained on sodium chloride solutions are shown in Figure 6. At areas per molecule greater than *ca.* 29 A² the curves show a marked dependence on ionic strength. However below that area, which corresponds to the formation of a close-packed monolayer, the curves at all the ionic strengths are coincident.

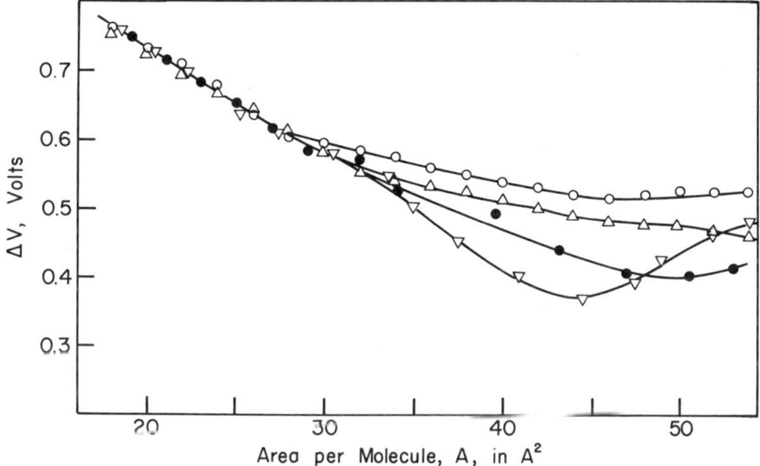

Figure 6. Curves of surface potential vs. *area per molecule for films of OMS on sodium chloride solutions:* (○) 10^{-4} *mole/dm³;* (△) 10^{-3} *mole/dm³;* (●) 10^{-2} *mole/dm³;* (▽) 10^{-1} *mole/dm³*

The surface pressure *vs.* area curves obtained on potassium thiocyanate solutions of different ionic strength are shown in Figure 7. In this case there is a clear influence of ionic strength on the packing of the molecules in the film, and the extrapolated areas per molecule for the first condensed region increase with increase in ionic strength as shown in Table I. The results seem to indicate an increase in repulsion between the head groups with increase in the concentration of electrolyte. The fact that this occurs with the thiocyanate ion and not with the chloride ion might indicate that some penetration of the thiocyanate ion between the head groups occurs possibly as a consequence of ion–dipole association between the thiocyanate ion and the sulfoxide dipole. This suggestion is supported by the ΔV *vs.* A data shown graphically in Figure 8. Unlike the results obtained on sodium chloride solutions, the curves show a definite trend with ionic strength at areas per molecule less than 29 A².

Discussion

Film Thickness. The measurement of the thickness of the horizontal foam films formed from DMS show that in the sodium chloride concentration region, 4×10^{-4}–10^{-2} mole/dm³, black films are formed with a

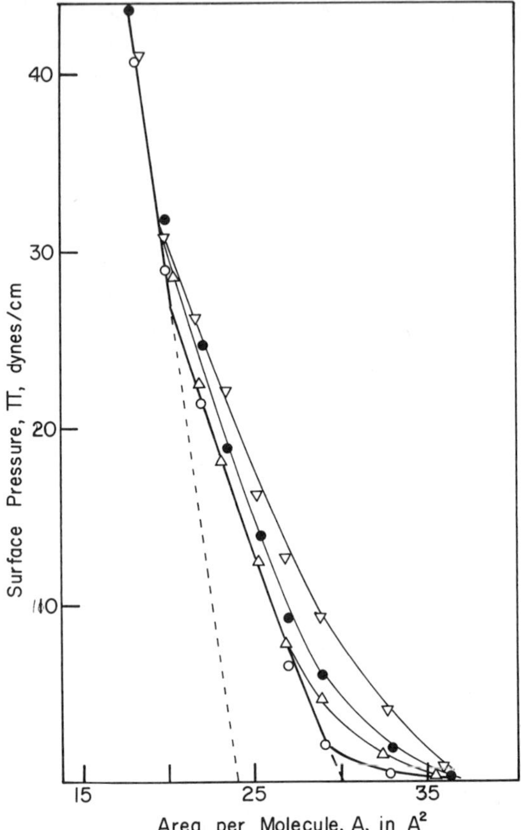

Figure 7. Curves of surface pressure vs. *area per molecule for films of OMS on potassium thiocyanate solutions:* (○) 10^{-4} *mole/dm*^{3}; (△) 10^{-3} *mole/dm*^{3}; (●) 10^{-2} *mole/dm*^{3}; (▽) 10^{-1} *mole/dm*^{3}

thickness of 60 ± 3 A. At 10^{-2} mole/dm^3 sodium chloride, a film of thickness 85 ± 3 A is also formed (*see* Figure 1). The film thickness decreases with an increase in sodium chloride concentration until *ca.* 0.5 mole/dm^3 salt, when the film again reaches a thickness of 60 A. The form of the curves is the same as that obtained by Clunie *et al.* (*4*) and Ingram (*5*). In agreement with these authors the lack of dependence of the 60-A film on salt concentration would suggest that this is a second black film. The formation of a thicker first black film at 10^{-2} mole/dm^3 salt suggests the onset of electrostatic repulsion within the film; the manner in which the thickness of the film decreases with increase in salt concentration also appears to confirm this.

Table I. Dependence of Close Packed Areas for OMS on Salt Concentration

Ionic Strength	*Limiting Area, A^2*
10^{-4}	30
10^{-3}	30
10^{-2}	31.5
10^{-1}	33

The films formed from potassium thiocyanate solutions behaved differently from those formed from sodium chloride solutions. At the lowest salt concentration examined, 4×10^{-4} mole/dm^3, the film had a thickness of 975 A, clearly a first black film. With increasing salt concentration the film thickness decreased and reached 60 A at a thiocyanate concentration of 5×10^{-1} mole/dm^3; it remained at this thickness as the salt concentration was increased to 1 mole/dm^3. This corresponded to the second black film thickness obtained with the films in sodium chloride solutions, and in this state the films were quite stable. The considerably thicker films formed in the presence of the thiocyanate ion indicate a stronger double layer repulsion effect than with chloride and hence a stronger adsorption of the thiocyanate ion to the film surface. These re-

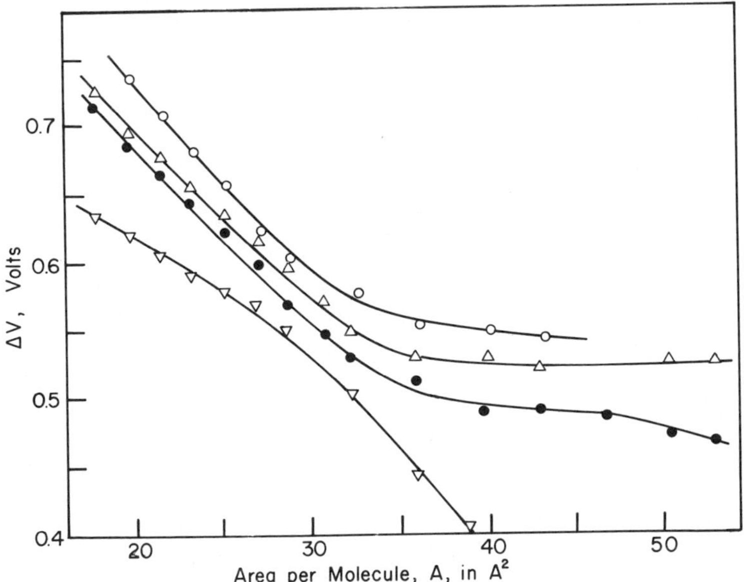

Figure 8. Curves of surface potential vs. *area per molecule for films of OMS on potassium thiocyanate solutions:* (○) *10^{-4} mole/dm^3;* (△) *10^{-3} mole/dm^3;* (●) *10^{-2} mole/dm^3;* (▽) *10^{-1} mole/dm^3*

sults indicate the dominance of the anion in determining the film thickness, an effect observed by Ingram (*5*) and Buscall (*6*).

The studies on the dependence of the film thickness on temperature (Figure 3) indicate that below 22°C in 10^{-3} mole/dm^3 sodium chloride solution, the film exists as a first black film and above this temperature as a second black film. The temperature sensitivity of the film thickness strongly suggests that the energy barrier between the first and second black film states is small and that a small decrease in temperature increases the ion binding of the chloride ion sufficiently to increase the electrostatic potential and consequently to increase the electrostatic repulsive energy. It also suggests that the energy of chloride ion binding to the film is small —*i.e.*, of the order of 1 kT.

Electrophoretic Mobility Studies. It is difficult but important to estimate the surface potential in the plane of the head group layers in a foam film as seen in Figures 1, 2, and 3. The influence of chloride and thiocyanate ions on the potential at the sulfoxide–water interface was studied by electrophoresis of dodecane droplets stabilized by DMS. The results in sodium chloride solutions (Figure 4) parallel those obtained on the foam films in that the zeta potential remained essentially zero until a salt concentration of 5×10^{-3} mole/dm^3 when it suddenly increased to -18 mV. The zeta potential then decreased gradually with increasing salt concentration. The potentials are quite low but adequate on the DLVO theory (*15, 16*) to ensure film stability. At a salt concentration of 5×10^{-3} mole/dm^3 the zeta potential (-18 mV) gives a charge density in the plane of shear of 2.96 $\mu C/cm^2$ corresponding to the presence of one anion in 540 A^2.

The results with the thiocyanate solutions indicate a fairly high zeta potential over the whole salt concentration range, the variation being between -40 and -60 mV. This is consistent with the considerably thicker foam films formed in the presence of thiocyanate ions and supports the idea that thiocyanate ions are more strongly bound to sulfoxide groups than chloride ions.

Monolayer Studies. The surface pressure *vs.* area curves indicate the formation of a condensed monolayer at low salt concentrations (10^{-4} mole/dm^3) with extrapolation to a close packed area per molecule of 30.4 A^2. At higher surface pressures, *ca.* 27 dynes/cm, a transition occurs to a closer packing of the head groups with the formation of a staggered array in the surface. With sodium chloride as the electrolyte the surface pressure *vs.* area curves appear independent of electrolyte concentration. With potassium thiocyanate in the substrate however an expansion of the film occurred with the close packed area per head group increasing from 29 A^2 in 10^{-4} mole/dm^3 to 32 A in 10^{-1} mole/dm^3 solution.

The surface potential *vs.* area curves obtained on sodium chloride solutions of different concentrations are independent of salt concentration for areas per head group less than 20 A². A dependence of surface potential on salt concentration is observed as soon as the area per head group becomes greater than that corresponding to a close packed monolayer.

The surface potential of a monolayer is given by (*17*)

$$\Delta V = 4\pi n\bar{\mu} + \psi \quad (3)$$

where ψ = the contribution of the electrical double layer, n = the number of dipoles per cm² of surface, and $\bar{\mu}$ = a composite dipole term for the molecules in the surface. The latter term is complicated because it includes contributions from the polar head group, the alkyl chain, water molecules, and ion pairs formed by association of ions from the electrolyte with the head group.

The results on sodium chloride solutions below $A = 29$ A² indicate that the arrangements of the dipolar parts of the molecules are the same at all chloride concentrations. The identity of the packing has already been inferred from the surface pressure curves. The zeta potential of the sulfoxide film at the oil–water interface was less than 20 mV at all electrolyte concentrations and thus, provided ψ has about the same magnitude, the variation of ψ would not easily be detected within the limits of the measurement of ΔV. The fact that ψ is small in the presence of chloride appears to be a reasonable assumption since the negative electrostatic potential in this case is not formed by the ionization of surface groups

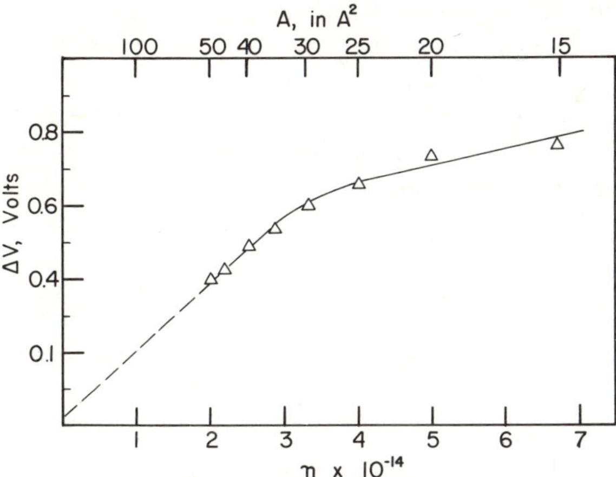

Figure 9. ΔV vs. n *for OMS films on* 10^{-1} *mole/dm³ sodium chloride solution*

but by the weak preferential adsorption of chloride ions on the sulfoxide head groups. A curve of ΔV *vs.* n for OMS on 0.1 mole/dm^3 sodium chloride shows two linear portions (Figure 9). If the ψ term is neglected, the initial slope for a film with the head groups occupying an area > 30 A^2 corresponds to $\bar{\mu} = 0.2 \times 10^{-18}$ esu, and the slope for $A < 30$ A^2 corresponds to $\bar{\mu} = 0.2 \times 10^{-18}$ esu. The reason for the change is not clear but could be connected with the rearrangement of both the surface active agent molecules and water molecules on forming a condensed monolayer. The dipole moments of dimethyl-, diethyl-, and dibutyl sulfoxide are all given as $3.9 \pm 0.1 \times 10^{-18}$ esu, and if OMS has a dipole of similar magnitude, it is clear that the composite surface dipole, $\bar{\mu}$, is appreciably smaller than the bulk value. The change in $\bar{\mu}$ at $A = 30$ A^2 could correspond also to a change in orientation of the dipole on forming a condensed film.

The surface potential *vs.* area curve obtained on 10^{-4} mole/dm^3 potassium thiocyanate solution appears almost identical to that obtained on 10^{-4} mole/dm^3 sodium chloride and has an identical slope in the low surface area region. The slopes of the curves in this region are also very similar for the results in 10^{-3} mole/dm^3 and 10^{-2} mole/dm^3 potassium thiocyanate but are displaced downward indicating an increase in negative potential ($-\psi$). The slope in 10^{-1} mole/dm^3 solution is smaller and is again moved in the negative direction. These data taken together with the expansion noted on the surface pressure curves indicate the interaction of the thiocyanate ion with the sulfoxide head groups, and the expansion would suggest some penetration of the thiocyanate ion between the head groups.

Both the surface pressure and the surface potential results confirm the electrophoretic measurements that binding of thiocyanate ions to the sulfoxide head groups is stronger than the binding of chloride ions. Approximate estimates of the binding energies give 1–2 kT for Cl^- and 6–7 kT for CNS^-. Consequently higher negative potentials are formed at the sulfoxide–water interface in the presence of thiocyanate ions than in the presence of chloride ions. As noted by Davies and Rideal (*17*) the order of polarization for anions at positively charged surfaces is,

$$CNS^- > I^- > Br^- > Cl^-$$

which suggests in the present case that the anions interact with the electropositive sulfur atom in the sulfoxide head group.

The overall results show that electrostatic factors lead to the formation of first black films from non-ionic surface active agents in electrolytes and that zeta-potentials as low as 18 mV are adequate to stabilize the film assuming the film potentials are comparable with those of the emulsion drop. As shown in more detail elsewhere, the DLVO theory can be used to give a good qualitative account of the effects of electrolyte concentrations on the thickness of first black foam films (*6*).

One of the interesting features of the films studies is the stability of the second black films at low electrolyte concentrations, at high electrolyte concentrations, and in the complete absence of the electrolyte. The most probable stabilizing factor is the so called "steric stabilization." However with a small head group the size of a methyl sulfoxide group configurational entropy terms cannot be invoked. Thus the stabilization must arise as a consequence of the solvation of the head groups; this suggests that the free energy required to remove the water from the close packed monolayers exceeds that obtained from van de Waals attraction. If the head groups in the film are staggered, as found for close packed OMS molecules on the trough, this could lead to a thicker solvated layer and hence aid the stability.

Acknowledgments

This work was carried out during the tenure of an S.R.C. CAPS studentship in conjunction with Procter and Gamble Ltd., and we are pleased to thank the Science Research Council for a maintenance grant to one of us (R.B.) during this period. We also thank B. Ingram and T. Walker for a number of useful discussions.

Literature Cited

1. Mysels, K. J., Shinoda, K., Frankel, S., "Soap Films," Pergamon Press Ltd., London, 1959.
2. Clunie, J. S., Goodman, J. F., Ingram, B. T., "Thin Liquid Films in Surface and Colloid Science," E. Matijević, Ed., Vol. 3, Wiley, New York, 1971.
3. Donaldson, R., Ph.D. thesis, University of Bristol, 1971.
4. Clunie, J. S., Corkill, J. M., Goodman, J. F., Ingram, B. T., *Spec. Discuss. Faraday Soc.* (1970) **1**, 30.
5. Ingram, B., *J. Chem. Soc. Faraday Trans. I* (1972) **68**, 2230.
6. Buscall, R., Ph.D. thesis, University of Bristol, 1973.
7. Adam, N. K., "Physics and Chemistry of Surfaces," Oxford, London, 1941.
8. Corkill, J. M., Goodman, J. F., Tate, J. R., *Trans. Faraday Soc.* (1969) **65**, 1743.
9. Corkill, J. M., Goodman, J. F., Ogden, C. P., *Trans. Faraday Soc.* (1965) **61**, 583.
10. Cabellero, D., *J. Opt. Soc. Amer.* (1947) **37**, 176.
11. Alexander, A. E., Saggers, L., *J. Sci. Instr.* (1948) **25**, 374.
12. Mattson, S., *J. Phys. Chem.* (1933) **37**, 223.
13. Henry, D. C., *J. Chem. Soc.* (1938) 997.
14. Wolstenholme, G. A., Schulman, J. H., *Trans. Faraday Soc.* (1950) **46**, 475.
15. Derjaguin, B. V., Landau, L., *Acta Physicochim.* (1941) **14**, 633.
16. Verwey, E. J. W., Overbeek, J. Th. G., "Theory of the Stability of Lyophobic Colloids," Elsevier, Amsterdam, 1948.
17. Davies, J. T., Rideal, E. K., "Interfacial Phenomena," Academic, London, 1961.

RECEIVED September 24, 1974.

8

A Model of the Electric Double Layer at a Completely Ionized Monolayer with Discreteness-of-Charge Effect

G. R. FÉAT

Department of Mathematics, University of Manchester, Manchester, England

S. LEVINE[1]

Department of Chemistry, University of Guelph, Guelph, Ontario, Canada

A model of the electric double layer at a completely ionized, insoluble monolayer is tested with the data of Mingins and Pethica for sodium octadecyl sulfate spread at the air/water interface. It is assumed that the Na^+ counterions adsorbed into the monolayer region can either form ion pairs with the sulfate head-group ions or remain mobile between these ions. The various physical quantities characterizing the double layer are determined by using the experimentally calculated variation with film molecule area A *of both surface potential* ΔV *and Esin–Markov coefficient* $(\partial \Delta V/\partial \log_{10} c)_{A,T}$ *where* c *is substrate electrolyte (NaCl) concentration and* T *the temperature. The apparent specific adsorption potential of the adsorbed Na^+ ions, particularly that of the mobile ions, varies strongly with* A. *Also the mean potential on the plane of the head-group ions has a maximum when* A *is varied at fixed* c *and* T. *Both properties are usually attributed to the discreteness-of-charge effect.*

The discreteness-of-charge effect (discrete-ion effect) is a general characteristic of electric double layers in aqueous media (*1*) and therefore should manifest itself in ionized monolayers. In a number of papers (*2, 3, 4, 5*), one of the authors and co-workers investigated the role of this

[1] On leave of absence from Department of Mathematics, University of Manchester.

effect in the physical properties of ionized insoluble monolayers at air–water (A–W) and oil–water (O–W) interfaces. In particular the variations in surface potential (ΔV) and surface pressure with area per film molecule A were studied theoretically. One property which is regarded as evidence for the discrete-ion effect is the observed maximum in the magnitude of the surface potential $|\Delta V|$ when plotted against A at a given substrate electrolyte concentration and temperature (*2*). The discrete-ion theory predicts that the electric double layer contribution $|\psi_0|$ to $|\Delta V|$ has a maximum and, provided any change in the dipole moment (μ) of the film molecule can be neglected and if ψ_0 has the same sign as μ, a maximum occurs if ΔV only reaches a limiting constant value with decrease in A. To show this consider the familiar relation between the surface potential ΔV, the double layer potential ψ_0, and the dipole moment μ:

$$\Delta V = \frac{4\pi\mu}{A} + \psi_0 \tag{1}$$

where, in the case of sodium octadecyl sulfate (SODS) for example both $\psi_0 < 0$ and $\mu < 0$. At high areas when $1/A \rightarrow 0$ we expect $\psi_0 \rightarrow 0$. Since $\psi_0 < 0$ for $1/A > 0$, $d\psi_0/d(1/A) < 0$ for small positive $1/A$. From Equation 1 the condition that ΔV tends to a constant value at low area, where $1/A \rightarrow \infty$, implies that

$$\frac{d\psi_0}{d(1/A)} \rightarrow -\ 4\pi\mu \left(1 + \frac{d\ \ln\ \mu}{d\ \ln\ A}\right) > 0 \tag{2}$$

assuming that μ does not decrease too rapidly with increase in A. Therefore at some intermediate $1/A$, $d\psi_0/d(1/A) = 0$—*i.e.*, $|\psi_0|$ has a maximum. Levine, Mingins, and Bell (L.M.B.) (*2*) collected several examples from the literature where $|\Delta V|$ either has a maximum or reaches a limiting value at $A > 60$ A^2, for which any dependence of μ on A should not alter the above conclusion.

A second property of the electric double layer, considered a cornerstone to demonstrate the discrete-ion effect, relates to the specific adsorption potential Φ of a counterion which is adsorbed into the so-called Stern inner region at a charged interface. If the discrete-ion effect for this ion is ignored, Φ unexplainably varies with the interfacial charge conditions (*1*). With an ionized monolayer there is evidence (*2*) that neglect of the discrete-ion effect causes a variation with film area A in the specific adsorption potential of a counterion which is oppositely charged to the ionized head-group of the film molecule and which has penetrated into the monolayer region.

Van Voorst Vader (*7*) and van den Tempel (*8*) proposed a modified Stern model of ionized monolayers in which the density of adsorption sites for such counterions in the monolayer region is much greater than the density of film molecules $1/A$. This assumption implies a small degree of specific binding of counterions to the charged head-groups of the film molecules. L.M.B. (*2*) used this model when incorporating the discrete-ion effect in a study of the ΔV–A relationship and found that the presence of mobile counterions, unpaired with the head-group ions, is necessary to explain the observed maximum in $|\Delta V|$–A plots mentioned above.

The theory of the ΔV–A behavior developed by L.M.B. (*2*) was based on earlier calculations by Mingins and Pethica (M.P.) (*9*) from their experimental work on monolayers of SODS at the A–W interface. Recently these authors (*10*) reported a numerical error in their earlier work; their conclusions question the model of the ionized monolayers used by L.M.B. (*2*) to explain the ΔV–A curves. The so-called Esin–Markov coefficient for adsorbed ions at the charged mercury/aqueous electrolyte has received considerable attention (*11, 12, 13*) particularly since it clearly demonstrates the discrete-ion effect. Its counterpart at ionized monolayers may be defined by the differential expression

$$(\partial \Delta V/\partial \log_{10} c)_{A,T} \tag{3}$$

where c is the electrolyte concentration in the substrate. M.P. (*9, 10*) studied this coefficient as a function of A. They suggest a key modification in the model of the monolayer region which is concerned with the degree of specific binding of the counterions (Na^+) with the primary head-group ions. They equate the number of adsorption sites for counterions to the number of film molecules, implying strong binding of counterions to the head-group ions, and calculate values of Φ at 5–7 kT for Na^+ on SO_4^- (k is Boltzmann's constant, and T the absolute temperature). Furthermore, the apparent variation with A in the specific adsorption potential (Φ) of a counterion (Na^+) adsorbed onto a head-group ion was calculated by M.P. as negligible, although the relevant discrete-ion effect has been ignored. Earlier works (*14, 15, 16*) seem to indicate that the Na^+ ion is not directly paired to the SO_4^- ion. However in Chapter 3 Mingins *et al.* no longer advocate an earlier suggestion (*2*) that entropies of compression of charged monolayers support weak specific binding. Difficulties in the model of M.P. are: (1). a significant variation in Φ with salt concentration which can be attributed to the discrete-ion effect for the bound Na^+ ions, and (2). no maximum in $|\psi_0|$ with change in A, if Φ is independent of A.

We propose a model of the electric double layer for ionized monolayers at A–W or O–W interfaces, which is a compromise between strong

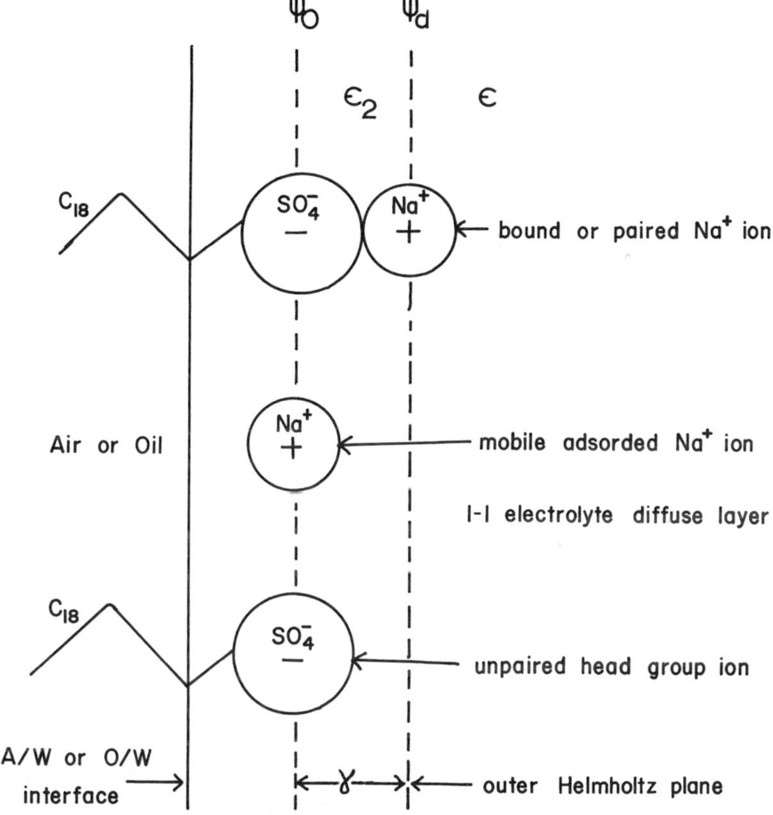

Figure 1. Model of electric double layer for a completely ionized monolayer

pairing of counterions to head-group ions and unpaired mobile counterions able to penetrate into the monolayer region between the head-group ions; this recognizes the discrete-ion effects which in our opinion are necessary. Such a modification inevitably introduces more parameters into the theory because of the two types of adsorbed counterions. We shall regard the monolayer molecule as sodium octadecyl sulfate, representative of a completely ionized, insoluble film, with SO_4^- as the head-group ion and Na^+ as the cation. In our model we consider two particular planes parallel to the interface: the plane on which the charge centers of the head-group ions are situated, denoted as the primary plane and the plane containing the charge center of the bound or paired off Na^+ ions, called the outer Helmholtz plane (ohp). It is convenient to choose the ohp as the boundary between the diffuse layer in the substrate and the monolayer region proper. The distance between the primary plane and ohp equals γ (Figure 1).

The centers of the adsorbed mobile or unpaired Na^+ counterions are located in the monolayer region between the ohp and the plane of nearest approach of such ions to the non-aqueous phase (air or oil). For simplicity we consider two idealized distributions of these Na^+ ions: (1) their centers are situated on the same plane as the charge center of the head-group ions—*i.e.*, on the primary plane; this model was used by L.M.B. (2) The mobile Na^+ ion centers are uniformly distributed in the "outer zone" situated between the primary plane and ohp. The first model would be reasonable if the potential energy of the Na^+ had a deep and/or narrow minimum on the primary plane. Such a minimum is the usual justification for models in which adsorbed ions are situated on the inner Helmholtz plane at other interfaces such as the mercury/electrolyte system. With the present system the mobile Na^+ ions will be attracted by the unpaired SO_4^- head group ions and may even form some loose Bjerrum type ion binding, a factor which would become more important as the area A is decreased. Otherwise it is not likely that any minimum exists in the potential energy of the Na^+ ion in the immediate vicinity of the non-aqueous phase. On the contrary image forces and hydration effects which will be felt as a specific adsorption potential tend to reduce the Na^+ ion concentration in the monolayer region. We expect that the Na^+ ions and some anions of the substrate electrolyte are distributed as a diffuse layer in the monolayer region. A diffuse layer which penetrates behind the centers of the head group ions was investigated by Haydon and Taylor (*18*) and Levine, Bell, and Pethica (*19*); we hope to extend this approach to the present problem in a later paper.

Currently no adequate quantitative theory of the discrete-ion potentials for adsorbed counterions at ionized monolayers exists although work on this problem is in progress. These potentials are more difficult to determine than those for the mercury/electrolyte interface because the non-aqueous phase is a dielectric medium and the distribution of counterions in the monolayer region is more complicated. However the physical nature of discrete-ion potentials for the adsorbed counterions can be described qualitatively. This paper investigates the experimental evidence for the discrete-ion effect at ionized monolayers by testing our model on the results of Mingins and Pethica (*9, 10*) for SODS. The simultaneous use of the Esin–Markov coefficient (Equation 3) and the surface potential ΔV as functions of A at the same electrolyte concentration c yields the specific adsorption potentials for both types of adsorbed Na^+ ions—bound and mobile. Two parameters which need to be chosen are the density of sites available to the adsorbed mobile Na^+ ions and the capacity per unit area of the monolayer region. The present work illustrates the value of using experimental ΔV–A plots at different electrolyte concentrations and temperatures to investigate the electric double layer properties at ionized monolayers.

Adsorption Isotherm for Counterions

If A is the area per film molecule, the surface charge density arising from the SO_4^- on these molecules is

$$\sigma_0 = -e_0/A \tag{4}$$

where e_0 is the proton charge. The surface charge density from the adsorbed counterions has two contributions: $\sigma_a^{(1)}$ from bound Na^+ ions and $\sigma_a^{(2)}$ from mobile Na^+ ions. If σ_d is the charge residing in the diffuse layer per unit interfacial area, the condition of electrical neutrality reads

$$\sigma_0 + \sigma_a^{(1)} + \sigma_a^{(2)} + \sigma_d = 0 \tag{5}$$

We use the Gouy-Chapman theory for the diffuse layer which is based on the Poisson-Boltzmann (P.B.) equation for the potential distribution. Although the different corrections to the P.B. equation in double-layer theory have been investigated (*20, 21, 22, 23*), it is difficult to state precisely the range of validity of this equation. In the present problem the P.B. equation seems a reasonable approximation at 0.1M of a 1-1 electrolyte to 50 mV for the mean electrostatic potential ψ_d at the ohp (*24*); this upper limit for ψ_d increases with a decrease in electrolyte concentration. All the values for ψ_d calculated in Tables I-IV are less than 50 mV—most of them are well below. If n is the volume density of each ion type of the 1-1 electrolyte in the substrate, ϵ the dielectric constant of the electrolyte medium, and

$$Y = -e_0 \psi_d/2kT \tag{6}$$

then the Gouy-Chapman theory yields

$$\sigma = -(\sigma_0 + \sigma_a^{(1)} + \sigma_a^{(2)}) = \left(\frac{2n\epsilon kT}{\pi}\right)^{1/2} \sinh Y \tag{7}$$

The surface charge density arising from the unpaired head-group ions is

$$\sigma_u = \sigma_0 + \sigma_a^{(1)} \tag{8}$$

Let the "outer" zone of the monolayer region between the primary plane and ohp, of thickness γ, have a uniform dielectric constant ϵ_2, so that its equivalent integral capacity per unit area is $K_2 = \epsilon_2/4\pi\gamma$. We consider first the model where the mobile Na^+ ions are on the primary plane. The potential drop across the outer zone is

$$\psi_0 - \psi_d = -\frac{1}{K_2}(\sigma_d + \sigma_a^{(1)}) = \frac{1}{K_2}(\sigma_0 + \sigma_a^{(2)}) \qquad (9)$$

The mean potential at the primary plane (ψ_0) should be identified with the ψ_0 in Equation 1 for ΔV.

The number of adsorption sites per unit area for bound Na^+ ions equals $1/A$. Assuming a simple Langmuir-type expression for the entropic term and making use of Equation 9, the adsorption isotherm of the bound counterion is given by the equilibrium relation:

$$kT \ln\left(\frac{\sigma_a^{(1)}}{-\sigma_u}\right) = \Phi_{Na}^{(1)} - e_0\psi_d + kT \ln\left(\frac{n}{n_0}\right) - \phi^{(1)} \qquad (10)$$

Table I. Calculated Parameters of Monolayer Model Using Experimental Na^+ Ions On

A	ΔV_{exp}	A_1	A_2
	T = 5°C, μ = −0.027 D		
100	−65	1.5×10^2	2.9×10^2
200	−53	$\simeq 3.4 \times 10^2$	$\simeq 5.9 \times 10^2$
500	−35	7.1×10^2	6.3×10^3
1000	−24[b] (−22.6)	1.8×10^3	1.7×10^4
	T = 10°C, μ = −0.13 D		
100	−76	1.5×10^2	2.4×10^2
200	−62	2.9×10^2	7.5×10^2
500	−43	7.3×10^2	6.9×10^3
1000	−31	1.9×10^3	5.8×10^5
	T = 14°C, μ = −0.27 D		
100	−84	2.7×10^2	1.3×10^2
200	−70	2.7×10^2	6.5×10^2
500	−49	8.1×10^2	2.3×10^3
1000	−35	1.9×10^3	1.9×10^4
	T = 18°C, μ = −0.50 D		
100	−90	—	—
200	−77	3.1×10^2	3.6×10^2
500	−53	7.6×10^2	2.0×10^3
1000	−39	1.8×10^3	6.8×10^3

[a] $N_2 = 10^{15}$ molecules/cm^{-2}; $\epsilon_2/\gamma = 20$; A, A_1, A_2 in A^2; ΔV_{exp}, ψ_d, ψ_0 in mV.

$\Phi_{Na}^{(1)}$ is the specific adsorption potential (energy), a function of T and pressure (p); n_0 is the volume density of water molecules in the aqueous substrate, and $\phi^{(1)}$ is the discrete-ion potential. ψ_d is the mean potential at a bound ion, assuming that the ion lies on the ohp. If we ignore pressure effects, $\phi^{(1)}$ depends on the three experimental variables (σ_0, n, and T), but at a given T we also regard $\phi^{(1)}$ as a function of the free or unpaired surface charge densities of head group ions and counterions, namely, σ_u and $\sigma_a^{(2)}$. An adsorption equation similar to Equation 10 is postulated for the adsorbed Na^+ ions situated on the primary plane on which there are N_2 adsorption sites, where N_2 is considered much larger than $1/A$.

$$kT \ln \left(\frac{\sigma_a^{(2)}}{N_2 e_0 - \sigma_a^{(2)}}\right) = \Phi_{Na}^{(2)} - e_0\psi_0 + kT \ln \left(\frac{n}{n_0}\right) - \phi^{(2)} \qquad (11)$$

where $\Phi_{Na}^{(2)}$ ($\neq \Phi_{Na}^{(1)}$) is the specific adsorption potential, and $\phi^{(2)}$ is the discrete-ion potential. Again $\phi^{(2)}$ is a function of σ_0, n, and T or

ΔV and Esin–Markov Coefficient of Mingins and Pethica. Mobile Primary Plane[a]

$\frac{\Phi_{Na}^{(1)}}{kT}$	$\frac{\Phi_{Na}^{(2)}}{kT}$	ψ_d	ψ_0
	$T = 5°C,\ \mu = -0.027\ D$		
8.2	1.7	1.5	−58
≃7	≃1.2	≃−19	≃−50
7.7	−0.3	−16	−32
7.1	−0.9	−14	−23
	$T = 10°C,\ \mu = -0.13\ D$		
9.1	3.1	24	−29
8	1.5	− 7	−40
7.5	−0.5	−18	−35
6.9	−4.6	−17	−26
	$T = 14°C,\ \mu = -0.27\ D$		
8.7	5.7	43	20
8.9	2.4	9	−23
7.5	1.0	−13	−27
6.9	−1.1	−16	−24
	$T = 18°C,\ \mu = -0.50\ D$		
—	—	—	—
9.6	4.6	38	18
7.9	1.4	− 7	−20
7.2	0.2	−12	−20

[b] Maximum theoretical $|\Delta V|$ (in brackets) less than experimental $|\Delta V|$.

Table II. Calculated

A	ΔV_{exp}	A_1	A_2
	$T = 5°C,\ \mu = -0.027\ D$		
100	−65	1.4×10^2	4.1×10^2
200	−53	3.4×10^2	8.6×10^2
500	−35	7.3×10^2	1.9×10^5
1000	−24[b] (−22.6)	1.9×10^3	4.6×10^5
	$T = 10°C,\ \mu = -0.13\ D$		
100	−76	1.5×10^2	3.1×10^2
200	−62	$\simeq 2.9 \times 10^2$	$\sim 1.0 \times 10^3$
500	−43	7.5×10^2	2.1×10^5
1000	−31[b] (−26)	1.9×10^3	4.3×10^5
	$T = 14°C,\ \mu = -0.27\ D$		
100	−84	2.9×10^2	1.4×10^2
200	−70	2.9×10^2	7.6×10^2
500	−49	8.4×10^2	3.3×10^3
1000	−35[b] (−33)	2.0×10^3	5.0×10^5
	$T = 18°C,\ \mu = -0.50\ D$		
100	−90	—	—
200	−77	3.3×10^2	3.7×10^2
500	−53	7.9×10^2	1.8×10^3
1000	−39	2.0×10^3	1.5×10^4

[a] $N_2 = 10^{15}$ molecules/cm²; $\varepsilon_2/\gamma = 40$; A, A_1, A_2 in A²; ΔV_{exp}, ψ_d, ψ_0 in mV.

equivalent of $\sigma_u, \sigma_a^{(2)}$, and T. Assuming $\sigma_a^{(2)} \ll N_2 e_0$ and substituting Equation 8, Equation 11 becomes

$$kT \ln\left(\frac{\sigma_a^{(2)}}{N_2 e_0}\right) = \Phi_{Na}^{(2)} - e_0\left(\psi_d + \frac{\sigma_0 + \sigma_a^{(2)}}{K_2}\right) + kT \ln\left(\frac{n}{n_0}\right) - \phi^{(2)} \quad (12)$$

If expressions for the discrete ion potentials $\phi^{(1)}$ and $\phi^{(2)}$ can be obtained, Equations 6, 10, and 12 can be solved for ψ_d, $\sigma_a^{(1)}$, and $\sigma_a^{(2)}$ in terms of the experimental variables n, σ_0 (*i.e.* A), and T, assuming values for N_2, $\Phi_{Na}^{(1)}$, $\Phi_{Na}^{(2)}$, and K_2.

Theories more recent than the simple Langmuir (random-mixing) adsorption statistics and usually regarded as more realistic can handle the entropic terms in Equations 10 and 12. Two of these, which have been used for adsorbed ions at the mercury/electrolyte interface, are based on Flory–Huggins (*12, 13, 25, 26*) and scaled-particle statistics (*31,*

Parameters as in Table I[a]

$\frac{\Phi_{Na}^{(1)}}{kT}$	$\frac{\Phi_{Na}^{(2)}}{kT}$	ψ_d	ψ_0
	T = 5°C, μ = −0.027 D		
7.5	1.5	−20	−54
6.5	1.0	−31	−49
7.3	−3.7	−22	−30
6.8	−4.2	−17	−22
	T = 10°C, μ = −0.13 D		
8.4	2.9	3	−28
≃7.5	≃1.3	≃−19	≃−42
7.2	−3.8	−24	−33
6.9	−4.1	−17	−21
	T = 14°C, μ = −0.27 D		
7.9	5.4	27	14
8.5	2.5	− 0.6	−17
7.1	0.6	−19	−26
6.7	−4.3	−18	−23
	T = 18°C, μ = −0.50 D		
—	—	—	—
9.0	4.5	28	17
7.7	1.7	− 7	−14
6.8	−0.7	−17	−21

[b] Maximum theoretical $|\Delta V|$ (in brackets) less than experimental $|\Delta V|$.

20, 26, 27, 28, 29). In the two-dimensional adaptation of the Flory–Huggins theory, we imagine a lattice of sites on a plane surface where one site can be occupied by a water molecule and p^1 sites by an adsorbed ion. If the area assigned to a head-group ion corresponds to p^1 sites, then the left side of Equation 10 is replaced by $kT\ [\ln\theta - p^1 \ln(1-\theta)]$, where $\theta = -\sigma_a^{(1)}/\sigma_0$ is the fraction of head-group ions paired off with bound Na^+. Equation 10 arose by assuming that the head-group ion provides only one site so that $p^1 = 1$. It is not clear how the more elaborate treatments of entropic terms apply to adsorbed Na^+ ions which are attached to head-group ions. Flory–Huggins or scaled-particle statistics seems most relevant to the adsorption isotherm of the mobile Na^+ ions, but here the modification to Equation 12 is a change in the value of $\Phi_{Na}^{(2)}$ if $\sigma_a^{(2)}$ is small compared with $N_2 e_0$, a condition which holds in Tables I-IV.

For the model where the mobile Na^+ ions are uniformly distributed in the outer zone situated between the primary plane and the ohp, Equation

Table III. Calculated Parameters as in Table I for Model

A	ΔV_{exp}	A_1	A_2
		$T = 5°C,\ \mu = -0.027\ D$	
100	−65	$\simeq 1.7 \times 10^2$	2.2×10^2
200	−53	4.0×10^2	4.6×10^2
500	−35	7.2×10^2	5.6×10^3
1000	−24	1.9×10^3	2.2×10^4
		$T = 10°C,\ \mu = -0.13\ D$	
200	−62	3.1×10^2	6.2×10^2
500	−43	7.4×10^2	6.1×10^3
1000	−31	1.9×10^3	4.3×10^5
		$T = 14°C,\ \mu = -0.27\ D$	
200	−70	2.9×10^2	5.3×10^2
500	−49	8.5×10^2	2.0×10^3
1000	−35	2.0×10^3	2.4×10^4
		$T = 18°C,\ \mu = -0.50\ D$	
200	−77	3.2×10^2	3.2×10^2
500	−53	7.9×10^2	1.7×10^3
1000	−39	1.9×10^3	5.0×10^3

[a] $N_2 = 10^{15}$ molecules/cm^{-2}; $\varepsilon_2/\gamma = 20$; A, A_1, A_2 in A^2; ΔV_{exp}, ψ_d, ψ_0 in mV.

9 is replaced by

$$\psi_0 - \psi_d = \frac{1}{K_2}\left(\sigma_0 + \tfrac{1}{2}\,\sigma_a^{(2)}\right) \tag{13}$$

Since the adsorption isotherm (Equation 12) needs modification, we replace the second term on the right side of Equation 12 by $-e_0\psi_m$, where ψ_m is the mean potential across the outer zone. Therefore,

$$\psi_m = \tfrac{1}{2}(\psi_0 + \psi_d) + \frac{1}{12}\frac{\sigma_a^{(2)}}{K_2} = \psi_d + \frac{1}{K_2}\left(\frac{1}{2}\,\sigma_0 + \frac{1}{3}\,\sigma_a^{(2)}\right) \tag{14}$$

using Equation 9. The right member of Equation 14 replaces the expression $\psi_0 + (1/K_2)(\sigma_0 + \sigma_a^{(2)})$ in Equation 12.

The Maximum in $|\psi_o|$ and the Discrete-Ion Potential

Using the monolayer model in which the mobile Na^+ ions are situated on the primary plane we examine a maximum in $|\psi_0|$ as a function of σ_0 (or A) at specified n and T. From Equation 9, at this maximum,

of Mobile Na⁺ Ions Uniformly Distributed in Outer Zone[a]

$\frac{\Phi_{Na}^{(1)}}{kT}$	$\frac{\Phi_{Na}^{(2)}}{kT}$	ψ_d	ψ_0
	$T = 5°C,\ \mu = -0.027\ D$		
8.4	3.6	14	−55
6.9	2.4	−13	−48
7.6	0.2	−15	−33
7.0	−1.0	−15	−24
	$T = 10°C,\ \mu = -0.13\ D$		
8.0	2.5	−4	−42
7.5	0	−17	−35
6.9	−4.1	−16	−26
	$T = 14°C,\ \mu = -0.27\ D$		
8.9	3.4	14	−22
7.4	1.4	−12	−28
6.9	−1.2	−16	−25
	$T = 18°C,\ \mu = -0.50\ D$		
9.6	5.1	41	9.7
7.8	1.8	−6	−21
7.2	0.7	−10	−18

$$0 = \frac{d\psi_0}{d\sigma_0} = \frac{d\psi_d}{d\sigma_0} + \frac{1}{K_2}\left(1 + \frac{d\sigma_a^{(2)}}{d\sigma_0}\right) - \frac{(\sigma_0 + \sigma_a^{(2)})}{K_2}\frac{d \ln K_2}{d\sigma_0} \quad (15)$$

Note that K_2 may depend on σ_0, n, and T. From Equations 6–7

$$\frac{d\sigma_a^{(1)}}{d\sigma_0} + \frac{d\sigma_a^{(2)}}{d\sigma_0} = -1 + \frac{e_0}{2kT}\left(\frac{2n\varepsilon kT}{\pi}\right)^{1/2} \cosh Y \frac{d\psi_d}{d\sigma_0} \quad (16)$$

Replacing the second term in the right side of Equation 12 by $-e_0\psi_0$ (according to Equation 9) and taking the derivative with respect to σ_0, Equation 12 yields at a maximum in $|\psi_0|$

$$\frac{1}{\sigma_a^{(2)}}\frac{d\sigma_a^{(2)}}{d\sigma_0} = -\frac{1}{kT}\frac{d\phi^{(2)}}{d\sigma_0} \quad (17)$$

The corresponding relation obtained from Equation 10 is

$$\frac{1}{\sigma_a^{(1)}}\frac{d\sigma_a^{(1)}}{d\sigma_0} - \frac{1}{\sigma_u}\left(1 + \frac{d\sigma_a^{(1)}}{d\sigma_0}\right) = -\frac{e_0}{kT}\frac{d\psi_d}{d\psi_0} - \frac{1}{kT}\frac{d\phi^{(1)}}{d\sigma_0} \quad (18)$$

In Equations 17 and 18 the discrete-ion potentials $\phi^{(1)}$ and $\phi^{(2)}$ are func-

Table IV. Calculated Monolayer Parameters with Various Fixed Values

A	ΔV_{exp}	A_1	A_2
$T = 5°C;\ \phi_{\ln n}^{(1)} = \tilde{\phi}_{\ln n}^{(1)} = 0.35$ kT, $\phi_{\ln n}^{(2)} = 0$			
100	−65	3.7×10^2	1.5×10^2
200	−53	9.6×10^2	3.5×10^2
500	−35	9.0×10^2	2.9×10^3
1000	−24	$\simeq 2.4 \times 10^3$	6.0×10^3
$T = 10°C;\ \phi_{\ln n}^{(1)} = \tilde{\phi}_{\ln n}^{(1)} = 0.27$ kT, $\phi_{\ln n}^{(2)} = 0$			
100	−76	2.2×10^2	1.7×10^2
200	−62	$\simeq 4.1 \times 10^2$	4.5×10^2
500	−43	8.8×10^2	2.7×10^3
1000	−31	2.5×10^3	1.9×10^4
$T = 10°C;\ \phi_{\ln n}^{(1)} = 0.1$ kT, $\phi_{\ln n}^{(2)} = 0$			
100	−76	1.7×10^2	2.2×10^2
200	−62	3.2×10^2	6.5×10^2
500	−43	7.8×10^2	5.1×10^3
1000	−31	2.0×10^3	5.9×10^4
$T = 10°C;\ \phi_{\ln n}^{(1)} = \phi_{\ln n}^{(2)} = 0.1\ kT$			
200	−62	1.5×10^3	2.8×10^2
500	−43	8.2×10^2	3.7×10^3
1000	−31	2.1×10^3	6.0×10^4

[a] $N_2 = 10^{15}$ molecules/cm^{-2}; $\varepsilon_2/\gamma = 20$; A, A_1, A_2 in A^2; ΔV_{exp}, ψ_d, ψ_0 in mV; $\phi^{(1)}_{\ln n} = 0.35\ kT$ at 5°C and $0.27\ kT$ at 10°C are the values of Mingins and Pethica.

tions of σ_0, n, and T. Equations 15–18 without $d\sigma_a^{(1)}/d\sigma_0$, $d\sigma_a^{(2)}/d\sigma_0$, and $d\psi_d/d\sigma_0$ satisfy the requirement that $|\psi_0|$ have a maximum,

$$\frac{1}{kT}\frac{d\phi^{(1)}}{d\sigma_0} + \frac{\sigma_a^{(2)}}{kT}\left[\Lambda\left(\frac{1}{\sigma_a^{(1)}} - \frac{1}{\sigma_u}\right) + \frac{e_0}{kTK_2}\right]\frac{d\phi^{(2)}}{d\sigma_0} \tag{19}$$
$$= \frac{1\text{-}\Lambda}{\sigma_u} + \frac{\Lambda}{\sigma_a^{(1)}} + (\sigma_0 + \sigma_a^{(2)})\left[(1\text{-}\Lambda)\left(\frac{1}{\sigma_a^{(1)}} - \frac{1}{\sigma_u}\right) - \frac{e_0}{kTK_2}\right]\frac{d\ln K_2}{d\sigma_0}$$

where

$$\Lambda = 1 + \frac{e_0}{kTK_2}\left(\frac{2n\varepsilon kT}{\pi}\right)^{1/2}\cosh Y \tag{20}$$

Since Equations 19 and 20 describe a fairly complicated relation, we consider four special cases:

(1). If the discrete-ion effect is ignored, then $\phi^{(1)}$ and $\phi^{(2)}$ are absent, and the left side of Equation 19 is identically zero. Now $\sigma_u < 0$, $\sigma_a^{(1)} > 0$, $\sigma_0 + \sigma_a^{(2)} < 0$, and $\Lambda > 1$. We expect a small variation of K_2

of $\phi_{\ln n}{}^{(1)}$ and $\phi_{\ln n}{}^{(2)}$ for Mobile Na^+ Ions on the Primary Plane[a]

$\Phi_{Na}{}^{(1)}/kT$	$\Phi_{Na}{}^{(2)}/kT$	ψ_d	ψ_0
$T = 5°C$; $\phi_{\ln n}{}^{(1)} = \phi_{\ln n}{}^{(1)} = 0.35$ kT, $\phi^{(2)}{}_{\ln n} = 0$			
5.4	2.4	−25	−56
4.6	1.6	−35	−55
6.9	0.4	−20	−35
6.5	0.2	−14	−20
$T = 10°C$; $\phi_{\ln n}{}^{(1)} = \phi_{\ln n}{}^{(1)} = 0.27$ kT, $\phi_{\ln n}{}^{(2)} = 0$			
7.5	3.3	7	−32
6.8	2.1	−14	−38
7.0	0.5	−18	−33
6.3	−1.2	−20	−28
$T = 10°C$; $\phi_{\ln n}{}^{(1)} = 0.1$ kT, $\phi_{\ln n}{}^{(2)} = 0$			
8.6	3.1	17	−32
7.6	1.6	−11	−42
7.3	−0.2	−19	−35
6.7	−2.3	−18	−27
$T = 10°C$; $\phi_{\ln n}{}^{(1)} = \phi_{\ln n}{}^{(2)} = 0.1$ kT			
4.5	2.5	−27	−40
7.1	0.2	−19	−34
6.7	−2.3	−18	−27

with σ_0 and that K_2 diminishes with an increase in $|\sigma_0|$. Thus all three terms on the right side of Equation 19 are positive—*i.e.*, there is no solution to Equation 19 and consequently no maximum in $|\psi_0|$.

(2). In the formulation of M.P., Equation 12 is entirely omitted, and the $\sigma_a{}^{(2)}$ is absent from all other equations. Thus Equation 19 simplifies to

$$\frac{1}{kT}\frac{d\phi^{(1)}}{d\sigma_0} = \frac{1\text{-}\Lambda}{\sigma_u} + \frac{\Lambda}{\sigma_a{}^{(1)}} + \left[(1\text{-}\Lambda)\left(\frac{1}{\sigma_a{}^{(1)}} - \frac{1}{\sigma_u}\right) - \frac{e_0}{kTK_2}\right]\frac{d\ln K_2}{d\ln\sigma_0} \quad (21)$$

We identify $\phi^{(1)} - \Phi_{Na}{}^{(1)}$ with M.P.'s specific adsorption potential Φ_{Na} which they calculated to be independent of A (or σ_0) at given n and T. This implies that the left side of Equation 21 vanishes, whereas the right side is the sum of three positive terms. Thus M.P.'s result means that there is no maximum in $|\psi_0|$. If $\phi^{(1)}$ is practically independent of σ_0, one necessary condition that the original relation (Equation 19) has a solution is that $d\phi^{(2)}/d\sigma_0 > 0$.

The discrete-ion energies $\phi^{(1)}$ and $\phi^{(2)}$ depend more closely on the free surface charge densities σ_u and $\sigma_a{}^{(2)}$ than on the equivalent variables σ_0 and n. Therefore it is instructive to consider the form of the condition (Equation 19) when $\phi^{(1)}$ and $\phi^{(2)}$ are functions of σ_u and $\sigma_a{}^{(2)}$. Since the dependence of K_2 on σ_u and $\sigma_a{}^{(2)}$ seems unimportant, for simplicity K_2 is assumed constant. We introduce into Equation 17 the relation:

$$\frac{d\phi^{(2)}}{d\sigma_0} = \frac{\partial\phi^{(2)}}{\partial\sigma_u}\left(1 + \frac{d\sigma_a{}^{(1)}}{d\sigma_0}\right) + \frac{\partial\phi^{(2)}}{\partial\sigma_a{}^{(2)}}\frac{d\sigma_a{}^{(2)}}{d\sigma_0}$$

A similar expression for $d\phi^{(1)}/d\sigma_0$ is substituted into Equation 18. Eliminating the same derivatives from Equations 15–18 which lead to Equation 19 now yields in place of Equation 19:

$$\left(\frac{\Lambda}{\sigma_a{}^{(1)}} + \frac{e_0}{kTK_2}\right)\frac{\partial\phi^{(2)}}{\partial\sigma_u} + \frac{(1\text{-}\Lambda)}{\sigma_a{}^{(2)}}\frac{\partial\phi^{(1)}}{\partial\sigma_u} - \left(\frac{\Lambda}{\sigma_a{}^{(1)}} + \frac{(1\text{-}\Lambda)}{\sigma_u} + \frac{e_0}{kTK_2}\right)\frac{\partial\phi^{(2)}}{\partial\sigma_a{}^{(2)}}$$
$$+ \frac{(1\text{-}\Lambda)}{kT}\left(\frac{\partial\phi^{(2)}}{\partial\sigma_a{}^{(2)}}\frac{\partial\phi^{(1)}}{\partial\sigma_u} - \frac{\partial\phi^{(2)}}{\partial\sigma_u}\frac{\partial\phi^{(1)}}{\partial\sigma^{(2)}}\right) = \frac{kT}{\sigma_a{}^{(2)}}\left(\frac{\Lambda}{\sigma_a{}^{(1)}} + \frac{(1\text{-}\Lambda)}{\sigma_u} + \frac{e_0}{kTK_2}\right) \quad (22)$$

with K_2 constant.

(*3*). If there are no bound ions, then $\sigma_a{}^{(1)} \rightarrow 0$, the function $\phi^{(1)}$ is absent, and Equation 22 simplifies to

$$\frac{\partial\phi^{(2)}}{\partial\sigma_u} - \frac{\partial\phi^{(2)}}{\partial\sigma_a{}^{(2)}} = \frac{kT}{\sigma_a{}^{(2)}} \quad (23)$$

which is the relation obtained by L.M.B. (*2*).

(**4**). Suppose that $\phi^{(1)}$ and $\phi^{(2)}$ depend on the net surface charge density only,

$$\sigma_t = \sigma_u + \sigma_a{}^{(2)} \quad (24)$$

Then

$$\frac{d\phi^{(1)}}{d\sigma_t} = \left(\frac{\partial\phi^{(1)}}{\partial\sigma_u}\right)_\sigma = \left(\frac{\partial\phi^{(1)}}{\partial\sigma_a{}^{(2)}}\right)_{\sigma_u} \quad (25)$$

similar relations hold for $\phi^{(2)}$. Then Equation 22 simplifies to

$$(1-\Lambda)\left[\frac{1}{\sigma_a{}^{(2)}}\frac{d\phi^{(1)}}{d\sigma_t} - \frac{1}{\sigma_u}\frac{d\phi^{(2)}}{d\sigma_t}\right] = \frac{kT}{\sigma_a{}^{(2)}}\left[\frac{\Lambda}{\sigma_a{}^{(1)}} + \frac{(1\text{-}\Lambda)}{\sigma_u} + \frac{e_0}{kTK_2}\right] \quad (26)$$

The right side of Equation 26 is positive. Then, if we assume that $d\phi^{(1)}/d\sigma_t$ and $d\phi^{(2)}/d\sigma_t$ have the same sign, they must be negative so that the left side of Equation 26 is also positive, a necessary (although not sufficient) condition for Equation 26 to have a solution.

The origins of the discrete-ion energies $\phi^{(1)}$ and $\phi^{(2)}$ follow. The potential $\phi^{(2)}$ for the mobile adsorbed Na^+ ions is described in the present simplified model where all surface charges, unpaired head-group ions, and mobile Na^+ ions are in the same (primary) plane. The term $e_0\psi_0$ in Equation 11 is the electrostatic work of transferring a Na^+ ion from the substrate interior to the primary plane under the hypothetical condition that no local redistribution of charge density occurs. The term $\phi^{(2)}$ in Equation 11 represents the additional work of Na^+ transfer arising from this redistribution. By adopting an approach used for the mercury/aqueous electrolyte interface (*30*), we regard $\phi^{(2)}$ to consist of three terms: $\phi_{self}^{(2)}$, $\phi_{disc}^{(2)}$, and $\phi_{outer}^{(2)}$. $\phi_{self}^{(2)}$ is the self-image potential energy arising from electrostatic imaging of a single ion at the interface where the dielectric constant is inhomogeneous. In the limiting case of high areas (A), $\phi_{self}^{(2)}$ becomes the only contribution to $\phi^{(2)}$ and depends on n, T, and p. The quantity $\phi_{self}^{(2)}$ will vary to some extent with σ_0 because of the dependence on σ_0 of the local electrolyte concentration in the diffuse layer. That part of $\phi_{self}^{(2)}$ which is independent of σ_0 and n can be absorbed into the specific adsorption potential, $\Phi_{Na}^{(2)}$. To discuss $\phi_{disc}^{(2)}$ and $\phi_{outer}^{(2)}$ we focus on the two-dimensional redistribution of charge on the primary plane—we imagine a Na^+ ion placed at some point on the plane. The term $\phi_{disc}^{(2)}$ arises from removal of the net charge density σ_t from a circular disc on the primary plane with the center at the adsorbed Na^+ ion. The radius a of this "exclusion" disc is the nearest mean distance of approach of other ions (SO_4^- or Na^+) situated on the primary plane. Note that this depletion of charge from the disc will be accompanied by the appropriate redistribution of the diffuse layer charge in the neighborhood. Usually with anionic monolayers, the total surface charge density (σ_t) will be negative, opposite in sign to the charge of the adsorbed Na^+ ion. Removing the charge density σ_t from the exclusion disc is equivalent to placing the adsorbed Na^+ ion at the center of a uniformly charged disc of radius a and total positive charge $-\pi a^2\sigma_t$ suitably screened by the local diffuse layer charge distribution. Thus the energy $\phi_{disc}^{(2)}$ is positive and roughly proportional to σ_t. We expect that $d\phi_{disc}^{(2)}/d\sigma_t$ is negative, bearing in mind that a positive increment $d\sigma_t$ implies a decrease in the magnitude of σ_t.

The energy term $\phi_{outer}^{(2)}$ is caused by a tendency for the unpaired head-group ions to be attracted to the adsorbed Na^+ ion and for other Na^+ ions on the primary plane to be repelled. Here again diffuse layer screening is taken into account; $\phi_{outer}^{(2)}$ will be negative, opposite in sign to $\phi_{disc}^{(2)}$. At a mercury surface the lateral electrostatic interaction energy between two adsorbed ions is short range (*31*) because of multiple electrostatic imaging by the mercury and diffuse layer phases, treating both as perfect conducting boundaries. Consequently the potential energy corresponding to $\phi_{outer}^{(2)}$ is generally very small. At the A-W or O-W interface however only one boundary, the diffuse layer, is effectively a perfect conductor, and the lateral interaction between two ions on the primary plane is no longer short-range but diminishes as the inverse third power of the separation at large separations. We expect therefore that $\phi_{outer}^{(2)}$ is appreciable and comparable with $\phi_{disc}^{(2)}$ in magnitude. The determination of $\phi_{outer}^{(2)}$ is more difficult than that of $\phi_{disc}^{(2)}$ principally because statistical mechanics is required to find the redistribution of charge on the primary plane beyond the exclusion disc.

The electrostatic part of the specific bonding energy of Na^+ to the head-group SO_4^- should contribute to the discrete-ion potential, $\phi^{(1)}$. This would depend slightly on σ_0 and n because the latter parameters determine the extent of local diffuse layer screening. The greater portion of this specific energy would depend on T and p and therefore may be incorporated into $\Phi_{Na}{}^{(1)}$. What remains of $\phi^{(1)}$ is divided into three terms: $\phi_{self}{}^{(1)}$, $\phi_{sphere}{}^{(1)}$, and $\phi_{outer}{}^{(1)}$, corresponding to the three contributions to $\phi^{(2)}$. The $\phi_{sphere}{}^{(1)}$ arises from the removal of the continuous volume distribution of charge, which has a "diffuse layer character," from the spherical exclusion volume assigned to the bound Na^+ ion. This corresponds to $\phi_{disc}{}^{(2)}$ which should be described as $\phi_{sphere}{}^{(2)}$. $\phi_{outer}{}^{(1)}$ is probably much smaller than $\phi_{outer}{}^{(2)}$ because ion-pair SO_4^-–Na^+ produces a local dipole field which is less intense and diminishes more rapidly with separation than the single charge field arising from a mobile adsorbed Na^+ ion. Once the main part of $\phi^{(1)}$ has been assigned to $\Phi_{Na}{}^{(1)}$, the remainder of $\phi^{(1)}$ should play a smaller role than $\phi^{(2)}$. This is confirmed by the results in Tables I-IV.

The Esin-Markov Coefficient

For the model where the mobile adsorbed Na^+ ions are situated on the primary plane, Equation 3 for the Esin-Markov coefficient may be written as

$$\left(\frac{\partial \Delta V}{\partial \log_{10} n}\right)_{\sigma_0, T} = 2.303 \left[\frac{\partial \psi_0}{\partial \ln n}\right]_{\sigma_0, T} = 2.303 \left[\frac{\partial \psi_d}{\partial \ln n} + \frac{1}{K_2}\frac{\partial \sigma_a{}^{(2)}}{\partial \ln n}\right]_{\sigma_0, T} \tag{27}$$

using Equations 1, 4, and 9. Assume here that at fixed σ_0 and T, μ, and K_2 are independent of electrolyte concentration n. [M.P. (9) considered that the second term in the square brackets of the right member was negligible]. To evaluate the derivatives in Equation 27 we differentiate Equations 7, 10, and 12, with respect to $\ln n$ at constant σ_0 and T and eliminate $\partial \sigma_a{}^{(1)}/\partial \ln n$ and $\partial \sigma_a{}^{(2)}/\partial \ln n$. For a comparison with the work of M.P., we introduce the areas A_1 and A_2, per bound and mobile adsorbed Na^+ ion. Then

$$\sigma_a{}^{(1)} = \frac{e_0}{A_1}, \quad \sigma_a{}^{(2)} = \frac{e_0}{A_2}, \quad \sigma_d = e_0\left(\frac{1}{A} - \frac{1}{A_1} - \frac{1}{A_2}\right), \tag{28}$$

where the last equation follows from Equation 4 and 7. We now define

$$\beta_1 = 1 + \frac{e_0^2}{A_2K_2kT}, \ \beta_2 = \frac{A_1^2}{A_2(A_1 - A)}, \ \beta_3 = \frac{1}{2}\left(\frac{A_1}{A} - \beta_2\right) \quad (29)$$

and write

$$\phi_{\ln n}^{(1)} = \left(\frac{\partial \phi^{(1)}}{\partial \ln n}\right)_{\sigma_0, T}, \ \phi_{\ln n}^{(2)} = \left(\frac{\partial \phi^{(2)}}{\partial \ln n}\right)_{\sigma_0, T} \quad (30)$$

We then obtain

$$\left(\frac{\partial \psi_d}{\partial \ln n}\right)_{\sigma_0, T} = \frac{kT}{e_0} \frac{\left[\beta_1\left\{1 + \beta_3 - \frac{\phi^{(1)}_{\ln n}}{kT}\right\} + \beta_2\left\{1 - \frac{\phi^{(2)}_{\ln n}}{kT}\right\}\right]}{\beta_1(1 + \beta_3 \coth Y) + \beta_2} \quad (31)$$

If we assume that the mobile adsorbed Na^+ ions are absent, then $A_2 \to \infty$, and Equation 31 simplifies to

$$\left(\frac{\partial \psi_d}{\partial \ln n}\right)_{\sigma_0, T} = \frac{kT}{e_0} \frac{\left[1 + \frac{A_1}{2A} - \frac{\phi_{\ln n}^{(1)}}{kT}\right]}{\left[1 + \frac{A_1}{2A} \coth Y\right]} \quad (32)$$

This is identical with Equation 3 of M.P. (9) if we identify their Φ_{Na} with our $\phi^{(1)} - \Phi_{Na}^{(1)}$. Thus, although M.P. did not explicitly introduce the function $\Phi^{(1)}$, they allowed for the discrete-ion effect by permitting their specific adsorption energy $\Phi_{Na}^{(1)}$ to vary with n. If it is assumed that the adsorbed mobile Na^+ ions are uniformly distributed in the "outer" zone, so that we use Equations 3 and 14, then in Equation 31

$$\beta_1 = 1 + \frac{e_0}{3A_2K_2kT} \quad (33)$$

If $\phi^{(1)}$ and $\phi^{(2)}$ are considered functions of $\sigma_a^{(1)}$ and σ_u, then after some algebraic manipulation, Equation 31 is identical wtih the relation

$$\left(\frac{\partial \psi_d}{\partial \ln n}\right)_{\sigma_0, T} = \frac{kT}{e_0} \frac{[\alpha_2 + \alpha_1 G(\frac{1}{2}\sigma_d)]}{[\alpha_2 + \alpha_1 G(\frac{1}{2}\sigma_d \coth Y)]} \quad (34)$$

where

$$\alpha_1 = \beta_1 + \frac{e_0}{kTA_2}\frac{\partial \phi^{(2)}}{\partial \sigma_a^{(1)}}, \ \alpha_2 = \beta_2 + \frac{e_0}{kTA_2}\left(\frac{\partial \phi^{(1)}}{\partial \sigma_a^{(1)}} - \frac{\partial \phi^{(1)}}{\partial \sigma_u}\right) \quad (35)$$

and

$$G(x) = \frac{\alpha_2 A_2 x}{e_0} + \left(1 + \frac{x}{kT}\frac{\partial \phi^{(1)}}{\partial \sigma_u}\right)\left(1 - \frac{e_0}{kTA_2\alpha_1}\frac{\partial \phi^{(2)}}{\partial \sigma^{(1)}}\right) \quad (36)$$

The derivatives of $\phi^{(1)}$ and $\phi^{(2)}$ are evaluated at constant σ_0 and T. If the discrete-ion effect is ignored, then

$$G(x) = 1 + \frac{\beta_2 A_2 x}{e_0} \quad (37)$$

and substituting Equation 37 and using Equation 7, Equation 34 simplifies to Equation 31 with $\phi_{\ln n}^{(1)} = \phi_{\ln n}^{(2)} = 0$. If we imagine that mobile adsorbed ions are absent, then in Equation 34 $\alpha_2 = 0$ and $G(x)$ simplifies to

$$G_0(x) = 1 + x\left[\frac{A_1^2}{e_0(A_1 - A)} + \frac{1}{kT}\left(\frac{\partial \phi^{(1)}}{\partial \sigma_a^{(1)}}\right)_{\sigma_0 T}\right] \quad (38)$$

Equation 34 becomes identical with Equation 32 of M.P.

M.P. calculated the Esin-Markov coefficient from their experimental ΔV–$1/A$ plots on SODS spread at the A-W interface. These plots at the three NaCl concentrations (0.1-, 0.01-, and 0.001M) indicate that ΔV is approximately linear in $\log_{10}c$ at a given A and T for values of $A <$ 500A^2. Therefore M.P. subtracted the experimental ΔV at $c = 0.1M$ from that at $c = 0.01M$ at selected A and T and interpreted the difference as the Esin-Markov coefficient ($\partial \Delta V/\partial \log_{10}c$) at 0.0316$M$. Also the mean of ΔV at 0.1- and 0.01M gives ΔV at 0.0316M, again at given A and T. Thus one can determine both the Esin-Maarkov coefficient and ΔV as a function of A at 0.0316M. This is available at the four temperatures: 5°, 10°, 14°, and 19°C, at which M.P. performed their experiments. The use of the experimental Esin-Markov coefficient and ΔV, plotted against $1/A$ at the *same* electrolyte concentration, is equivalent to applying the experimental ΔV-$1/A$ data at two different concentrations provided ΔV is linear in $\log_{10}c$.

M.P. describe how the film molecule dipole moment μ can be obtained from the intercept of the plot of $\lim_{1/A \to 0} \partial(\Delta V)/\partial(1/A)$ *vs.* $n^{-1/2}$. Using their values for ΔV, $(\partial(\Delta V)/\partial \log_{10} c)$, μ at specified T, film molecule A, and electrolyte concentration c and choosing the number of adsorption sites (N_2) and monolayer capacity (K_2), we determine the specific adsorption potentials $\Phi_{Na}^{(1)}$ and $\Phi_{Na}^{(2)}$, areas A_1 and A_2, and potentials ψ_0 and ψ_d. In the absence of sufficient information about discrete-ion potentials $\phi^{(1)}$ and $\phi^{(2)}$, we regard the two derivatives $\phi_{\ln n}^{(1)}$ and $\phi_{\ln n}^{(2)}$ in Equation 31 as adjustable constants. However we assume that the discrete-ion potentials $\phi^{(1)}$ and $\phi^{(2)}$ will vary with A or σ_0; we then incorporate $-\phi^{(1)}$ and $-\phi^{(2)}$ into the adsorption potentials $\Phi_{Na}{}^{(1)}$ and $\Phi_{Na}{}^{(2)}$ in Equations 10 and 12. This is equivalent to omitting $\phi^{(1)}$ and $\phi^{(2)}$ from Equations 10 and 12 and to recognizing the discrete-ion effect through the dependence on A of $\Phi_{Na}{}^{(1)}$ and $\Phi_{Na}{}^{(2)}$.

The following is a convenient procedure for calculating the various parameters in our model when the mobile Na^+ ions are on the primary plane. The potential ψ_0 is determined from Equation 1 knowing ΔV and μ. We write Equation 9 as

$$\psi_d = \psi_0 - \frac{e_0}{K_2}\left(\frac{1}{A_2} - \frac{1}{A}\right) \tag{39}$$

and, from Equations 4, 6, 7, and 28

$$\frac{1}{A_1} = \frac{1}{A} - \frac{1}{A_2} - \frac{\sqrt{c}}{\kappa^*}\sinh \mathrm{Y} \tag{40}$$

where, if c is the electrolyte concentration in moles/l and N is Avogadro's number, $\kappa^* = e_0\,(1000\pi/2\epsilon NkT)^{1/2}$. Equations 39–40 express ψ_d and A_1 in terms of A_2. Equation 27 is written as

$$\left(\frac{\partial \psi_d}{\partial \ln n}\right)_{\sigma_0, T} = \frac{1}{2.303}\left(1 + \frac{e_0{}^2}{kTA_2K_2}\right)\left(\frac{\partial \Delta V}{\partial \log_{10} c}\right)_{\sigma_0, T} - \frac{e_0}{A_2K_2}\left(1 - \frac{1}{kT}\,\phi^{(2)}{}_{\ln n}\right) \tag{41}$$

making use of Equations 9, 12, and the second relation in Equation 28; this is substituted for the left side of Equation 31. Choosing constant values for $\phi_{\ln n}{}^{(1)}$ and $\phi_{\ln n}{}^{(2)}$ (*see* next section) and recalling that A_1 and ψ_d are expressed in terms of A_2, Equation 31 has only one unknown, A_2; it is solved by a suitable root-finding method. The adsorption potentials are then obtained from Equations 10 and 12, which are written as

$$\Phi_{Na}{}^{(1)} = -kT\left[2.303 \log_{10}\left\{\frac{(55.6/c)}{(A_1/A - 1)}\right\} - 2\mathrm{Y}\right] \tag{42}$$

$$\Phi_{Na}{}^{(2)} = -kT\left[2.303 \log_{10}\left(\frac{55.6}{c\,A_2N_2}\right) - 2\mathrm{Y}\right] + \frac{e^2{}_0}{K_2}\left(\frac{1}{A_2} - \frac{1}{A}\right), \tag{43}$$

with $\phi^{(1)}$ and $\phi^{(2)}$ absorbed into $\Phi_{Na}{}^{(1)}$ and $\Phi_{Na}{}^{(2)}$.

If it is assumed that the adsorbed mobile ions are uniformly distributed in the outer-zone, so that we make use of Equations 13 and 14, then Equation 39 is replaced by

$$\psi_d = \psi_0 - \frac{e_0}{K_2}\left(\frac{1}{2A_2} - \frac{1}{A}\right), \tag{44}$$

and Equation 41 by

$$\left(\frac{\partial\psi_d}{\partial \ln n}\right)_{\sigma_0,T} = \left(1 - \frac{e^2{}_0}{6kTA_2K_2}\right)^{-1} \times \tag{45}$$

$$\left[\left(1 + \frac{e^2{}_0}{3kTA_2K_2}\right)\frac{1}{2.303}\frac{\partial\Delta V}{\partial \log_{10}c} - \frac{e_0}{2A_2K_2}\left(1 - \frac{1}{kT}\phi^{(2)}{}_{\ln n}\right)\right]$$

Also in Equation 43 the last term on the right side is replaced by

$$\frac{e^2{}_0}{K_2}\left(\frac{1}{3A_2} - \frac{1}{2A}\right)$$

Numerical Results and Discussion

Results of the comparison between our calculations and the experimental data of M.P. (9) are shown in Tables I–IV at 5°, 10°, 14°, and 18°C for $A = 100$, 200, 500, and 1000 A^2. Corresponding values of the dipole moment μ were chosen as follows. According to M.P., we may write

$$\Delta V = \chi_F - \chi_W \tag{46}$$

where χ_W and χ_F denote respectively the χ potentials of the clean and film-covered interface. Figure 1 of M.P. shows plots against $n^{-1/2}$ of

$$\lim_{1/A\to 0}\,[\partial\Delta V/\partial(1/A)]_{n,T} = \lim_{1/A\to 0}\,[\partial\chi_F/\partial(1/A)]_{n,T} \tag{47}$$

noting that χ_W is independent of A. With the Gouy–Chapman model of the double layer, in which $\sigma_0 = -\sigma_d$ and $\psi_0 = \psi_d$,

$$\Delta V = \frac{4\pi\mu}{A} - \frac{2kT}{A}\left(\frac{\pi}{2\varepsilon kTn}\right)^{1/2} \tag{48}$$

at large A and therefore the plot of 15.2 there is a straight line with intercept $4\pi\mu$ against $n^{-1/2}$, where $n^{-1/2} \to 0$. Figure 10 of M.P. shows deviations from the Gouy-Chapman theory, but if the potential ψ_0 in Equation 1 goes to zero more rapidly than $1/A$ at large A and n, the above intercept will still equal $4\pi\mu$. The value of μ obtained in this way strongly increases in magnitude with temperature yielding −0.027-, −0.13-, −0.27-, and −0.50 Debye at the four temperatures listed above. We used these values for μ, ignoring any dependence on A or n. An increase in the magnitude of μ with temperature suggests that the contributions to μ from the film molecule and from the immediately adjacent water molecules are opposite in sign. Although both contributions individually may be decreasing with an increase in temperature, the smaller one need only decrease more rapidly to give the observed dependence of μ on temperature. Contrary to the discussion by M.P. (*10*), no information about χ_W is available from their Figure 10.

The range of the ratio ϵ_2/γ examined was 10(10)50 and Tables I and II have the results for $\epsilon_2/\gamma = 20$ and 40 when the mobile Na^+ ions are on the primary plane; also we assumed $\phi_{\ln n}{}^{(1)} = \phi_{\ln n}{}^{(2)} = 0$ and $N_2 = 10^{15}\,cm^2$. The calculated values of A_1, A_2, $\Phi_{Na}{}^{(1)}$, $\Phi_{Na}{}^{(2)}$, ψ_d, and ψ_0 are given. At the high and low values of A no solution which would fit both the experimental ΔV, and the Esin–Markov coefficient could be found. At the high A a solution was obtained by choosing a $|\Delta V|$ smaller than the experimental value. In such cases (written in brackets) we equated $|\Delta V_{exp}|$ to the maximum $|\Delta V|$ which yields a solution for the given A and determined the various parameters listed on that basis. It is difficult to reach any conclusion about this apparent discrepancy since the experimental results are less accurate at high A; M.P. had similar problems in fitting their model. In two cases at $A = 100$ A^2, a solution existed only for $|\Delta V|$ exceeding $|V_{exp})$, and these have been left blank in the Tables.

We observe from Tables I and II that A_1 is less than A_2 and that A_1 increases more rapidly than A_1 with increase in A. This is an unexpected result because with increase in A, the density of adsorption sites for bound ions is being reduced whereas that for mobile ions remains unchanged. To explain this behavior, we divide Equation 10 by Equation 12 to obtain

$$\frac{A_2}{A_1} = \frac{(1-A/A_1)}{N_2 A} \exp\left[\frac{1}{kT}\left(\Phi_{Na}{}^{(1)} - \Phi_{Na}{}^{(2)} + \psi_0 - \psi_d\right)\right] \qquad (49)$$

recalling that $\phi^{(1)}$ and $\phi^{(2)}$ have been absorbed into $\Phi_{Na}{}^{(1)}$ and $\Phi_{Na}{}^{(2)}$. If the specific adsorption potentials had remained constant, we would have obtained the expected decrease in A_2/A_1 with increase in A. It is seen from Tables I and II that the source of the increase in A_2/A_1 with A is the dependence on A of the difference $\Phi_{Na}{}^{(1)} - \Phi_{Na}{}^{(2)}$ and $\psi_0 - \psi_d$, particularly the former. We note that $\Phi_{Na}{}^{(1)}$ is large and positive, indicating strong binding between the Na^+ and SO_4^-, in general agreement with the conclusions of M.P. However our $\Phi_{Na}{}^{(1)}$ increases somewhat with decrease in A, a characteristic of the discrete-ion effect. $\Phi_{Na}{}^{(2)}$ for the mobile ions has a much stronger dependence on A than $\Phi_{Na}{}^{(1)}$ for the bound ions, indicating a marked discrete-ion effect. It is the property that $\Phi_{Na}^{(1)} - \Phi_{Na}^{(2)}$ become more positive as A increases that compensates for the diminution in adsorption sites available to the bound ions. Results corresponding to those in Table I are shown in Table III for the model in which the mobile Na^+ ions are uniformly distributed in the outer zone. Comparison of Tables I and III indicate a similar behavior of the various parameters, suggesting that the manner in which the mobile Na^+ ions are distributed in the monolayer region does not change the general features described above. At $A = 100\ A^2$ and $\epsilon_2/\gamma = 20$ a solution was obtained only at $T = 5°C$.

Earlier we described qualitatively the nature of the discrete-ion terms with the present model of the ionized monolayer. The maximum in $|\psi_0|$ with change in A, which characterizes the discrete-ion effect but is not predicted by the model of M.P., is demonstrated in Tables I–III. However preliminary calculations on $\phi^{(2)}$ indicate that a change in $\Phi_{Na}^{(2)}$ of several kT over a range in A of 100–1000 A^2, shown in the tables, is too large. A consequence of the high ion-pair binding energies $\Phi_{Na}^{(1)}$ and of this large variation in $\Phi_{Na}^{(2)}$ with A is that the potential at the ohp, ψ_d, may become positive as A is decreased. This implies that the density of absorbed Na^+ ions exceeds that of the head group SO_4^- ions. This behavior of $\Phi_{Na}^{(1)}$ and $\Phi_{Na}^{(2)}$ is consistent with the unexpectedly low concentration of mobile adsorbed ions at large A. We wish to investigate whether the variation in $\Phi_{Na}^{(2)}$ with A can be reduced by retaining in Equation 31 non-zero values for the derivatives $\phi_{\ln n}^{(1)}$ and $\phi_{\ln n}^{(2)}$. Indeed the assumption $\phi_{\ln n}^{(1)} = \phi_{\ln n}^{(2)} = 0$ makes the comparison of our theory with experiments not entirely self-consistent because the dependence of the discrete-ion potentials $\phi^{(1)}$ and $\phi^{(2)}$ on σ_0 has been considered but not that on n.

Since in fact the forms of $\phi_{\ln n}^{(1)}$ and $\phi_{\ln n}^{(2)}$ cannot be established without an adequate theory of the discrete-ion potentials $\phi^{(1)}$ and $\phi^{(2)}$, a few typical results at $\epsilon_2/\gamma = 20$ and $T = 5°C$ and $10°C$ (*see* Table IV), are sufficient. At the larger A, where $A_1 \ll A_2$, the density of adsorbed mobile ions is so small that our results should be identical with those of M.P. (*10*). From their Figure 9 we estimate that the derivative $\phi_{\ln n}^{(1)}$ in Equation 32 equals 0.35 kT at 5°C and 0.27 kT at 10°C at $A = 1000\ A^2$. These particular values of $\phi_{\ln n}^{(1)}$ (denoted by $\tilde{\phi}_{\ln n}^{(1)}$), with $\phi_{\ln n}^{(2)} = 0$ are used in the first two sets of calculations in Table IV. Both $\Phi_{Na}^{(1)}$ and $\Phi_{Na}^{(2)}$ vary with A although not to the same extent as in the case where $\phi_{\ln n}^{(1)} = \phi_{\ln n}^{(2)} = 0$. If $\phi_{\ln n}^{(1)} = \phi_{\ln n}^{(2)} = \tilde{\phi}_{\ln n}^{(1)}$, no solutions are obtained at 10° for $A \lesssim 500$ A and 5°C for any A. The pair of values $\phi_{\ln n}^{(1)} = \tilde{\phi}_{\ln n}^{(1)}$, $\phi_{\ln n}^{(2)} = 0.5$ also have no solutions at 5° or 10°C. Where other combinations gave solutions (*see* Table IV for two examples) $\Phi_{Na}^{(1)}$ and in particular $\Phi_{Na}^{(2)}$ can still vary considerably with A. For all cases considered $\Phi_{Na}{}^{(2)}$ always decreases towards negative values with increase in A, but the reverse trend is sometimes observed with $\Phi_{Na}{}^{(2)}$.

These preliminary calculations strongly suggest that when the discrete-ion potentials are incorporated into $\Phi_{Na}{}^{(1)}$ and $\Phi_{Na}{}^{(2)}$, the latter have a significant dependence on A; this would be attributed to the discrete-ion effect. However no reliable estimates of this effect are possible until the potentials $\phi^{(1)}$ and $\phi^{(2)}$ have been determined. Such a task would require a more careful analysis of the distribution in the monolayer region of mobile ions, possibly both cations and anions. Also configurational entropy effects associated with ion-size in the monolayer region merit further study. Further work along these lines is in progress.

Acknowledgments

We are indebted to the Science Research Council of the United Kingdom for a research assistantship to G.R.F. The senior author (S.L.) has a visiting science lectureship in Canada sponsored by the Nuffield Foundation of the United Kingdom and the National Research Council of Canada.

Literature Cited

1. Levine, S., Mingins, J., Bell, G. M., *J. Electroanal. Chem.* (1967) **13,** 280.
2. Levine, S., Mingins, J., Bell, G. M., *J. Phys. Chem.* (1963) **67,** 2095.
3. Bell, G. M., Levine, S., Pethica, B. A., *Trans. Faraday Soc.* (1962) **58,** 904.
4. Bell, G. M., Levine, S., *Z. Phys. Chem. (Leipzig)* (1966) **231,** 289.
5. Bell, G. M., Levine, S., Pethica, B. A., Stephens, D., *J. Colloid Interface Sci.* (1970) **33,** 482.
6. Bell, G. M., Levine, S., Stephens, D., *J. Colloid Interface Sci.* (1972) **38,** 609.
7. van Voorst Vader, G., *Proc. Intern. Conf. Surface Activity, 3rd,* (University of Mainz Press) 1960, **2,** 276.
8. van den Tempel, M., *Recl. Trav. Chim.* (1953) **72,** 419.
9. Mingins, J., Pethica, B. A., *Trans. Faraday Soc.* (1963) **59,** 1892.
10. Mingins, J., Pethica, B. A., *Faraday Trans. I* (1973) **69,** 500.
11. Parsons, R., *Proc. Intern. Congr. Surface Activity, 2nd, London,* 1957, **3,** 38.
12. Levine, S., Robinson, K., *Proc. Intern. Congr. Surface Active Agents, 6th, Hanser, Munich,* 1973, 615.
13. Levine, S., Robinson, K., *J. Electroanal. Chem. Interface Electrochem.* (1975) **58,** 19.
14. Kotin, L., Nagasawa, H., *J. Amer. Chem. Soc.* (1961) **83,** 1026.
15. Lapanje, S., Rice, S. A., *J. Amer. Chem. Soc.* (1961) **83,** 496.
16. Mukerjee, P., *J. Phys. Chem.* (1962) **66,** 843.
17. Mingins, J., Owens, N. F., Taylor, J. A. G., Brooks, J. H., Pethica, B. A., ADVAN. CHEM. SER. (1975) **144,** 14.
18. Haydon, D. A., Taylor, F. H., *Philos. Trans. Roy. Soc. London* (1960) **A252,** 225, 255.
19. Levine, S., Bell, G. M., Pethica, B. A., *J. Chem. Phys.* (1964) **40,** 2304.
20. Buff, F. P., Stillinger, F. H., *J. Chem. Phys.* (1963) **39,** 1963.
21. Bell, G. M., Levine, S., "Chemical Physics of Ionic Solutions," Conway, B. E., Barradas, R. G., Eds., Wiley, New York, 1966.

22. Outhwaite, C. W., *Chem. Phys. Lett.* (1970) **7**, 636.
23. *Ibid., Mol. Phys.* (1974) **27**, 561.
24. Levine, S., Bell, G. M., *Discuss. Faraday Soc.* (1966) **42**, 69.
25. Levine, S., Bell, G. M., Calvert, D., *Can. J. Chem.* (1962) **40**, 518.
26. Levine, S., *J. Colloid Interface Sci.* (1971) **37**, 619.
27. Helfand, E., Frisch, H. L., Lebowitz, J. L., *J. Chem. Phys.* (1961) **34**, 1037.
28. Lebowitz, J. L., Helfand, E., Praestgaard, E., *J. Chem. Phys.* (1965) **43**, 774.
29. Plesner, I. W., Michaeli, I., *J. Chem. Phys.* (1974) **60**, 3016.
30. Levine, S., Robinson, K., Bell, G. M., Mingins, J., *J. Electroanal. Chem. Interface Electrochem.* (1972) **38**, 253.
31. Levine, S., Robinson, K., *J. Electroanal. Chem. Interface Electrochem.* (1973) **41**, 159.

RECEIVED September 17, 1974.

9

The Significance of Volta and Compensation States and the Measurement of Surface Potentials of Monolayers

B. A. PETHICA, M. M. STANDISH, J. MINGINS, C. SMART, D. H. ILES, M. E. FEINSTEIN, S. A. HOSSAIN, and J. B. PETHICA

Unilever Research, Port Sunlight Laboratory, Port Sunlight, Wirral, Cheshire L62 4XN, England

The thermodynamic analysis of the Volta effect by Koenig was studied for insoluble monolayers at air/water interfaces. Capacity measurements with a frequency bridge and with vibrating plate electrodes were used to test the basic assumptions of Koenig. The compensation effect was studied, and the conditions to verify the compensation method for measuring changes in χ *potential were established for different types of spread monolayers. The field effect we reported for zwitterionic molecules is shown now to be an artefact produced by changes in the contact angle at the edge of the working interface at high monolayer pressures, with consequent changes in the geometry of the interfacial condenser. Within experimental limits, Kelvin's assumption that the Volta and compensation potentials are equal and opposite is correct for the systems studied.*

Surface potential measurements have broadened our understanding of both solid and liquid surfaces. The early works of Frumkin (*1, 2, 3*) and Guyot (*4*) demonstrated that adsorbed molecules can change the electrical potential at a liquid/vapor interface. This change for adsorbed and spread monolayers at aqueous interfaces has been studied by many surface chemists, including N. K. Adam (*5*), whose contributions to surface science we honor in this volume. The method was used by Schulman, Rideal, and Alexander to investigate ionic charge distributions and the orientations of surface dipoles (*6, 7*) and more recently to study the ionization of monolayer molecules (*8, 9*).

In most of this work at the air/water interface the surface potentials were measured with an ionizing electrode in the air above the aqueous phase. The theory of the ionizing electrode is not well understood, and it cannot be regarded as an absolute method. The vibrating plate method, as developed by Zisman (*10*) and others, rests on a more secure theoretical understanding. It depends on the relatively simple determination of the potential that is applied for null current (compensation potential) in a circuit containing an oscillating capacitor made of a reference electrode and the surface under study. Where both the Zisman and radioactive electrode methods were compared directly, the same results were found (*11, 12*). To this extent the radioactive electrode method is reliable, although further detailed studies are desirable (*13*).

The other method used for liquid surfaces is the flow method of Kenrick (*14*) in which a jet of one solution is passed down the center of a tube whose walls carry a flowing layer of a second solution. The potentials between the flowing liquids are monitored with a quadrant or other electrometer. This method has been used with good results by Randles (*15*) and Parsons (*16*). Case and Parsons (*17*) compared the Kenrick and radioactive electrode methods for methanol–water mixtures. They found good agreement except at elevated methanol concentrations where methanol adsorption at the air electrode probably occurs. Measurement of the null current (compensation) potential in the Kenrick method is suitable for determining the surface potentials of solutions where rapid surface equilibrium occurs, but it is not convenient for spread monolayers or adsorbed films that have slow time effects.

From the measurements with a form of the Zisman apparatus, we establish experimentally several features of the surface electrical states described by Koenig in his paper on the thermodynamic analysis of the Volta effect (*18*). Also we wish to establish conditions for reproducible and defined measurements of compensation and surface potentials. We first define the relevant potentials and give a brief summary of Koenig's findings so that we can clarify the form in which we will test some of the assumptions he isolated. We confine the discussion to condenser states where temperature, pressure, and solution composition are constant but where the composition of the working interface may be changed by spreading an insoluble monolayer. The electrical circuit includes a half-cell, sometimes with a salt bridge. The relatively minor considerations thus introduced are fully discussed by Koenig, and for present purposes we can neglect them.

Figure 1 gives the essential features for describing the Volta and compensation condenser states. In Figure 1a two conducting phases, α and β, face each other at a distance x across a non-conducting gap of,

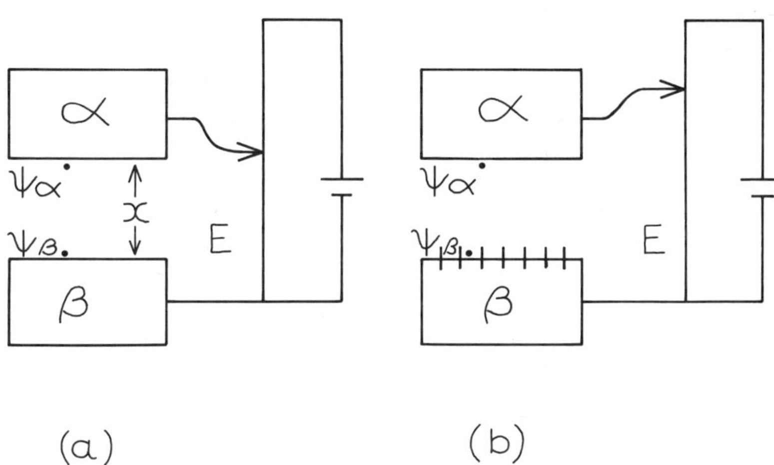

Figure 1. Representation of the outer potentials (ψ) in the non-conducting gap between the plane surfaces of two conducting phases, α and β, separated by a distance, x, *and connected to a potentiometer at potential* E

(a) Clean surfaces, (b) surface of phase β covered by an insoluble monolayer

for example, air or paraffin oil. Phase α is a metal reference electrode, and phase β may be another metal, a semiconductor, or (in our case) an aqueous salt solution. These two phases are coupled through an appropriate circuit to a potentiometer, and a potential E may be applied. The surfaces of the two condenser plates carry charges $\pm q$; their outer potentials are denoted by ψ_α and ψ_β. The outer potential is the work done in bringing unit charge from infinity to a point just outside a surface where it is beyond the range of electrical double layers and other surface dipoles on the phase in question. It is associated solely with the free charge q on the surface and as such depends on E. The Volta state corresponds to zero applied potential ($E = 0$); the quantity $(\Delta\psi)_{E=0}$, where $\Delta\psi$ is the difference in ψ_α and ψ_β, is the classical Volta or contact potential. Outer potentials have sometimes been referred to as Volta potentials, but here, as in Koenig's discussion, the Volta potential is a *difference* in outer potential. The compensation state is given by $\psi_\alpha = \psi_\beta$, and therefore by $q = 0$. The value of E in the compensation state is denoted by E_0, the compensation potential.

In Figure 1b the same two conducting phases are opposed, but the surface of phase β now includes a defined surface density of molecules of an insoluble monolayer. The reference electrode, phase α, is assumed unchanged by the spreading of the monolayer on β. In practice this means waiting until the spreading solvents have fully evaporated from the region of study. The Volta and compensation states are defined in the same way as before. The value of E_0 is changed experimentally by the presence of the monolayer. If we denote the respective systems with and without a monolayer by (1) and (2), the Volta and compensation potentials for the two systems can be written $(\Delta\psi_1)_{E=0}$, $(\Delta\psi_2)_{E=0}$, $(E_0)_1$, and $(E_0)_2$. The surface potential (ΔV) for the insoluble monolayer is then usually defined as $\Delta V = (E_0)_2 - (E_0)_1$. We write χ as the drop in electrostatic potential across the dipole layer at an interface and denote the absence and presence of a monolayer by suffixes (1) and (2). For this system $\Delta V = (\chi_0)_2 - (\chi_0)_2$, where the subscript zero means that the values of χ are those in the compensated state (*23*).

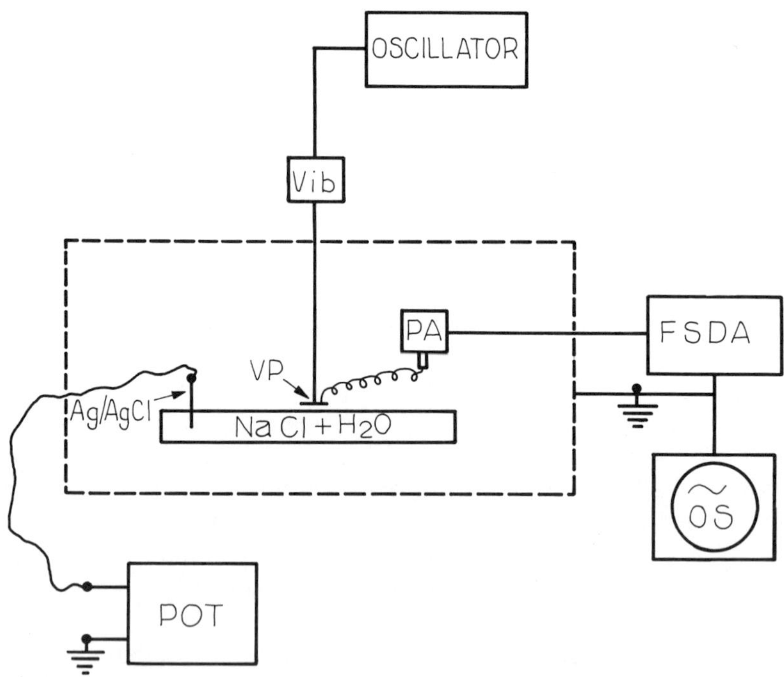

Figure 2. Schematic for measuring the potentials between a vibrating plate electrode and an aqueous electrolyte.

VP—vibrating plate; Vib—vibrator; PA—preamplifier; FSDA—frequency selective detector amplifier; POT—potentiometer; OS—oscilloscope. Dotted line represents a Faraday cage.

Koenig (*18*) has analyzed the thermodynamic relations between the potentials described above. He points out that certain long-standing assumptions attributed to Kelvin (*19*), Bridgman (*20*) and Lorentz (*21*) are strictly non-thermodynamic and that these assumptions have not been experimentally tested directly. It is sufficient to state that the assumptions of Bridgman and Lorentz concerning Volta and compensation potentials can be shown to contain the Kelvin assumption—namely that $(\Delta\psi)_{E=0}$ is equal to $-E_0$. In this communication we test the assumption by noting that it requires that for a given surface χ should not change with variation of surface charge (*18*). The χ potential itself is not measurable (*23*), but the change in χ at a given surface caused by spreading an insoluble monolayer is measurable. Hence the Kelvin assumption will require that in general $\partial\Delta_\chi/\partial q = 0$ for the surface under study, and we tested it in this form. We also tested the simple requirement of the Koenig analysis that the compensation potential should not change with the distance between the opposing surfaces, irrespective of the Kelvin assumption.

Experimental

Apart from some studies of the behavior of an ionizing air electrode in non-compensated states (*24*) our measurements of the potentials at liquid interfaces were made with the vibrating plate method (*10*) (*see* Figure 2). The reference electrode (usually silver) is vibrated sinusoidally through a small amplitude with a loud-speaker coil unit. In many experiments the frequency and amplitude of the vibration were varied. The amplitude was measured, as required, from the output of a fixed cylindrical coil partially surrounding another coil wound on the vibrator drive shaft. In early experiments the plate gap was measured with a cathetometer. This was inaccurate; later the position of the water surface was monitored with a microscope graticule. The plate was often located by measuring the gap capacity. Strictly speaking, since derivation of the gap distance from the capacity involves using the Kelvin assumption, the capacity method is only to be used to give precise relocation of the vibrating plate between experiments. Note in Figure 2 that the area of the vibrating plate is small compared with the area of the liquid surface. This introduces possible complications referred to later. In one series of experiments, plates of various sizes, shapes, and metal components were studied. No effects relevant to this report were found. Likewise, perforating the plates to reduce air drag had no effect.

In the vibrating plate studies the liquid surface was formed in a waxed Langmuir–Adam trough in which the surface could be cleaned by waxed slides. Details of how cleanliness was controlled, how monolayers were spread, and how surface pressures were measured are described elsewhere (*25*). We studied numerous monolayers but only report sodium octadecyl sulfate and distearoyl lecithin monolayers.

We also measured the capacity of a reference electrode–liquid surface system with a Wayne–Kerr Universal bridge. The capacitor was formed between a large aluminum plate (20 cm diameter) which slightly overlapped a hydrophobed dish (18 cm diameter) which held an electrolyte solution. The measuring circuits included a potentiometer to dc bias the electrode system.

We measured the surface tension of clean and monolayer covered electrolyte solutions with and without an applied field using a Wilhelmy

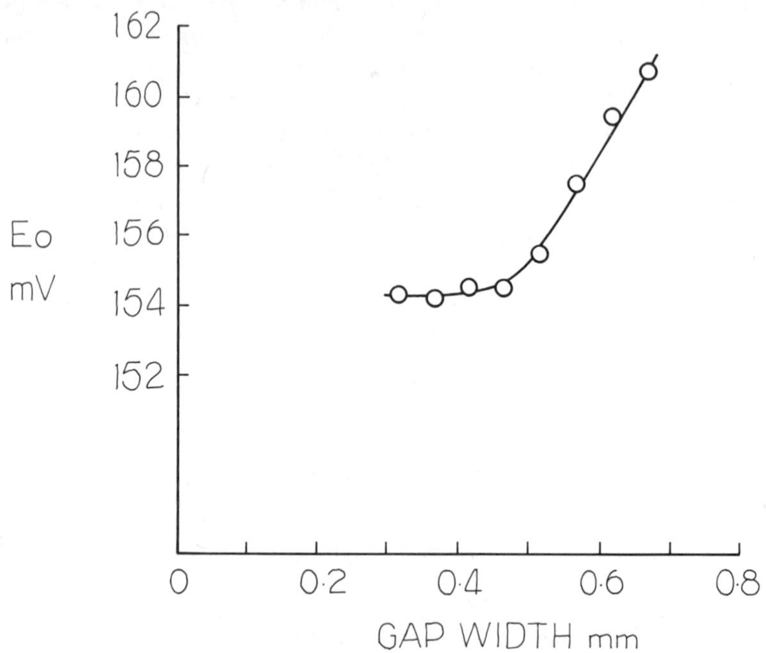

Figure 3. Variation of E_0 *with gap width.*
Trough dimensions—ca. 15 × 25 cm; plate diameter—ca. 4 cm; vibration frequency—170 Hz; amplitude—3.6 × 10^{-2}mm. The trough was filled with 0.1M NaCl at 20°C.

plate that passed through a 4 cm diameter hole in a reference electrode which covered the whole surface of the trough. No effect of field was found for a clean electrolyte or for a surface containing a monolayer of lecithin at several coverages.

Results and Discussion

The Compensation State. The compensation potentials for vibrating electrodes over clean and monolayer covered solutions were measured as a function of gap width in numerous experiments. A typical result is shown in Figure 3. In this experiment, a circular plate *ca.* 4 cm diameter was vibrating above the center of a trough *ca.* 15 × 25 cm. For a condenser well separated from other matter and with surface dimensions very large compared with the gap width, the compensation potential should not vary with separation (*18*). In our experiments these conditions apply only at very small gap widths where E_0 does become constant. For larger gaps calculation confirms that the apparent change in E_0 is primarily the result of the change in the solid angle with distance between the opposing surfaces. Effects arising from the movement of the plate with respect to the metal frame of the trough assembly etc. are also significant. If the plate is not parallel to the water surface, the range of gap widths where E_0 is constant is diminished (*22*). In measuring ΔV for a monolayer the gap width should be fixed as small as possible. The plate will, for example, detect the potential of a clean surface on the other side of a waxed slide if the gap is large, and the apparent ΔV will then depend on the plate position. This fact may be one of the causes of the variability in published values of ΔV.

Non-Compensated States. The vibration of the reference electrode produces an alternating current in the circuit (*see* Figure 2). The magnitude of the current depends at a given temperature on the geometry (including the gap width) of the electrode/surface condenser, on the magnitude of the outer potential difference between the opposing surfaces, on the amplitude of the vibration at a given frequency, and on the detailed design of the circuitry (which need not be considered here). Thus for a given frequency, amplitude, and geometry, the current measures $\Delta\psi$. If the Kelvin assumption is correct, all terms $\partial\chi/\partial q$ are zero, and the value of $-\Delta\psi$ will be given precisely by $(E - E_0)$, the bias potential applied to the vibrating capacitor. The form of the current-bias relationship for a given frequency, amplitude, and geometry will not depend on the presence or absence of a monolayer. If the Kelvin assumption is incorrect, $-\Delta\psi$ will be different from $(E - E_0)$ by an amount dependent on $\partial\chi/\partial q$. In other words, polarization effects produced in the surface dipoles by the imposed charge will appear as a back e.m.f. and will reduce $\Delta\psi$ accordingly. The form of the current-bias relationship will be different for clean and monolayer covered surfaces unless $\partial\chi/\partial q$ is fortuitously the same in both cases. The test of the Kelvin assumption is thus to measure $\Delta\psi$ as a function of $(E - E_0)$ for clean surfaces and for a variety of monolayers under identical condenser geometry and constant circuit parameters.

The monolayers chosen included cephalins and lecithins as examples of molecules expected to have a high polarizability normal to the interface (*9, 26*). Long chain sulfate and quaternary ammonium ions were studied as examples of monolayers with diffuse ionic double layers. Other experiments were made with protein films, with a long chain β-alanine, and with monolayers of equimolar mixtures of long chain sulfates and quaternary ammonium ions. The various results can be explained by the effects illustrated below for long chain sulfates and lecithin.

The first problem was to ensure that at the chosen frequency and amplitude the geometry remained constant as the field was varied. The electrostatic force across the condenser will act to pull the plate down on its mounting and raise the liquid surface below the plate in the center of the trough. The first effect was negligible, and microscopic movements of

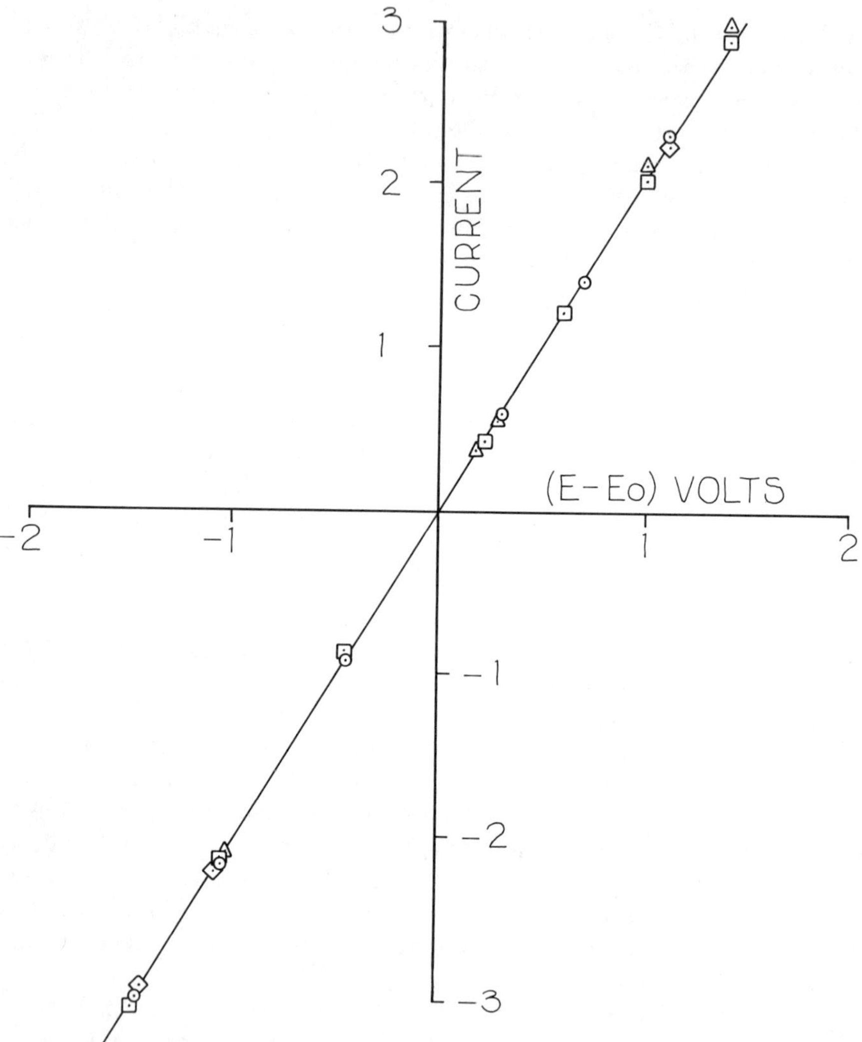

Figure 4. The ac current in the vibrating capacitor as a function of applied bias (E–E_0).

Current is given in arbitray units (peak-to-peak displacement on an oscilloscope) for a fixed frequency, amplitude, and plate size for a clean 0.1M NaCl surface—○; and for the surface covered with a monolayer of $C_{18}H_{37}SO_4Na$ at 0.5 dynes/cm—△, 1.5 dynes/cm—□, and 10 dynes/cm—◇. The monolayers are spread and compressed without moving the plate.

the water surface were only observed, with or without a monolayer, at field strengths well above those used in most experiments. Capacity measurements, as a function of applied field, revealed slight movements of the liquid surface at moderate fields. These movements were in accord with approximate calculation, on the assumption that gravity and surface tension were acting together as restoring forces. These minute movements of the liquid surface under a small electrode should give an apparent change in $\Delta\psi$ in a vibrator experiment when the field is applied. The sensitivity of our vibrator measurements was, however, too low to detect these shifts. If the lift only depends on electrode and trough geometry and on surface tension, future experiments with more sensitive equipment should show apparent changes in $\Delta\psi$ for different monolayers which will be the same at identical surface pressures if the Kelvin assumption is correct.

Many of the early experiments gave the type of result illustrated in Figure 4. The graph shows the current (measured as the peak-to-peak displacement on the oscilloscope) as a function of $(E - E_0)$ for a clean electrolyte surface and for a monolayer of sodium octadecyl sulfate at surface pressures up to 10 dynes/cm. The results for the two surface conditions are identical within experimental error. When studies began with the phospholipids, however, remarkable differences between clean and monolayer covered surfaces were soon revealed. The result shown in Figure 5 is one of the largest differences of any of the monolayers we examined. These results seemed to show that for lecithins at high surface pressures, as had been hoped, $-\Delta\psi$ was not equal to $(E - E_0)$, and the Kelvin assumption had been disproved. We discussed this result in our earlier report (*27*). However, the magnitude of the apparent changes in $\Delta\chi$ was often too large to be explained by any simple model—as for example that the zwitterions of the monolayer molecules were rotating normal to the surface; obviously some uncontrolled experimental factor was involved. One hypothesis which we tested depended on the fact that since the vibrating plate was small in comparison with the size of the trough, the region under the plate was almost at constant surface pressure. If χ is altered materially in the applied field, the surface density depends on the field at constant pressure. The large shifts in apparent $\Delta\psi$ might therefore arise from a change in molecular density in the monolayer under the plate when the field was applied. This hypothesis was disproved since the surface pressure of a lecithin monolayer under a large plate which covered the whole of a trough did not depend, within experimental error, on the field up to nearly 1000 V/cm.

At that stage, we had not tested the possibility that compressing the monolayer (with the plate *in situ*) might itself alter the gap width. This effect was the required explanation; as a monolayer of a phospholipid was compressed, the contact angles at the waxed slides and edge of the trough changed at higher surface pressures, with the liquid surface increasingly wetting the waxed surfaces. It is normal to have a small meniscus above the ground sides of a trough, and the change in contact angle causes the liquid to wet up the slides and to spread slightly over the sides of the trough, thus altering the level in the trough. Calculation showed that this effect verified our results such as those shown in Figure 5. To confirm the point, the level in the trough was observed with a microscope, and the change on compressing the film was measured. The vibrating plate was then lowered by the same amount using a vernier head. The current-bias curve was identical with that for the clean surface at the same gap width. The magnitude of the shift in the current-bias curve, illustrated in Figure 5, could also be varied by altering the meniscus height. Also, contact angle changes were found with monolayers

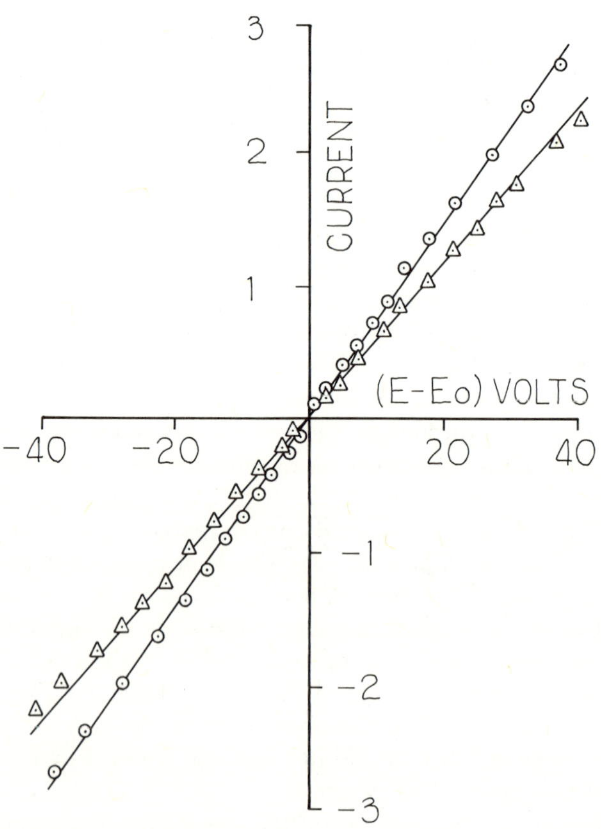

Figure 5. Current bias (E–E$_0$) *relation for a clean 0.1*M *NaCl surface—○ and for a surface with a monolayer of distearoyl lecithin at a surface pressure of 30.4 dynes/cm at 20°C—(△).*

Gap width—0.61 mm, frequency—170 Hz, amplitude—4.45 × 10^{-2}mm. The plate was rectangular, 1 × 3 cm. The current is represented in arbitrary oscilloscope readings.

other than phospholipids provided that they could be compressed to 30 dynes/cm, as for example with docosyl sulfate.

The vibrator experiments were followed by a series of tests of the Kelvin assumption using capacity measurements directly. The liquid surface was made level with the flat hydrophobic sides of the dish, with the static reference electrode slightly overlapping the dish. In this way changes in liquid height and surface density were reduced to very small proportions. After cleaning the surface and positioning the plate, the capacity of the clean system was measured as a function of field up to approximately 1000 V/cm. Lecithin monolayers were then spread without moving the plate, and the experiment was repeated. The minimum gap widths were set by two factors: the presence of ripples and the difficulty in arranging the large metal plate exactly horizontal. The final experiments were made on a vibration-free bench, but small ripples caused by spreading etc., limited the method, particularly as ripples were pulled up in the field. If the plate was not exactly horizontal, the difference in field strength across the gap caused the liquid to flow and to wet the plate. At the limit of these experiments, therefore, we were able to show that the capacity of the electrode-liquid surface system at gap widths down to 1 m*M* was not altered to an accuracy of one part in 20,000 by the presence of a monolayer of phospholipid or by a field in the presence or absence of the film. To this limit, therefore, Kelvin's assumption is confirmed.

As Koenig (*18*) notes, there is no reason to suppose that the interrelated assumptions of Kelvin, Bridgman, and Lorentz are correct, except as close approximations in most systems. From our experiments the approximations seem good. The deviations from these assumptions are worth pursuing since they offer the possibility of observing field-dependent surface effects. Our next series of experiments on liquid surfaces will require an improved control of gap geometry and of surface transverse waves. The search will also include solid semiconductors.

We conclude by expressing our pleasure in pursuing this study of Lord Kelvin's work with methods that owe so much to N. K. Adam's classic studies on insoluble monolayers.

Literature Cited

1. Frumkin, A. N., *Z. Phys. Chem.* (1924) **109**, 34.
2. *Ibid.* (1925) **111**, 190.
3. *Ibid.* (1925) **116**, 485.
4. Guyot, J., *Ann. Phys. Paris* (1924) **10**, 506.
5. Harding, J. B., Adam, N. K., *Trans. Faraday Soc. Ser. A* (1933) **29**, 837.
6. Alexander, A. E., Schulman, J. H., *Proc. Roy. Soc. Ser. A* (1937) **161**, 115.
7. Schulman, J. H., Rideal, E. K., *Proc. Roy. Soc. Ser. A* (1931) **130**, 259.
8. Betts, J. J., Pethica, B. A., *Trans. Faraday Soc.* (1956) **52**, 1581.
9. Standish, M. M., Pethica, B. A., *Trans. Faraday Soc.* (1968) **64**, 1113.

10. Yamins, H. G., Zisman, W. A., *J. Chem. Phys.* (1933) **1,** 656.
11. Phillips, J. N., Rideal, E. K., *Proc. Roy. Soc. Ser. A* (1955) **232,** 159.
12. Mingins, J., unpublished data.
13. Guastalla, J., *C. R. Acad. Sci. Paris* (1947) **224,** 1498.
14. Kenrick, C. B., *Z. Phys. Chem.* (1896) **19,** 265.
15. Randles, J. E. B., *Trans. Faraday Soc.* (1956) **52,** 1573.
16. Parsons, R., Rubin, B. T., *J. Chem. Soc. Faraday Trans. I* (1974) **70,** 1636.
17. Case, B., Parsons, R., *Trans. Faraday Soc.* (1967) **63,** 1224.
18. Koenig, F. O., *C.R. Com. Intern. Thermodyn. Cinet. Electrochem., 3rd* (1951), 299.
19. Lord Kelvin, *Proc. Roy. Soc., Edin.* (1897–98) **22,** 118.
20. Bridgman, P. W., *Phys. Rev.* (1919) **14,** 306.
21. Lorentz, H. A., *Wied. Ann.* (1889) **36,** 593.
22. Uhlig, H. H., *J. Appl. Phys.* (1951) **22,** 1399.
23. Parsons, R., "Modern Aspects of Electrochemistry," Bockris, J. O. M. and Conway, B. E., Eds., Butterworths, London, 1954.
24. Pethica, J. B., Saraga, L. Ter Minassian, unpublished data.
25. Mingins, J., Owens, N. F., Iles, D. H., *J. Phys. Chem.* (1969) **73,** 2118.
26. Pethica, B. A., "Surface Activity and the Microbial Cell," Society of Chemical Industry, Monograph No. 19, p. 85 (1964).
27. Pethica, B. A., Standish, M. M., Mingins, J., Iles, D. H., *Nature* (1965) **205,** 348.

RECEIVED November 4, 1974.

10

The Effect of Ethylenic and Acetylenic Groups on the Properties of Fatty Acid Monolayers

HARRY L. WELLES[1] and GEORGE ZOGRAFI

School of Pharmacy, University of Wisconsin, Madison, Wis. 53706

CHARLES M. SCRIMGEOUR and FRANK D. GUNSTONE

St. Andrews University, St. Andrews, Scotland

The monolayer properties of some cis- *and* trans-*octadecenoic acids and a series of octadecynoic acids have been measured at 25°C. Within the cis series identical film behavior is noted for all isomers from Δ4 to Δ14, and greater film condensation occurs as the double bond is moved closer to each end of the chain. The same is noted for the trans series, except that identical behavior occurs from Δ5 to Δ13. The similar pattern of change for cis and trans isomers suggests that chain configuration is not the major factor in determining their film properties. Octadecynoic acid films are more expanded than those of the ethylenic series, with progressively greater expansion going from Δ2 to Δ15. This suggests the importance of chain–substrate interactions.*

An important constituent of all fatty acids and their derivatives is the alkyl chain. These chains vary in length (number of carbons) and in the extent of branching and polar group substitution. Chains also vary in the type, number, position, and geometric configuration of unsaturated carbon–carbon bonds. Such differences in alkyl chain structure are largely responsible for the broad variation in the biological properties of

[1] Present address: College of Pharmacy, Dalhousie University, Halifax, Nova Scotia, Canada.

lipids. Because these lipids accumulate and orient at interfaces where water is one of the phases, the study of their interfacial properties has been of interest for many years. The insoluble monomolecular film technique has often been utilized for this purpose since it allows, among other things, precise control of the surface concentration of the film-forming molecules (*1*).

Since the work of Langmuir (*2*), it has been known that an unsaturated fatty acid, such as *cis*-Δ9-octadecenoic acid (oleic acid), has a greater tendency to spread on water as a monomolecular film than the corresponding saturated acid—octadecanoic acid (stearic acid). This observation, coupled with the fact that saturated acids with shorter alkyl chain lengths also have a greater spreading tendency, led to the conclusion that a lower tendency for interaction between alkyl chains in the film leads to greater spreading tendency (*3*). A few studies reported the influence that cis and trans isomerization and position of the double bond have on fatty acid monolayer properties (*3, 4, 5, 6, 7, 8, 26*). Comparisons of oleic acid with its trans isomer, elaidic acid, and of the C_{22} isomers, erucic and brassidic acids, show that cis isomers produce more expanded films than trans isomers, presumably because of a configuration which inhibits close packing and strong chain–chain interaction. The influence of bond position was studied previously by comparing *trans*-Δ2-octadecenoic acid with its trans-Δ6, trans-Δ9, and trans-Δ11 isomers, and *cis*-Δ6-octadecenoic acid with its cis-Δ9 and cis-Δ11 isomers, whereas the Δ2-isomer is more condensed than the Δ9-isomer, *i.e.*, more like stearic acid; little difference is observed between cis-Δ6, cis-Δ9 and cis-Δ11 and between trans-Δ6, trans-Δ9 and trans-Δ11. One study (*5*) also reported that Δ9-octadecynoic acid formed a more expanded film than oleic acid, possibly because of a triple bond-water interaction which reduces the tendency of alkyl chains to interact with one another.

Recent syntheses of unsaturated fatty acids that differ systematically in alkyl chain structure (*9, 10*) make it possible to study the influence of unsaturation on a number of chemical and biochemical processes (*11*). It seemed to us that many questions concerning the influence of alkyl chain structure on fatty acid interfacial properties could be answered better by studying the monomolecular film properties of these compounds. This paper examines the general surface film properties of some cis and trans ethylenic and acetylenic 18-carbon acids.

Experimental

Materials. High purity saturated fatty acids, *cis-* and *trans*-Δ9-octadecenoic acids (oleic and elaidic acids) and 12-hydroxystearic acid, were obtained from Applied Science Laboratories. *Trans*-3-octadecenoic acid was a gift from Charles Sih (School of Pharmacy, University of Wisconsin). The remaining unsaturated acids were synthesized and purified as previously reported (*9, 10*). For convenience, the unsaturated acids are abbreviated in a form which indicates the position of the unsaturated bond and whether it is *cis*-ethylenic, *trans*-ethylenic, or acetylenic. For example, oleic, elaidic, and Δ9-octadecynoic acids are written as Δ9*c*, Δ9*t*, and Δ9*a*. The spreading solvents were Lipopure hexane and chloroform (Applied Science Laboratories). The aqueous subphase for all studies was 0.01*N* HCl; it maintained the acids in an undissociated form. Water was triple distilled from a permanganate solution.

Procedure. Surface pressure, π, and surface potential, ΔV, were measured at 25° ± 0.2°C on a Teflon-coated trough, 30 cm × 10 cm × 2 cm. Surface pressure was determined by the Wilhelmy plate technique using a 1-cm platinum plate attached to a Cahn model RG electrobalance. Surface potential was measured with a Keithley model 610B electrometer using a 1-mc ^{241}americium air electrode (Amersham-Searle) and a calomel reference electrode. The apparatus was designed to produce constant flow of nitrogen which would preclude the possibility of fatty acid oxidation. Since no such problem was encountered, nitrogen was not used.

Films were spread as solutions in hexane, except for Δ2*a*, Δ16*a*, and Δ17*a* which were more soluble in chloroform. An Agla micrometer syringe (Burroughs–Wellcome) was used to administer the desired quantities. Films were compressed continuously at rates from 0.2–0.4 A^2/molecule/min, depending on the amount of fatty acid on the surface. Prior investigation indicated that below the equilibrium spreading pressure of each film, properties were independent of compression rate. Equilibrium spreading pressures (*ESP*) were measured for each compound by placing a small sample of the pure fatty acid on the surface and measuring the surface pressure which resulted from the equilibrium between the monolayer and bulk compound. It took from 1 min to 12 hr for the various films to reach an apparent equilibrium.

Results

Equilibrium Spreading Pressures. The *ESP* values at 25°C for the various unsaturated fatty acids are presented in Table I. They represent the average of at least two measurements which did not differ by more than 0.5 dyne/cm. Since the *ESP* reflects the tendency of molecules to leave the bulk phase and to spread over the aqueous phase (*12*), we expect that it is related to a combination of the intermolecular forces in the bulk and the hydrophilic properties of the molecule, working against each other. Comparison of these values with the reported melting points (*9, 10*) indicates no exact correlation except that cis isomers with lower melting points than trans isomers generally exhibit larger *ESP* values. However the acetylenic derivatives, with melting points equivalent to the trans isomers, have *ESP* values similar to those for the cis isomers. Likewise, within any of the three series, despite a tendency for an increase in melting point when the unsaturated group is situated near

Table I. Equilibrium Spreading Pressures (dynes/cm) for Various Compounds at 25°C

Compound	*cis*	*trans*	*Acetylene*
Δ2	21.0	—	44.0
Δ3	44.7	36.1	—
Δ4	32.5	9.0	30.3
Δ5	31.0	19.0	26.4
Δ6	35.0	—	34.9
Δ7	30.2	—	30.4
Δ8	30.0	—	31.9
Δ9	30.0	21.4	29.6
Δ10	30.0	24.7	27.3
Δ11	30.5	—	23.8
Δ12	31.4	22.8	31.7
Δ13	33.8	24.3	11.4
Δ14	27.5	10.0	12.4
Δ15	30.0	19.0	28.4
Δ16	21.4	—	9.5
Δ17	26.1	—	15.7

either end of the chain, no systematic decrease in *ESP* occurs. Note the rather large *ESP* values for Δ2*a*, Δ3*c*, and Δ3*t*. Factors, other than those that determine chain–chain interaction, seem to play an important role in spreading; the hydrophilic nature of the chain is one of these factors. Note also that the *ESP* is the highest surface pressure to which a film can be compressed without introducing the possibility of time-dependent behavior and film instability (*13*).

Surface Pressure *vs.* Area Per Molecule. Figure 1 shows the π–A curves for stearic acid and the three Δ9-unsaturated acids used. It illustrates, as noted by others (*3, 7*), that the order of film expansion for the different types of fatty acids is: saturated < *trans*-ethylenic < *cis*-ethylenic < acetylenic. Figures 2–4 indicate the results of compressing the unsaturated acids that differ in the configuration of the double bond and in the position of the double or triple bond on the alkyl chain.

A comparison of the cis acids reveals no measurable differences in film behavior when the double bond is situated between the Δ4 and Δ14 positions (Figure 2). Moving the double bond to either end of the molecule, however, increases film condensation. Film condensation, or a reduced tendency to cover the surface, is reflected by the area under the π–A curve; the greater the area, the greater the expansion of film or the less the extent of condensation. In Table II we tabulated the area under the curve from 50 to 26 A^2 per molecule for some *cis*- and *trans*-ethylenic acids and some saturated acids. These areas could be expressed as the free energy of compression, but, as suggested by Gershfeld and Pagano (*14*), this is only thermodynamically valid if surface pressure

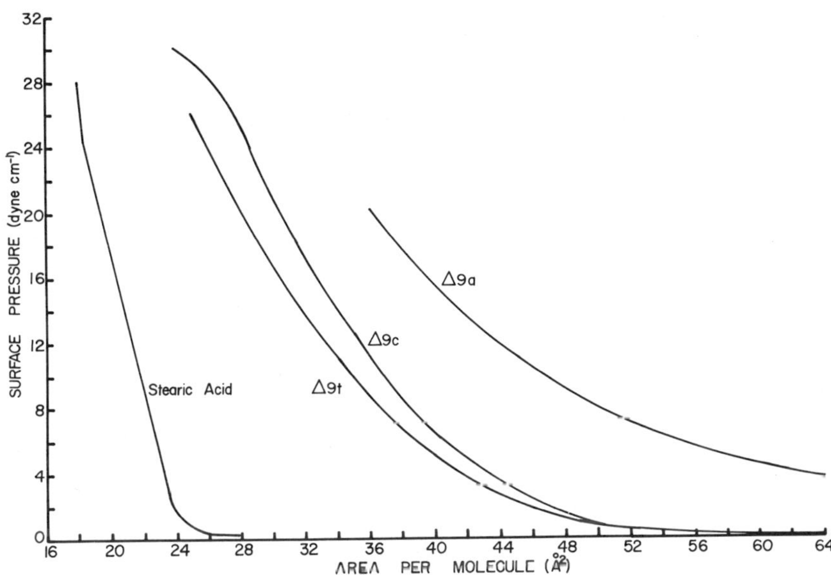

Figure 1. Comparison of surface pressure vs. area per molecule of some 18-carbon fatty acids at 25°C

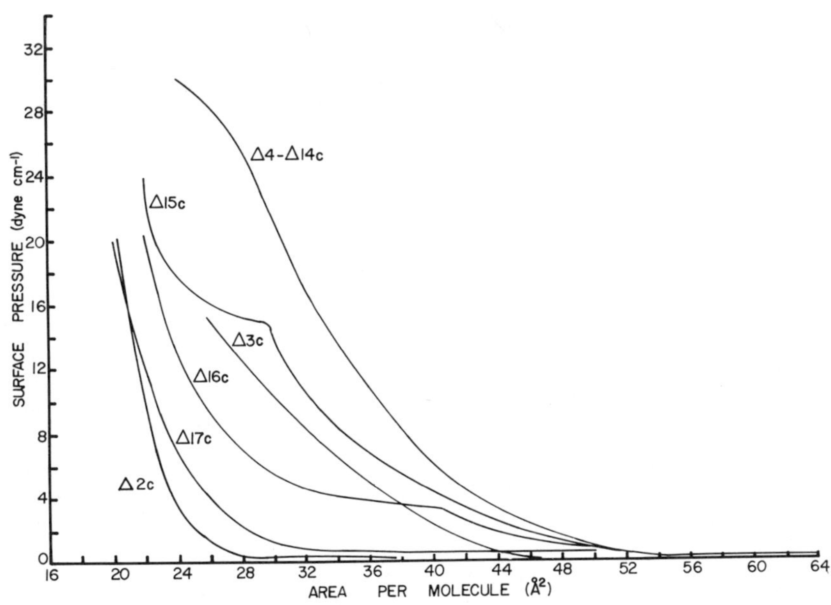

Figure 2. Surface pressure vs. *area per molecule for* cis-*octadecenoic acid isomers at 25°C*

measurements are made to where the film behaves as a two-dimensional gas. Thus, these values only give a convenient means of comparing various films over the same range of area per molecule.

Comparisons of the various trans compounds in Table I, and shown in Figure 3 and in Table II, reveal a pattern similar to the cis series except that the point at which film changes occur shifts one bond closer to the middle of the chain. As expected, trans compounds always give more condensed films than their cis isomers; however, the difference between cis and trans does not remain the same for different bond posi-

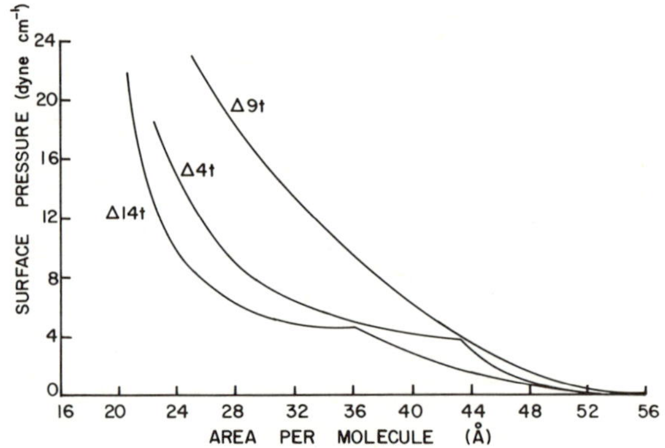

Figure 3. Surface pressure vs. *area per molecule for some* trans-*octadecenoic acid isomers at 25°C*

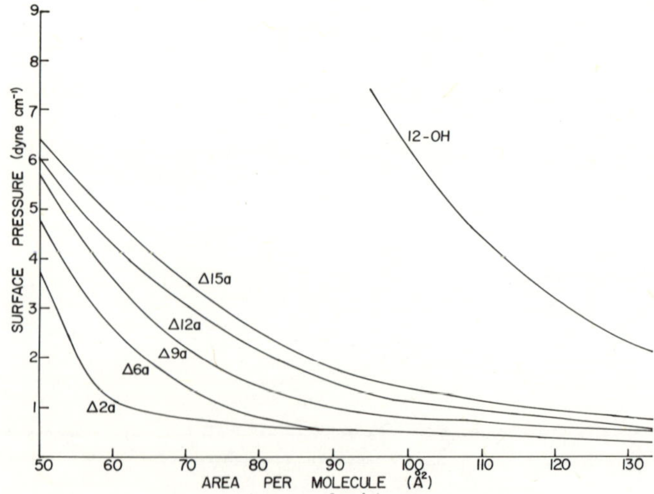

Figure 4. Surface pressure vs. *area per molecule for some octadecynoic acid isomers at 25°C*

Table II. Area Under the π–A Curve Between 26 and 50 A² per Molecule for Some Saturated and Ethylenic Acids[a] at 25°C

Compound	*Area*[b]
Δ2c	5
Δ3c	116
Δ9c	268
Δ15c	166
Δ16c	87
Δ17c	23
Δ3t	15
Δ4t	107
Δ9t	206
Δ14t	81
Δ15t	52
18:0	3
16:0	13
15:0	112
14:0	142

[a] Ethylenics not listed have values equal to their Δ9 isomer.
[b] Units of area are A²-dyne/molecule/cm.

tions. For example, the ratio of areas under the curve for cis to trans is 7.7, 1.3, and 3.2, for the Δ3, Δ9, and Δ15 position.

The results for the complete series of acetylenic acids, except Δ3*a*, are given in Figure 5 by a plot of the area under the curve *vs.* triple bond position (*see* Figure 4 for some of the π–A curves). All of the compounds form more expanded films than the corresponding ethylenic acids; consequently, the range of areas used for the comparisons in Figure 5 was chosen arbitrarily as 140 to 50 A²/molecule. Here, the pattern of change is quite different from that with the ethylenic series. The Δ2*a* and Δ4*a* are relatively condensed and similar to each other.

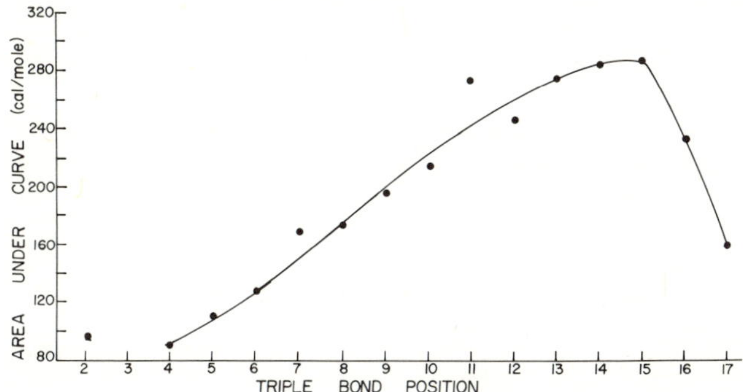

Figure 5. Area under the π–A curve from 50–140 A²/ molecule as a function of triple bond position for the octadecynoic acid isomers at 25°C

When the triple bond is moved further from the carboxyl group, the film is more expanded. With the triple bond in the Δ15 position, expansion is maximum, and the Δ16 and Δ17 compounds are more condensed. Assuming that expanded film behavior might arise partially from a water–triple bond interaction, (*5*), we studied 12-hydroxystearic acid. As Figure 4 shows, the hydroxy group causes greater expansion of the film than the corresponding acetylenic acid, but its equilibrium spreading pressure is only 9.7 dynes/cm.

Surface Potential Studies. Surface potential measurements were made on all compounds at the same time surface pressure was measured. Since it is preferable to have a parameter which reflects only changes in the fatty acid and surface water dipole moments and not changes resulting from merely increasing the number of dipoles per unit area, surface potential values, ΔV, were converted to an apparent surface dipole moment, $\mu_\perp$ (*1*), where

$$\mu_\perp = \frac{\Delta V}{12\pi n}$$

and n is molecules/A^2, π is 3.14, and $\mu_\perp$ has millidebye units (mD). Typical plots of ΔV and $\mu_\perp$ *vs.* area/molecule for an ethylenic and acetylenic acid are given in Figure 6. The value of $\mu_\perp$ increases as the film is compressed and reaches a maximum when the surface pressure

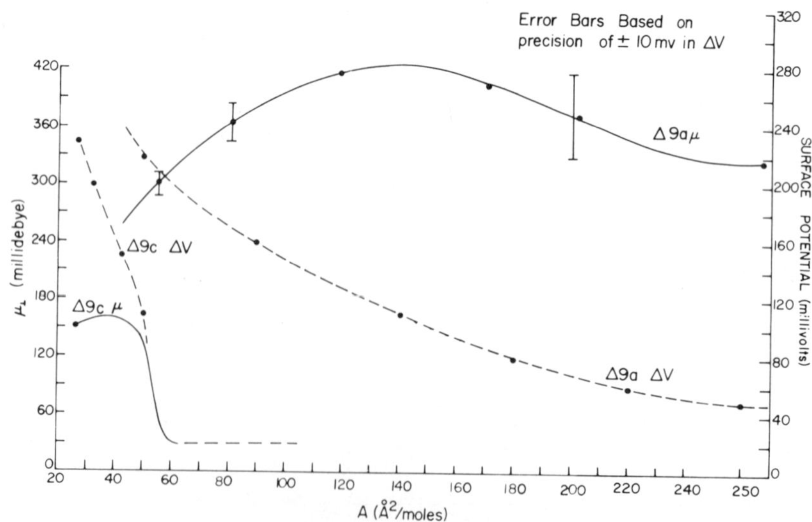

Figure 6. Surface potential and surface dipole moment vs. *area per molecule for* cis-*Δ9-octadecenoic (oleic) acid and Δ9-octadecynoic acid at* 25°C

Table III. Surface Dipole Moments, $\mu_\perp$, for Various Octadecenoic Acids[a] at 26 A^2 per Molecule

Compound	$\mu_\perp (mD)$	*Compound*	$\mu_\perp (mD)$
Δ2c	240	Δ3t	300
Δ3c	230	Δ4t	270
Δ4c	160	Δ5t	220
Δ9c	160	Δ9t	190
Δ14c	160	Δ12t	210
Δ15c	180	Δ13t	220
Δ16c	190	Δ14t	210
Δ17c	100	Δ15t	200

[a] Ethylenics not listed have values essentially the same as their Δ9 isomer.

begins to increase significantly. At lower areas, $\mu_\perp$ remains fairly constant or decreases gradually until collapse.

Table III lists values for the various cis and trans isomers at 26 A^2/molecule—the lowest area where all ethylenic compounds could be compared. In all cases these values are within 20 mD of the maximum observed value of $\mu_\perp$. Compounds not listed in Table III give values essentially the same as their Δ9 isomers. At 26 A^2/molecule, a slightly higher $\mu_\perp$ is observed with trans compounds relative to cis compounds, but a similar pattern is followed. Table IV includes the maximum values for the various acetylenic compounds and the value at 45 A^2 to show how $\mu_\perp$ decreases with compression. When it was possible to compress the acetylenics to lower areas, values similar to the ethylenics and satu-

Table IV. Surface Dipole Moments, $\mu_\perp$, in mD for Various Octadecynoic Acids

Compound	*45 A^2*	*Maximum Value*
Δ2a	270	280
Δ4a	280	280
Δ5a	290	300
Δ6a	290	340
Δ7a	330	420
Δ8a	290	440
Δ9a	280	420
Δ10a	240	440
Δ11a	260	490
Δ12a	260	440
Δ13a	250	420
Δ14a	280	420
Δ15a	280	390
Δ16a	260	370
Δ17a	140	340
12-OH	collapse	500

rated fatty acids were observed. For example, Δ16*a*, which can be compressed to 26 A^2/molecule, has a surface dipole moment of 370 mD at its maximum and 160 mD at 26 A^2/ molecule. This indicates that the acetylenics can assume configurations similar to the saturated and ethylenic acids at low areas, but a major difference in overall dipole moment exists at the higher areas.

Previously, it was reported (*4, 5*) that Δ2*t* had a greater surface dipole moment than Δ9*t* because of the proximity of the double bond to the carboxyl group. We also observed that Δ2*c,* Δ3*c,* Δ3*t,* and Δ4*t* gave higher dipole moments at 26 A^2/molecule than the other ethylenics. Likewise, note that Δ2*a* has a higher value at 26A^2/molecule than any of the other acetylenics which can be compressed to that pressure. Whether this effect arises from a change in the dipole moment of the carboxyl group or a change in the water dipole contribution, or both, is unknown; however, the effect apparently can extend as high as the Δ4 position in the case of the trans isomer.

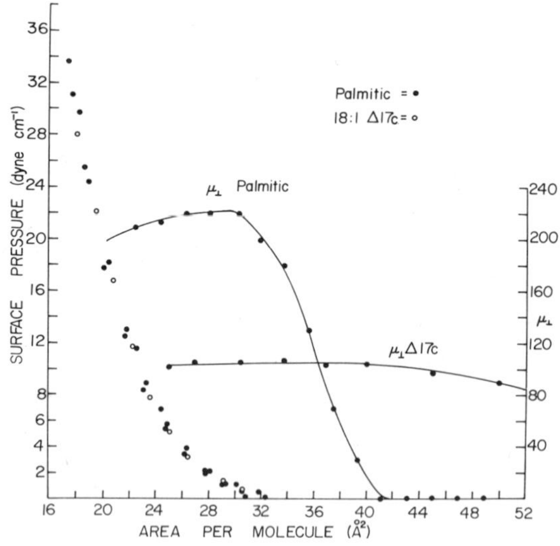

Figure 7. Surface pressure and surface dipole moment vs. *area per molecule for* cis-Δ*17-octadecenoic acid and hexadecanoic (palmitic) acid at 25°C*

The two Δ17 acids exhibit lower $\mu_{\perp}$ values than the other members of their series (Tables III and IV). This is seen also in Figure 7, where despite identical π–*A* curves for Δ17*c* and palmitic acid, $\mu_{\perp}$ for Δ17*c* is much lower. These results appear to arise from the different contributions to the overall dipole moment from the terminal groups. The dipole moment of the terminal carbon–hydrogen bond in a saturated molecule is about 400 mD with the hydrogen atoms positive (*15*) while the moments for a R—CH═CH_2 and a R—C≡CH group are about 360 and 750 mD, with the R group positive (*16*). Consequently, the presence of a double or triple bond at the end of the fatty acid acyl chain results in considerable lowering of the net molecular dipole moment while its presence near the carboxyl group raises $\mu_\perp$. For all other compounds the values of $\mu_\perp$ do not appear to be very different at the lowest areas of compression.

Discussion

To put these results into better perspective we consider what occurs when fatty acids are spread as monolayers on water. The presence of a polar group facilitates the spreading of the fatty acid molecules over the surface. At very high areas/molecule, the fatty acid molecules are relatively independent of one another. However, the alkyl chains, as well as the polar groups, must be interacting with water molecules. As compression takes place, association into surface micelles or islands occurs because of chain–chain interaction and because the chains are hydrophobic. The longer the chain or the less the polarity of the head group, the lower the surface pressure and the higher the area/molecule where this occurs (*17*). A reduction in temperature will significantly promote such behavior. The tendency of chains to associate is determined by the relative strength of chain–chain and chain–water interactions, with the latter interaction working to offset the self-association of chains. For example, Davies (*18*) indicated that the energy of adhesion per CH_2 group per mole (to water) is about 1700 cal whereas the introduction of a hydroxy group on the chain contributes an additional 3400 cal/mole. Unsaturated groups should produce energies of adhesion which lie between these values. The importance of such interaction is seen when one compares fatty acid film behavior on water with that on mercury (*19*), glycerol (*20*), and aqueous urea solutions (*21*). In each case, compared with water, there is a marked tendency for film expansion. Increases in chain–subphase attraction caused by increased dispersion forces explain the effect of mercury and, perhaps, glycerol; however, the effect of urea suggests that changes in water structure at the surface most likely contribute by providing an entropy change which favors chain–subphase contact.

We can now examine the influence of unsaturation on alkyl chain behavior. This significant influence is seen by comparing the total enthalpy and entropy of compression of oleic acid with those for stearic acid (*14*). The values at 25°C are: −6,537 cal/mole and −29 e.u. for oleic acid and −15,684 cal/mole and −57 e.u. for stearic acid. To understand this behavior, we determine what effect introducing a double or triple bond has on the interaction between two alkyl chains, and later, we consider chain–substrate interactions. Salem (*22*) estimated that the net attractive force between two parallel saturated chains, 18 carbons long, at a distance corresponding to 20 A^2/molecule is about 8.4 kcal/mole. Using this datum, we made a similar calculation for comparison, substituting one double bond in the chain and fixing the separation at 20 A^2/molecule. When the bond is in the middle of the chain, the net attractive force is less than that for stearic acid by about 0.5 kcal/mole.

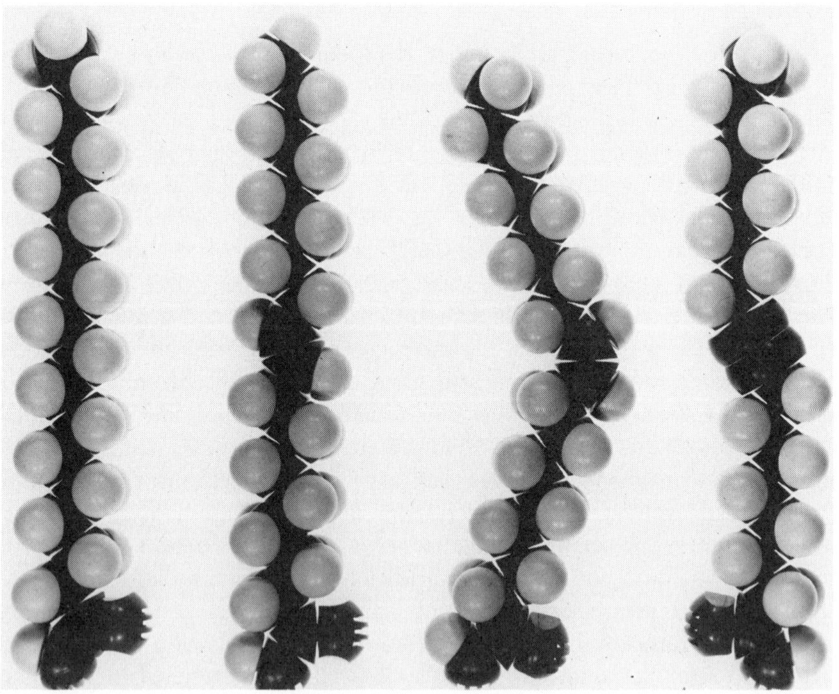

Figure 8. Molecular models of octadecanoic acid, trans-Δ9-*octadecenoic acid,* cis-Δ9-*octadecenoic acid, and* Δ9-*octadecynoic acid; left to right*

This is equivalent to about one less CH_2 group. As the bond is moved toward either end, the value decreases to about 0.3 kcal/mole; the loss of two hydrogens offsets the increased polarizability of the double bond. Similar data which would allow this calculation for the triple bond are not available, but based upon the small differences in melting points for acetylenics and trans-ethylenics and their lower melting points relative to saturated acids, we assume that a similar small reduction in chain–

chain interaction energy occurs. Thus, without considering any configurational effects, which might alter the possible distance of approach for unsaturated molecules, we conclude that changes in the intrinsic interaction energy between alkyl chains are insufficient to produce the observed film expansion.

In Figure 8, molecular models of saturated, cis- and trans-ethylenic, and acetylenic chains are shown. Here, the cis isomer with the double bond toward the middle of the chain, more than any other, should produce configurational effects that deter chain–chain interaction. Calculation of interaction energy based upon packing in a crystal for oleic acid (22.5 A^2/molecule) and stearic acid (20.5 A^2/molecule) indicates that there should be a decrease in the attractive energy between the chains of about 1.8 kcal/mole, or roughly 20%. Liquid, crystalline, fatty acid bilayers (23) from x-ray and spin labeling studies suggest that the great area occupied by a $\Delta 9c$ chain relative to the saturated acid leads to a reduction in molar attraction of 300 cal per CH_2 group. This adds up to more than a 5 kcal/mole decrease rather than 1.8 kcal/mole, presumably because the distance of approach in the bilayer is much greater than it is in the pure crystal.

Current data are insufficient to quantitate, in thermodynamic terms, changes that take place in the monolayer. However, it appears from the greater expansion of cis films relative to the trans and saturated acids, that configurational factors are important, particularly when the double bond is positioned in the middle of the chain. The cis isomer in Figure 9 shows that as the double bond moves toward either end of the chain, many uninterrupted CH_2 groups are produced along with greater chain linearity. When the number of uninterrupted groups reaches 13 to 14 (Table II), the films appear to give areas under the curve closely related to saturated fatty acids that have a similar number of methylene groups. Thus, at these positions cis configurational effects on chain–chain interaction are no longer predominant.

Since shifting the double bond position in the trans series apparently does not change the configuration of an alkyl chain (*see* Figure 8), this cannot be the factor contributing to a lack of surface pressure *vs.* area/molecule difference when the double bond is between the $\Delta 5$ and $\Delta 13$ positions. This effect, likewise, cannot be the result of reduced intrinsic interaction energy, as indicated by the previously described calculations since the expansion of trans films relative to stearic acid is generally equivalent to more than one CH_2 group. For example, $\Delta 9t$ has an area under the curve corresponding to a 12- or 13-carbon saturated acid. In addition, the dissimilarities over a wide range of double bond positions in both the cis and trans series and the relatively small differences between cis and trans isomers compared with differences between saturated and ethylenic acids suggest that factors other than configurational predominate in determining the monolayer properties of both cis and trans compounds. Since the double bond is more hydrophilic than a CH_2 group (*24*), it is conceivable that upon compression it would tend to remain associated with water and thereby cause the molecule to be more expanded. Although this was originally suggested by Langmuir (*2*) as an explanation for the expanded behavior of oleic acid, Gaines (*25*) strongly attributed the effects of unsaturation to packing and not to attraction between double bonds and water.

In view of the present results, however, one is forced to conclude that there must be another more general factor which causes ethylenic fatty acids to be expanded, and this factor must be related to a chain–subphase interaction.

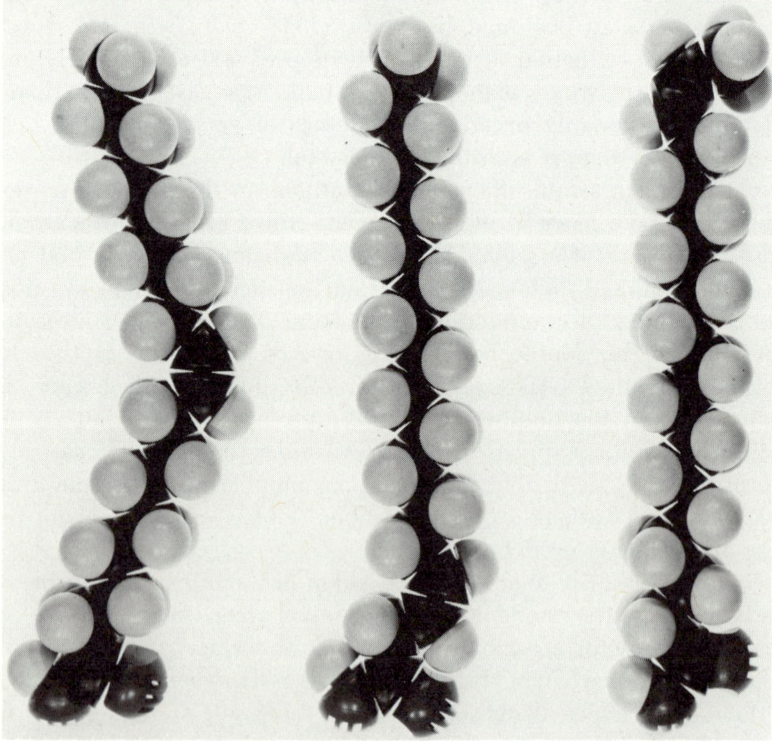

Figure 9. Molecular models of cis-Δ9-*octadecenoic acid,* cis-Δ2-*octadecenoic acid, and* cis-Δ15-*octadecenoic acid; left to right*

To gain insight into the specific interactions between double bonds and water, compare the results from the acetylenic series, where triple bond–water interactions are more likely (*5*), with the results from the ethylenic series. The view that water–triple bond interactions are significant is strengthened by the high surface dipole moments first observed upon compression, followed by a rapid decrease to values equal to those obtained with the ethylenic and saturated acids. This latter result reflects the lack of any contribution by the triple or double bonds to the overall dipole moment when the alkyl chain has assumed a more vertical orientation. Similar surface dipole moments of acetylenics with 12-hydroxystearic acid support this idea. Similar values of surface dipole moment have been reported for other situations where the alkyl chain is most likely bound to the aqueous surface at high areas, *i.e.*, dihydroxy stearic acid isomers (*26*) and α-ω, dicarboxylic acids (*27*).

At high areas alkyl chains with a triple bond can lie fully extended on the surface, giving rise to higher surface pressures and significant increases in surface dipole moment. As illustrated with the molecular models in Figure 10, we expect that upon compression the chain might first 'lift off" from the aqueous surface to the point where the triple bond is positioned. Such a mechanism was suggested for unsaturated fatty acid film behavior on mercury where such chain–substrate interactions are quite strong (*19*). If this occurs in the present case, closer packing and more film condensation should occur when the triple bond is closer to the carboxyl group, as observed. However, when the triple bond is positioned beyond carbon 15 many uninterrupted methylene groups are in the alkyl chain. If these predominate, the film then would be expected to revert to a more condensed state. Obviously, the triple bond is less polar than a hydroxy group, and hence we see much greater expansion for 12-hydroxystearic acid. We wonder what would happen if the hydroxy group were placed on the terminal chain position and whether behavior parallel to the acetylenics would occur. So far we know that going from position 2 to position 9, using the methyl ester of hydroxystearic acid, produces more expansion (*13*).

In contrasting the ethylenic with the acetylenic series, a comparison of surface dipole moments at high areas per molecule (Table III, Table IV) indicates no real differences with saturated acids. Likewise, there is no change in $\mu_\perp$ upon further compression, yet we might expect some change with "lift off" if the chain–water interaction was significant. Also, changing the triple bond position produces different π–A curves, whereas cis and trans acids have no film differences for positions covering a large portion of the alkyl chain. Thus, double bond–water interactions appear to have minor effects, based upon these observations.

A factor which has not been considered is the entropy change associated with the compression of a monolayer. In a manner analogous to micelle formation in solution, we would expect both "lift off" and the association of oriented chains to involve a negative entropy change; the removal of the alkyl chain from contact with water should contribute

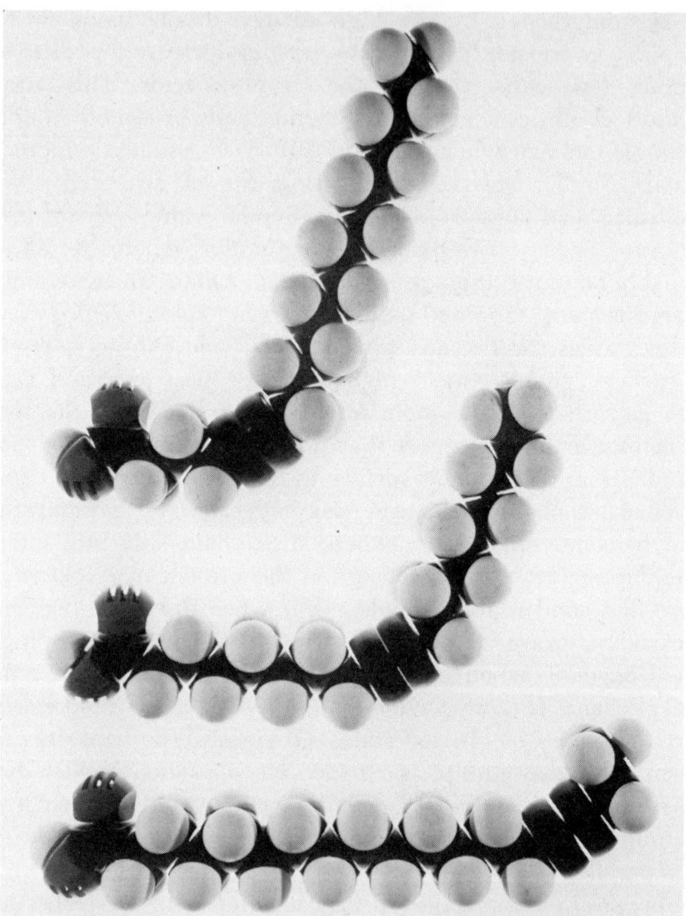

Figure 10. Molecular models showing possible effects of monolayer compression on the configuration of (top to bottom) Δ4-octadecynoic acid, Δ8-octadecynoic acid, and Δ14-octadecynoic acid

a positive entropy change because of changes in water structure. Gershfeld and Pagano (*14*) concluded that changes in water structure are likely to occur in the formation of each monolayer state; this is supported by the tendency of urea–water solutions to favor more expanded monolayer properties of fatty acids relative to a pure water subphase (*21*). Thus, we suggest that the presence of a double bond in contact with water could reduce the favorable portion of the entropy change associated with the lift off of alkyl chains; this favors continued contact with water and initiates more expanded film behavior relative to saturated chains. Differences between cis and trans isomers would reflect more of an effect caused by packing while differences between ethylenics and acetylenics would be primarily caused by stronger water–triple bond interactions for the latter compounds.

Conclusions

The influence of unsaturation on alkyl chain behavior at interfaces involving water is significant. At higher areas chain–substrate interactions appear to predominate for the acetylenic series, but their role with the ethylenics is not as clear. At lower areas, after chain lift off, the acetylenics act as if chain–chain interactions occur easily. Only the cis-ethylenic series suggests a possibility for a configurational effect which prevents close packing and interaction. However, the trans acids, less expanded than the cis series, are much more expanded than saturated acids and depend on a double bond position which is very similar to the cis series. This suggests a more general effect by the double bond, possibly an entropy difference associated with chain–water interactions.

Acknowledgment

The authors thank Ibrahim Jalal for assisting with the equilibrium spreading pressure measurements.

Literature Cited

1. Gaines, G. L., "Insoluble Monolayers at Liquid-Gas Interfaces," Wiley-Interscience, New York, 1966.
2. Langmuir, I., *J. Amer. Chem. Soc.* (1917) **39,** 1848.
3. Adam, N. K., *Proc. Roy. Soc.* (1922) **A101,** 516.
4. Hughes, A. H., Rideal, E. K., *Proc. Roy. Soc.* (1933) **A140,** 253.
5. Hughes, A. H., *J. Chem. Soc.* (1933) 338.
6. Marsden, J., Rideal, E. K., *J. Chem. Soc.* (1938) 1163.
7. Schneider, V. L., Holman, R. T., Burr, G. O., *J. Phys. Chem.* (1949) **53,** 1016.
8. Glazer, J., Goddard, E. D., *J. Chem. Soc.* (1950) 3406.
9. Gunstone, F. D., Ismail, I. A., *Chem. Phys. Lipids* (1967) **1,** 209, 264.
10. Barve, J. A., Gunstone, F. D., *Chem. Phys. Lipids* (1971) **7,** 311.
11. Gunstone, F. D., *J. Amer. Oil Chem. Soc.* (1973) **50,** 486A.
12. Cary, A., Rideal, E. K., *Proc. Roy. Soc.* (1925) **A109,** 301.
13. Sims, B., Zografi, G., *J. Colloid Interface Sci.* (1972) **41,** 35.
14. Gershfeld, N. L., Pagano, R. E., *J. Phys. Chem.* (1972) **76,** 1231.
15. Demchak, R. J., Fort, T., *J. Colloid Interface Sci.* (1974) **46,** 191.
16. Dale, J., in "Chemistry of Acetylenes," H. G. Viehe, Ed., p. 30, Marcel Dekker, New York, 1969.
17. Langmuir, I., *J. Chem. Phys.* (1933) **1,** 756.
18. Davies, J. T., *Trans. Faraday Soc.* (1953) **49,** 949.
19. Smith, T., *Advan. Colloid Interface Sci.* (1962) **3,** 161.
20. Cadenhead, D. A., Demchak, R. J., *Biochem. Biophys. Acta* (1969) **176,** 849.

21. Cadenhead, D. A., Bean, K. E., *Biochem. Biophys. Acta* (1962) **290**, 43.
22. Salem, L., *J. Chem. Phys.* (1962) **37**, 2100.
23. Axel, F., Seelig, J., *J. Amer. Chem. Soc.* (1973) **95**, 7972.
24. Shafrin, E. G., Zisman, W. A., *J. Phys. Chem.* (1967) **71**, 1309.
25. Gaines, G. L., "Insoluble Monolayers at Liquid-Gas Interfaces," p. 185, Wiley-Interscience, New York, 1966.
26. Goddard, E. D., Alexander, A. E., *Biochem. J.* (1951) **47**, 331.
27. Jeffers, P. M., Daen, J., *J. Phys. Chem.* (1965) **69**, 2368.

RECEIVED September 23, 1974. Work supported by grant from the Research Committee, Graduate School, University of Wisconsin—Madison.

11

The Chain-Length Compatibility and Molecular Area in Mixed Alcohol Monolayers

D. O. SHAH and S. Y. SHIAO

Departments of Chemical Engineering and Anesthesiology, University of Florida, Gainesville, Fla. 32611

The average area per molecule in mixed monolayers of alkyl alcohols depends on the chain-length compatibility of the surfactant molecules. The minimum area/molecule was observed when both components possessed the same chain length. The results were interpreted in terms of the thermal motion of surfactant molecules in the monolayer. There was no significant effect of film compression rate on surface pressure–area curves of pure component films, but for mixed monolayers an increased compression rate caused significant expansion (excess area/molecule). The thermal motion of additional methylene groups in higher molecular weight alkyl alcohols increases the intermolecular spacing by 0.12 A at 20 dynes/cm. This suggests that such small changes in the intermolecular spacing affect the properties of systems containing mixed surfactants.

The chain-length compatibility strikingly influences the properties of systems containing mixed surfactants or a surfactant and a paraffinic oil. Schick and Fowkes (*1*) showed that the maximum lowering of CMC (critical micelle concentration) and maximum foam stability occurred when both the anionic and nonionic surfactants had the same chain length. Lawrence (*2*) reported that the maximum emulsion stability and maximum surface viscosity of mixed monolayers occurred when both the ionic and nonionic surfactants had the same chain length. Fort (*3*) reported that the coefficient of friction between polymeric surfaces decreased strikingly as the chain length of paraffinic oils approached that of

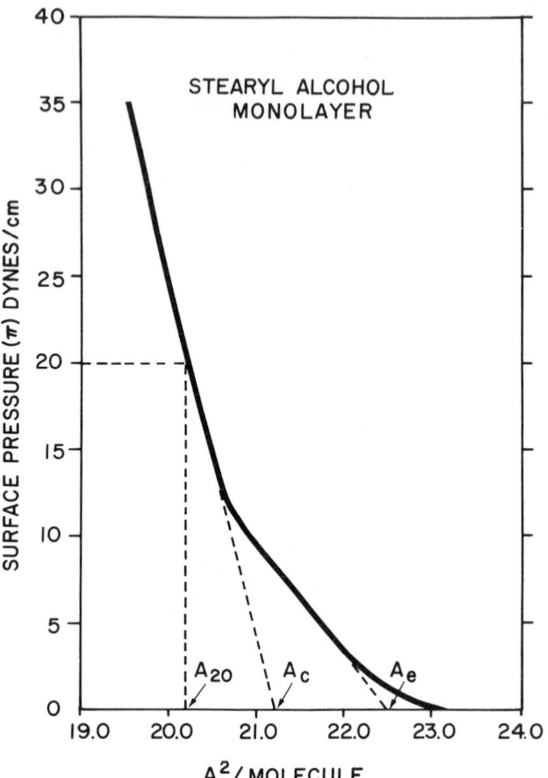

Figure 1. Surface pressure–area curve of stearyl alcohol monolayer at pH 2.0, 25°C. Compression rate: 6.5 A²/molecule/min.

stearic acid which was used as a lubricant additive. Cameron and Crouch (*4*) studied the effect of chain length of fatty acids on lubrication by measuring scuff load. The maximum scuff load was observed when the fatty acid had the same chain length as that of the hydrocarbon oil. Askwith *et al.* (*5*) studied the viscosity of very thin layers (3.5×10^{-4} inch thickness) of oil containing fatty acids of various chain length. Here also, the maximum viscosity of the thin layers was observed when fatty acid and oil had the same chain length. Similar effects of chain-length compatibility on rust prevention and dielectric absorption were reported by others (*6, 7*). However, no explanation based on molecular properties has been offered to explain the chain-length compatibility effect in these systems.

The present study was designed to obtain a better understanding of the chain-length compatibility effect in terms of molecular areas of surfactants. Although several systems, used in the chain-length compatibility studies mentioned previously (*1–7*), were mixtures of ionic and nonionic surfactants, we have used the surfactants with the same polar group in our mixed monolayer studies to attribute any change in the molecular area to the chain-length compatibility. In the mixed monolayers reported here, the chain length of one component was kept constant, and that of the other component was varied from C_{16} to C_{22}.

Materials and Methods

The experiments on mixed monolayers were carried out using alkyl alcohols (C_{16}, C_{18}, C_{20}, C_{22}). These surfactants were obtained from Supelco, Inc. (Bellefonte, Pa.) with purity greater than 99+%. The C_{22} alcohol was purchased from the Applied Science Laboratories, Inc. (State College, Pa.). The solutions of all pure alkyl alcohols were prepared (concentration 0.004 *M*) in a mixture of chloroform, methanol, and

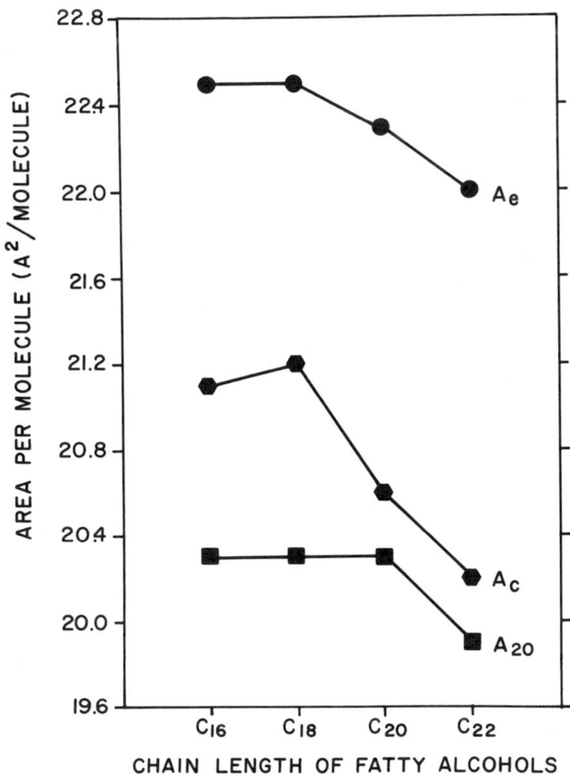

Figure 2. Average area per molecule (A_e, A_c, and A_{20}) of various alkyl alcohols

n-hexane (1:1:3 v/v/v). First, the single component films of alkyl alcohols were studied on the subsolution of 0.01*M* HCl. Next, the mixed monolayers of alkyl alcohols were studied on the subsolution of 0.01*M* HCl. Reagent grade chemicals and double-distilled water were used in all experiments.

The Wilhelmy plate connected to a Honeywell transducer was used to measure the surface tension which was fed directly to Y axis of an X–Y recorder. The monolayer tray was made from Plexiglass with dimensions of 25 $\times$ 12 $\times$ 1.5 cm. The temperature of the subsolution was controlled by pumping the water from a thermostated water bath through a glass coil which was immersed in the subsolution. The temperature was controlled at 25° $\pm$ 0.1°C. The subsolution was poured in the trough about 2 to 3 mm above the rim of the trough. The surface of the subsolution was cleaned by moving a wax coated glass slide from one side of the trough to the other and then sucking the impurities by a capillary con-

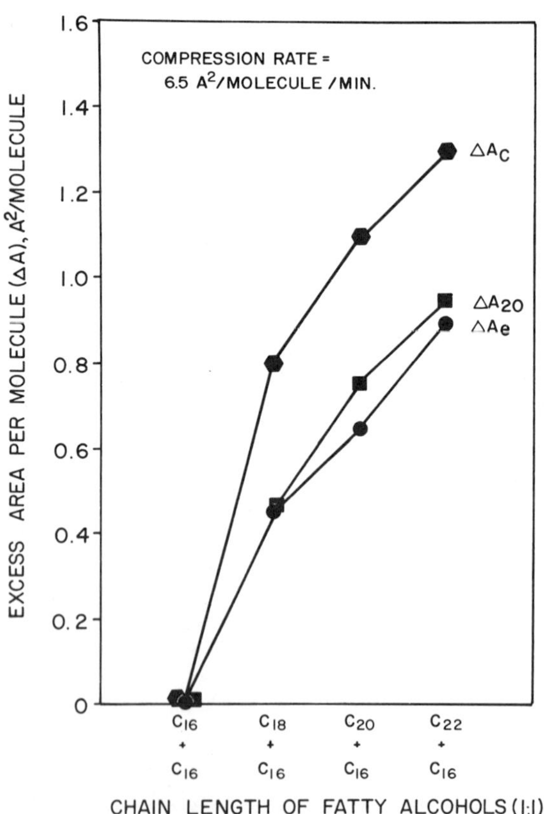

Figure 3. Excess area per molecule of A_e, A_c, *and* A_{20} vs. *various alkyl alcohol mixtures (1:1 molar ratio) with* C_{16} *alcohol as the common component at pH 2.0, 25°C*

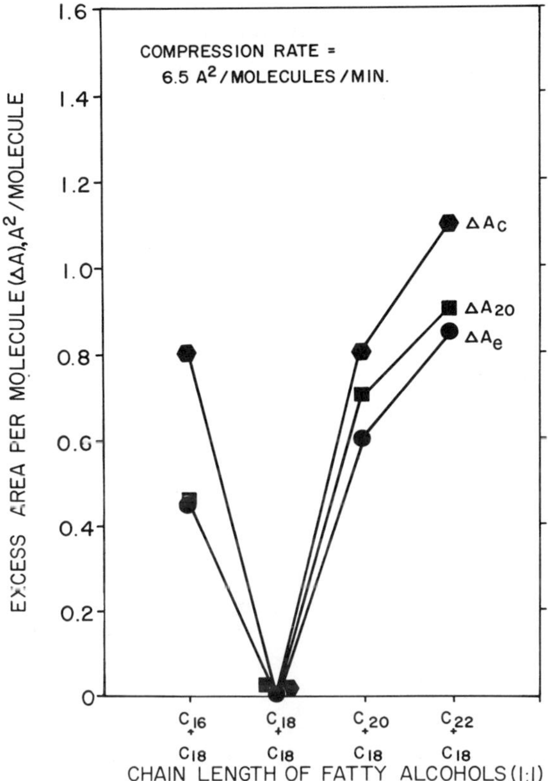

Figure 4. Excess area per molecule of A_e, A_c, *and* A_{20} vs. *various alkyl alcohol mixtures (1:1 molar ratio) with* C_{18} *alcohol as the common component at pH 2.0, 25°C*

nected to a vacuum pump. This was done several times to make sure no impurities remained on the surface. The monolayers were spread evenly on the surface by using an Agla microsyringe. The spreading volume of the solution was 0.025 ml for each experiment. A time interval of 5 minutes was allowed for spreading solvents to evaporate or diffuse in the subsolution from the monolayers. The monolayer was compressed at 6.5, 2.7, and 1.0 A²/molecule/min. The motor driving the compression bar was connected to a 10-turn potentiometer which provided the potential difference proportional to the movement of the compression bar. This was connected to the X axis of the X–Y recorder. In this way, the surface pressure–area curve for a monolayer was automatically plotted on the X–Y recorder for each surfactant mixture. Three to five monolayers of each mixture were studied, and the results reported are average values. Because of the standardization of the experimental procedure such as the method of spreading, the evaporation time of solvents, and the rate of compression, the standard error of the surface pressure–area curves is ±0.15 A²/molecule.

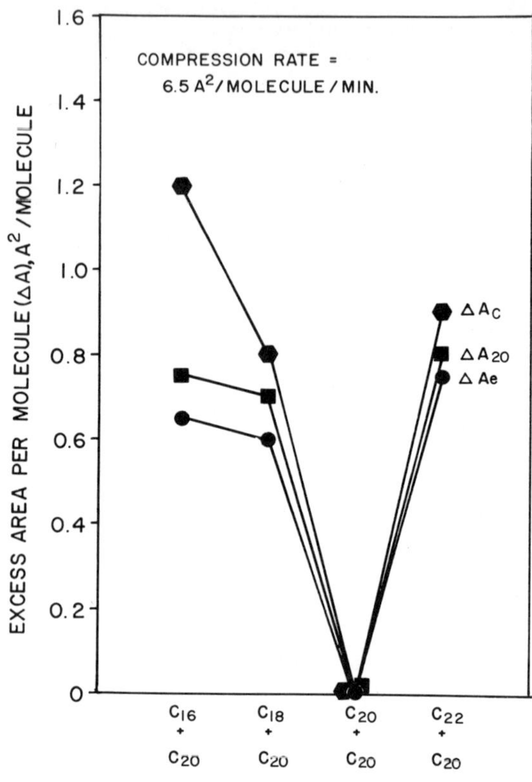

Figure 5. Excess area per molecule of A_e, A_c, *and* A_{20} vs. *various alkyl alcohol mixtures (1:1 molar ratio) with* C_{20} *alcohol as the common component at pH 2.0, 25°C*

Theory

If at a surface pressure, π, the average area per molecule of two surfactants in their individual monolayers is A_1 and A_2, then in the mixed monolayer (1:1 molar ratio) of these two surfactants, the average area per molecule should be $(A_1 + A_2)/2$ at the same surface pressure provided the surfactant molecules occupy the same area in the mixed monolayer as they do in their individual monolayers (*8, 9*). However, in many cases, the average area per molecule in a mixed monolayer is greater or smaller than that expected from the simple additivity rule (*10, 11, 12*). A reduction in the average area/molecule in a mixed monolayer can be attributed to the molecular attraction between the surfactants or to the "intermolecular cavity effect" (*9, 13*). An expansion in the average area/

molecule in a mixed monolayer can be attributed to molecular repulsion or to greater disorder in the mixed monolayer. In the present study, we have calculated excess molecular area for mixed monolayers using the following expression:

$$\text{Excess molecular area} = \left(\frac{\text{av. area}}{\text{molecule}}\right)\text{Expt.} - \left(\frac{\text{av. area}}{\text{molecule}}\right)\text{Ideal}$$

$$= (A_{\text{m}})_{\pi} - \left(\frac{A_1 + A_2}{2}\right)_{\pi}$$

where A_{m}, is the area/molecule in the mixed monolayer and A_1, A_2 are

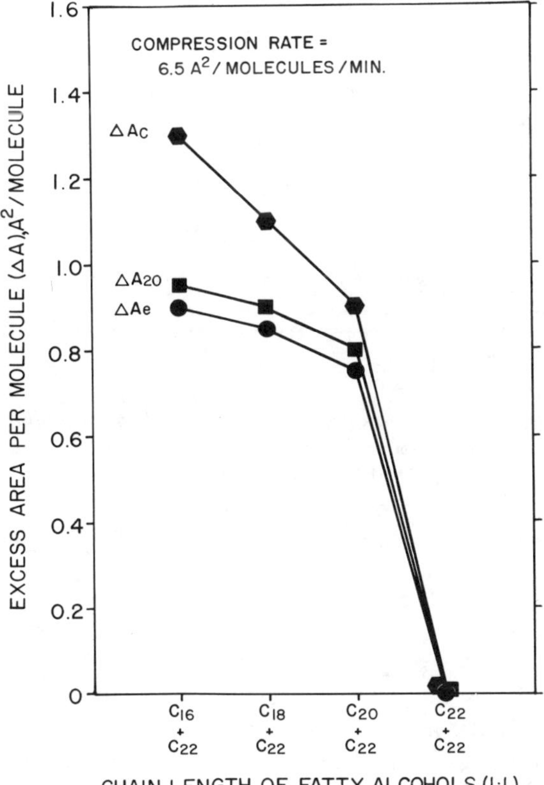

Figure 6. Excess area per molecule of A_e, A_c, *and* A_{20} vs. *various alkyl alcohol mixtures (1:1 molar ratio) with* C_{22} *alcohol as the common component at pH 2.0, 25°C*

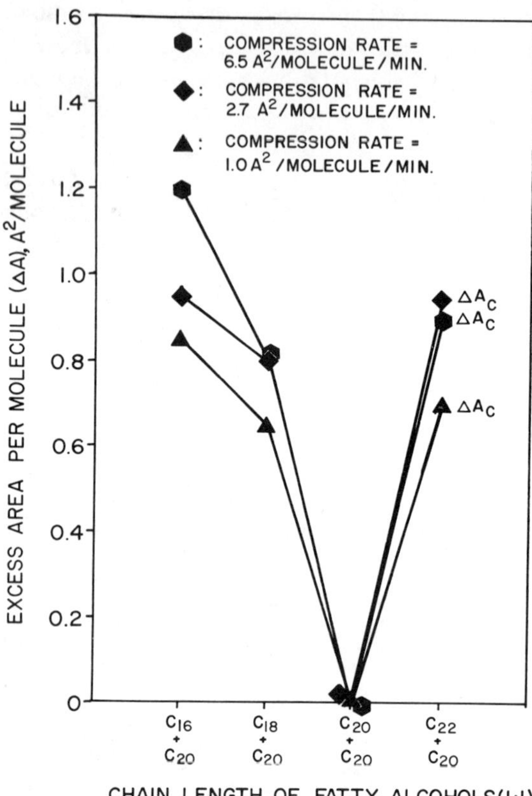

Figure 7. Effect of the rate of compression on the excess area/molecule of A_c *for various alkyl alcohol mixtures with* C_{20} *alcohol as a common component at pH 2.0, 25°C*

the area/molecule of the components in their individual monolayers at a specific surface pressure π.

Results

Figure 1 shows the surface pressure–area curve of stearyl alcohol monolayer at pH 2.0 and 25°C, where A_e is the area/molecule in the expanded state at zero surface pressure, A_c is the area/molecule in the condensed state at zero surface pressure, and A_{20} is the area/molecule at surface pressure of 20 dynes/cm. Figure 2 shows the values of A_e, A_c, and A_{20} for various alkyl alcohols. From this diagram the average area per molecule in mixed monolayers was calculated using simple additivity rule. Figure 3 shows the excess molecular area when C_{16} alkyl alcohol was mixed with alkyl alcohols of various chain lengths. This result indicates that there is an expansion of the mixed monolayers compared with the molecular area calculated using the simple additivity rule. It is also evident that the greater the difference between the chain length of the components, the greater is the excess molecular area in the mixed monolayers. It is also evident that A_c is most strikingly influenced by the differences in chain length. Figures 4, 5, and 6 show the results on the excess molecular area when the common components in the mixed monolayers were C_{18}, C_{20}, and C_{22} alkyl alcohols, respectively. Note that the extremely small values of excess area/molecule shown in Figures 3 to 6 were not measured directly. The area/molecule values in the mixed monolayers were measured experimentally and showed considerable expansion when compared with the surface pressure–area curves of pure components. The excess area/molecule is the difference between the experimentally obtained area/molecule in the mixed monolayers and that expected from the simple additivity rule.

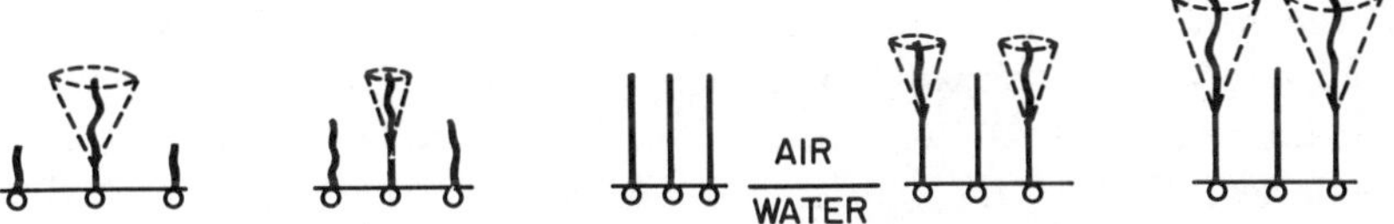

Figure 8. Schematic of the thermal motion of the upper segments of hydrocarbon chains at the air–water interface. The cones shown by broken lines represent the time-average space occupied by the thermal motion of the segments.

Figure 7 shows the effect of the rate of compression on the excess area/molecule in A_c for various alkyl alcohol mixtures with C_{20} as a common component. There seems to be a small increase in excess area/molecule at a faster rate of compression. Similar rate-of-compression effects were observed also with the mixed monolayers containing C_{16}, C_{18}, or C_{22} alcohol as common component. The area/molecule of pure components did not exhibit any significant dependence on the compression rates used in this present study. The area/molecule values of pure components at various compression rates were within experimental error (± 0.15 A^2/molecule) at any given surface pressure.

Discussion

The results on the surface pressure–area curve of stearyl alcohol monolayer shown in Figure 1 and the area/molecule of other alkyl alcohols shown in Figure 2 agree with the results reported for some of these alkyl alcohols by others (*14, 15*). Figure 3 shows that the values for the excess molecular area of A_c are 1.3, 1.1, and 0.8 A^2 per molecule for the difference of six, four, and two methylene groups between the common component and the longer alkyl alcohols.

In the previous studies on the chain-length compatibility (*1–7*), the only explanation suggested for the observed effects was the term "the best fit" of surfactant molecules. Our explanation for these chain-length compatibility effects is shown schematically in Figure 8 based upon the monolayer studies. When the chain lengths of the two components are equal, the mixed film will be very condensed. However, when the chain length is different, the portion of molecules above the height of the adjacent molecules will exhibit thermal motion (vibrational, oscillational, and rotational modes). If this thermal disturbance were limited to the portion above the height of the adjacent molecules, it would not change the average area per molecule. However, this disturbance presumably propagates along the chain at a considerable length towards the polar group of the molecule. Therefore, the mixed monolayers will be more expanded and hence exhibit a greater average area per molecule. As the difference in the chain length increases, the corresponding thermal disturbance and expansion will also increase because the length of the segment undergoing rotation, vibration, and oscillation increases. Figure 8 schematically shows the thermal motion of these segments and its effect propagating considerably below the height of the adjacent molecules. The time–average space occupied by this segment of methylene groups is shown by a cone with dashed lines. The concept of thermal motion of hydrocarbon chains at the interface was advanced by N. K. Adam (*16*) in his classical book on the physics and chemistry of surfaces. We have also used this concept to propose the existence of intermolecular cavities in mixed monolayers which may be responsible for an apparent condensation in lecithin–cholesterol monolayers (*8, 9, 10, 13*). A similar concept was also used by Bruun (*17*) to explain his results on mixed monolayers of isodextropimaric acid and straight chain fatty acids.

Philips *et al.* (*18*) reported that in mixed monolayers of dioleoyllecithin–distearoyllecithin, the area/molecule showed expansion from the additivity rule because the unsaturated fatty acid chains increased the kinetic motion of saturated chains. However, in contrast to the studies here on $C_{16} + C_{18}$ alcohols, they found no expansion in the mixed monolayers of dipalmitoyllecithin–distearoyllecithin.

It is interesting that the rate of compression did not influence significantly the area/molecule in the monolayers of pure components. However, for mixed monolayers, there was a small but significant increase in the excess area/molecule with increase in the rate of compression (Figure 7). This suggests that perhaps the disordered segments of hydrocarbon chains may begin to orientate themselves in a more ordered state, resulting in smaller excess area/molecule as one decreases the rate of compression.

We have shown (*10, 19, 20*) that a change of 0.3 to 1.5 A in the intermolecular spacing between lipid molecules in the monolayer strikingly influences the interaction of metal ions and enzymic hydrolysis in monolayers. One can calculate the intermolecular spacing in monolayers by assuming the average area per molecule to be a circle and then calculating the radius R. The distance $2R$ will be the intermolecular spacing in the monolayer which is the distance between the centers of the adjacent molecules. For example, in mixed monolayers of $C_{16} + C_{22}$ alkyl alcohols, (Figures 3 and 6) where an excess molecular area of 0.95 A^2 was observed at 20 dynes/cm, we can calculate the increase in the intermolecular spacing caused by mixing. For the mixed monolayers of $C_{16} + C_{22}$ alkyl alcohols, the average area per molecule based upon simple additivity rule at a surface pressure of 20 dynes/cm is 20.10 A^2 per molecule. The experimentally observed area per molecule is 21.05 A^2 per molecule. The corresponding radii for the circles of these areas are calculated to be 2.53 and 2.59 A respectively. Hence, the intermolecular spacing ($2R$) for these two molecular areas would be 5.06 and 5.18 A. Therefore, the difference between these two intermolecular spacings represents the expansion in the intermolecular spacing caused by the excess molecular area which is induced by the thermal motion of chain segments in the mixed monolayer. The increase in the intermolecular spacing caused by expansion of the monolayer would be 0.12 A ($5.18 - 5.06 = 0.12$ A). These calculations suggest that a change in the intermolecular spacing of the order of 0.12 A must be of significant importance in determining the properties of mixed surfactant systems and phenomena such as foams, emulsions, as well as lubricants, and dielectric absorption by paraffins.

Summary

The exess area per molecule in the mixed monolayers of alkyl alcohols of dissimilar chain length is a minimum when the two components have the same chain length. The expansion of mixed monolayers arising from unequal chain length is presumably caused by the thermal motion of the hydrocarbon chains. The thermal motion increases the intermolecular spacing in the monlayer by about 0.12 A at a surface pressure of 20 dynes/cm. In view of the observed effects of chain-length compatibility in foams, emulsions, and lubrication it is proposed that such small changes in intermolecular spacing are important in determining the properties of the above systems.

The authors gratefully acknowledge the support of the Departments of Anesthesiology and Chemical Engineering for carrying out this research. The authors also acknowledge their deep appreciation of the classical book "Physics and Chemistry of Surfaces," by N. K. Adam, which stimulated the imagination of so many young minds that entered into the two-dimensional world of surfaces.

Literature Cited

1. Schick, M. J., Fowkes, F. M., *J. Phys. Chem.* (1957) **61,** 1062.
2. Lawrence, A. S. C., *Soap, Perfumery Cosmetics* (1960) **33,** 1180.
3. Fort, T., Jr., *J. Phys. Chem.* (1962) **66,** 1136.
4. Cameron, A., Crouch, R. F., *Nature* (1963) **198,** 475.
5. Askwith, T. C., Cameron, A., Crouch, R. F., *Proc. Roy. Soc.* **A291** (1966), 500.
6. Ries, H. E., Jr., Gabor, J., *Chem. Ind.* (1967) 1561.
7. Meakins, R. J., *Chem. Ind.* (1968) 1768.
8. Shah, D. O., Schulman, J. H., *Advan. Chem. Ser.* (1968) **84,** 189.
9. Shah, D. O., *Advan. Lipid Res.* (1970) **8,** 347.
10. Shah, D. O., Schulman, J. H., *J. Lipid Res.* (1967) **8,** 215.
11. Shah, D. O., *J. Colloid Interface Sci.* (1970) **32,** 577.
12. Dervichian, D. G., in "Surface Phenomena in Chemistry and Biology," J. F. Danielli, K. G. A. Parkhurst, and O. C. Riddiford, Eds., p. 70, Pergamon Press, New York, 1958.
13. Shah, D. O., in "Biological Horizons in Surface Science," L. M. Prince and D. F. Sears, Eds., p. 69, Academic Press, New York, 1973.
14. Harkins, W. D., "Physical Chemistry of Surface Films," pp. 118-119, Reinhold, New York, 1954.
15. Nutting, G. C., Harkins, W. D., *J. Amer. Chem. Soc.* (1939) **61,** 1180.
16. Adam, N. K., "The Physics and Chemistry of Surfaces," p. 67, Oxford University Press, London, 1938.
17. Bruun, H. H., *Acta Chem. Scand.* (1955) **9,** 342.
18. Phillips, M. C., Ladbrooke, B. D., Chapman, D., *Biochim. Biophys. Acta* (1970) **196,** 35.
19. Shah, D. O., Schulman, J. H., *J. Colloid Interface Sci.* (1967) **25,** 107.
20. Shah, D. O., Schulman, J. H., *J. Lipid Res.* (1967) **8,** 227.

RECEIVED October 25, 1974.

12

Thermodynamics of Monolayer Solutions of Lecithin and Cholesterol Mixtures by the Surface Vapor Pressure Method

KAZUO TAJIMA[1] and N. L. GERSHFELD

Laboratory of Physical Biology, National Institute of Arthritis, Metabolism, and Digestive Diseases, National Institutes of Health, Bethesda, Md. 20014

Thermodynamic parameters for the mixing of dimyristoyl lecithin (DML) and dioleoyl lecithin (DOL) with cholesterol (CHOL) in monolayers at the air–water interface were obtained by using equilibrium surface vapor pressures π_v*, a method first proposed by Adam and Jessop. Typically,* π_v *was measured where the condensed film is in equilibrium with surface vapor (*$\pi < 0.1 \pm 0.001$ *dyne/cm) at 24.5°C; this exceeded the transition temperature of gel* $\rightleftharpoons$ *liquid crystal for both DOL and DML. Surface solutions of DOL–CHOL and DML–CHOL are completely miscible over the entire range of mole fractions at these low surface pressures, but positive deviations from ideal solution behavior were observed. Activity coefficients of the components in the condensed surface solutions were greater than 1. The results indicate that at some elevated surface pressure, phase separation may occur. In studies of equilibrium spreading pressures with saturated aqueous solutions of DML, DOL, and CHOL only the phospholipid is present in the surface film. Thus at intermediate surface pressures, under equilibrium conditions (*$40 > \pi > 0.1$ *dyne/cm), surface phase separation must occur.*

Since the initial observation by Leathes (*1*) that cholesterol exerts a condensing effect on spread monomolecular films of natural lecithins,

[1] Present address: Tokyo Metropolitan University, Tokyo, Japan.

extensive investigations of the condensing phenomenon with synthetic lecithins have been reported (*2*). These studies have been criticized recently on the basis that the films may be neither homogeneous nor at equilibrium (*3*).

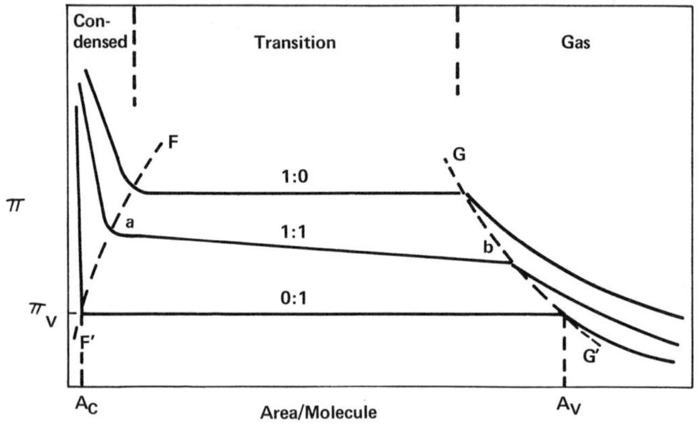

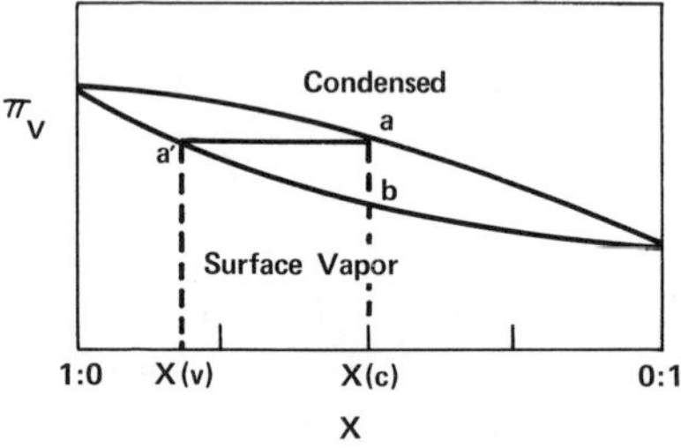

Figure 1. Upper: Schematic of π–A isotherms in the transition region: F-F' represents region where bulk of film material is in the condensed monolayer state; G-G' represents the region where virtually all of the film is in the gaseous monolayer state. Area per molecule is the total area occupied by the sum of all lipid molecules in the surface.
Lower: π_v-x surface phase diagram where data for upper curve containing point a are obtained from F-F' of the π–A isotherm; lower curve containing point b is obtained from G-G' of the π–A isotherm. At any value of π_v, condensed film at point a (x = x(c)) is in equilibrium with vapor film at point a' (x-x(v)).

To study the interactions in mixed films of lecithin and cholesterol under conditions of homogeneity and equilibrium, the surface vapor pressure method, first explored by Adam and Jessop (*4*), was used. This method uses spread films in the transition region of the isotherms, where the condensed films are in equilibrium with two-dimensional vapor (*5, 6*). Unlike the high pressure region of the isotherm, where virtually all other studies have been carried out, equilibrium can be established; the phase rule correctly predicts the variance of the system depending on whether two or three surface phases are present (condensed monolayer(s) and surface vapor) (*7*). The miscibility in the condensed surface phases can be readily specified by this procedure; application of the method to various monolayer systems has been reported (*8*). Preliminary studies with dipalmitoyl lecithin (DPL) at temperatures below the liquid-crystalline transition temperature, T_c, showed that cholesterol is not miscible with the gel state of this lecithin (*9*). We now use the method to analyze the two-component mixtures of cholesterol with DOL and DML at temperatures above T_c. We demonstrate that large positive deviations from ideal mixing result from mixing cholesterol with these lecithins, and we predict and verify experimentally that at elevated surface pressures, phase separation in the surface film will occur.

Experimental

Chemical Activities by the Surface Vapor Pressure Method. Surface pressure measurements in the transition region between the condensed and gaseous monolayer states of a single lipid component spread as a monolayer on water yield a value of π which is independent of the surface area. This value—the surface vapor pressure, π_v—is analogous to the vapor pressure of a liquid in equilibrium with its vapor. When a second lipid component is in the surface, the limits of miscibility in the condensed phase may be determined on the basis of the surface vapor pressure dependence on the mole fraction in the condensed phase (*8*).

Figure 1 represents the isotherms for two lipid components which are miscible in the condensed monolayer state. The major feature of the isotherms for the pure components (1:0, 0:1) is the transition region in which the surface pressure is independent of surface area; here the limits of the transition region are at the low area end, A_c, and at the high area end, A_v. These areas are characteristic of each lipid and represent the area per molecule of the lipid in the condensed and vapor states (*10*). For an equimolar mixture of the two components (1:1), the surface pressure in the transition region depends on the surface area; according to the phase rule (*11, 12, 13, 14*), two surface phases coexist here: a condensed phase of lipids and the surface vapor phase. To obtain the activity coefficient γ_i of the ith component in the condensed phase the following relation may be used:

$$\gamma_i = \frac{x_i(v)\pi_v}{x_i(c)\pi_v^{i,o}} \tag{1}$$

where the mole fractions (x) in the surface vapor (v) and condensed (c) monolayer states and the surface vapor pressures of the pure $\pi_v^{i,o}$ component and of the mixture π_v must be measured.

Equation 1 assumes that Dalton's law of partial pressures applies to the mixture of lipid molecules in the surface vapor state; studies with various lipid mixtures in this state support this assumption (*8*).

Correction terms for non-ideal behavior of the lipid molecules in the surface vapor state were calculated from the second virial coefficients of the mixtures using procedures described by Prausnitz (*15*). The contributions to the activity coefficients in the vapor were negligible; thus Equation 1 was unchanged. Each term in Equation 1 is experimentally available. From Figure 1, points along the dotted line FF′ represent the surface vapor pressure in equilibrium with the mixture in the condensed film state that is composed of the material deposited on the surface; most of the material is in the condensed state, and only a small amount is present in the vapor state. To obtain the composition of the equilibrium vapor phase one must examine the transition region of the isotherms where the bulk of the lipid is in the vapor state—*i.e.*, along the line GG′ which joins values of A_v (Figure 1). Values of π_v along the line GG′ represent π_v–$x_i(v)$ data. The data can be represented by a conventional π_v–x phase diagram (Figure 1). The upper curve represents the π_v data as a function of the composition of the condensed state; the lower curve is for the composition of the vapor state. For the 1:1 mixture the values of π_v on lines FF′ and GG′ are the points a and b; these are shown in the π_v–x phase diagram. The composition of the surface vapor in equilibrium with the condensed mixture $[x = x(c)]$ is at point a′ where $x = x(v)$.

The millidyne film balance used in these studies has been described in detail (*7, 16*). With proper shock mounting, the precision of the balance is ±0.2 mdyne/cm below 30°C; above this temperature the precision is poorer because of the thermal disturbances caused by evaporation from the water surface. We restricted these studies to below 30°C. The temperature of the trough was controlled to ±0.2°C by circulating

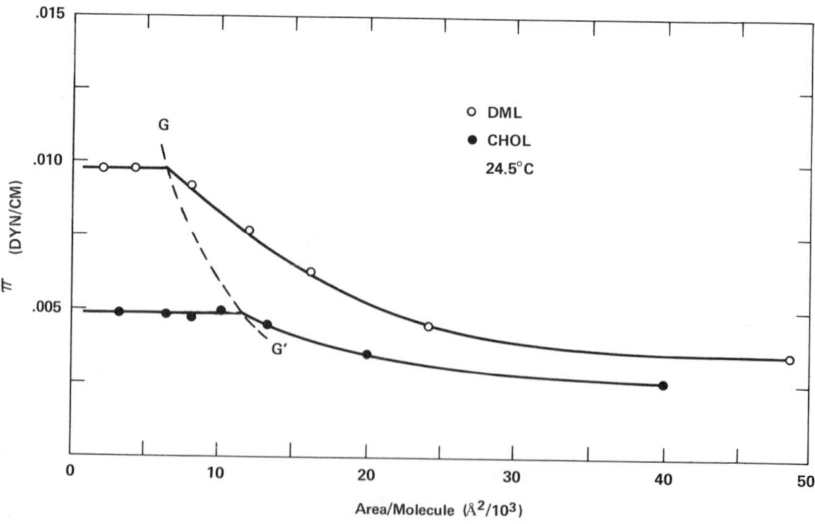

Figure 2. π–A isotherms of DML and CHOL on water, pH 5.8 at 24.5°C

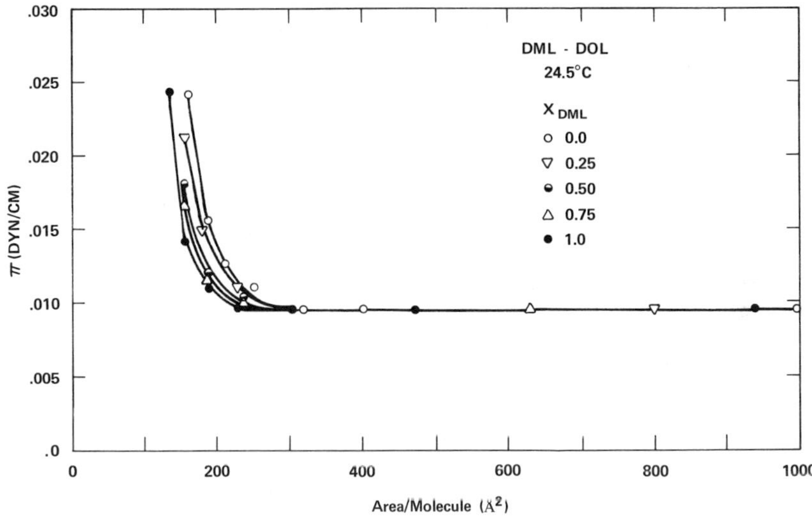

Figure 3. π–A isotherms for mixtures of DML and DOL on water, pH 5.8 at 24.5°C

water from a thermostated bath through coils in the base of the trough. The temperature was monitored by thermistor probes placed at different points in the water surface. The procedure for measuring the very low surface pressures and the evidence for equilibrium in these systems has been discussed at length (*7*).

Materials. Dimyristoyl lecithin (DML) from Nutritional Biochemical Corp. was purified by column chromatography (Unisil, activated silica gel); solvent elution gradients of chloroform and methanol mixtures were used (*17*). Pure dioleoyl lecithin (DOL) was synthesized and donated by R. E. Pagano. The final purity of the lecithins was confirmed by thin layer chromatography with a mixed solvent system of chloroform:methanol:water (65:25:4 v/v) (*18*). The fatty acid composition and purity of the lecithins were verified by gas chromatography. We estimate the purity of these samples to be >99 mole %. Cholesterol from Applied Science Laboratories was used without further purification because only the cholesterol peak was observed in the gas chromatogram; the melting point (148.5°C) and melting range (±0.10°C) were characteristic of purity > 99 mole %.

Chloroform and hexane were the spreading solvents and were purified by being passed through an activated silica gel column (Florosil). Surface active impurities were detected by spreading 0.1 cc of the solvent on the surface of the millidyne film balance and monitoring the surface pressure after the solvent had evaporated (1–5 min). No residue was detected with the purified solvents as indicated by the absence of surface pressure after evaporation. If non-purified solvents were used, appreciable amounts of surface active impurities were detected. All solvents were stored under nitrogen to avoid oxidation; the lipid solutions were stored at −20°C but for periods not exceeding three days. Water was purified as described previously (*19*). All studies were at pH 5.8, without the addition of buffer or electrolyte to avoid extraneous contaminations. Isotherms were determined point by point.

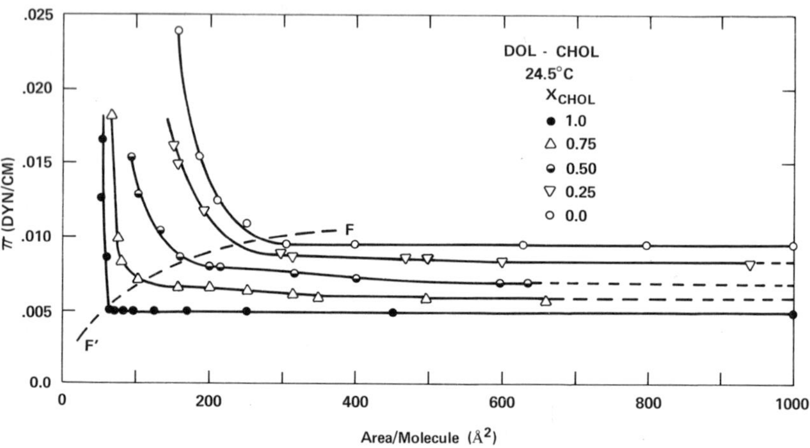

Figure 4. π–A isotherms for mixtures of DOL and CHOL on water, pH 5.8 at 24.5°C

Results

Surface Pressure–Area (*A*). Isotherms for CHOL, DML, DOL and their mixtures are shown in Figures 2–5; their molecular areas were calculated by dividing the total surface area by the total number of lipid molecules which had been deposited on the surface. The isotherms for the mixtures extend to approximately 1000 A^2; for pure CHOL and DML the isotherms are shown into the gaseous region. The isotherm for DOL in the gaseous region at the same temperature was indistinguishable from that for DML and is not shown.

For the mixtures of DOL and DML with cholesterol (Figures 4 and 5) the transition to the liquid-expanded film is represented by line FF′; at each point along FF′ the surface vapor pressure is in equilibrium with the condensed phase whose mole fraction is primarily that of the condensed monolayer. The surface vapor pressures in the region of A_v, where virtually all of the spread film is in the surface vapor state, were obtained by extrapolating the data obtained at about 1000 A^2 to the region of A_v (about 5000–10,000 A^2) assuming a linear decrease in π_v. This extrapolation is justified on the grounds that it is relatively short (*see* Figure 2) and that the slopes of the curves are small.

The dependence of the π_v values on the composition of the vapor and condensed states for DML–CHOL, DOL–CHOL, and DOL–DML mixtures is shown in Figure 6. The upper curve is the surface vapor pressure as a function of the mole fraction of the liquid-expanded film; the lower curve is for the dependence of π_v on the composition of the gaseous phase. Ideal mixing behavior is given by the linear dotted line which joins the π_v° points for each of the pure compounds. In all cases there was complete miscibility of the components as represented by the continuous function of π_v with x. In the cholesterol mixtures positive deviations from Raoult's law are observed; for the mixture of lecithins, ideal mixing is observed. These results confirm those obtained with lipid mixtures—*i.e.*, cholesterol mixed with liquid-expanded lipid films forms non-ideal mixtures with positive deviations; for mixtures of lipids which are in the same monolayer state, as in the case of the liquid-expanded DOL–DML mixtures, ideal mixing results (8).

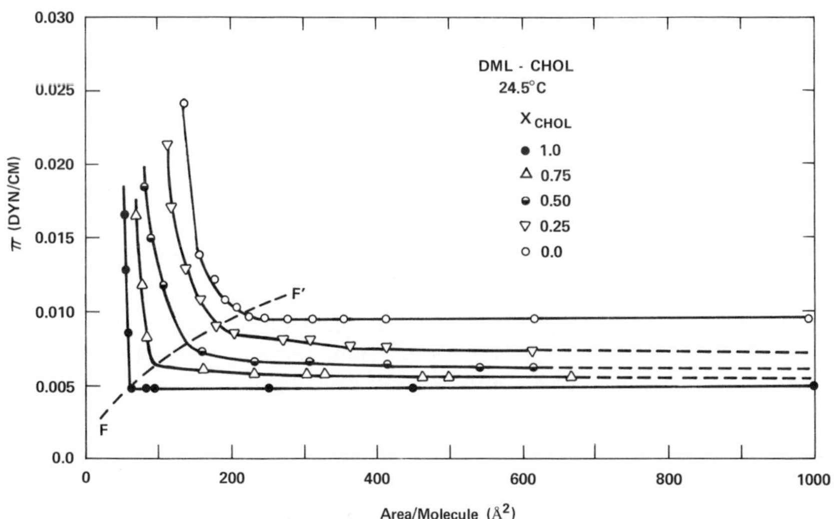

Figure 5. π–A *isotherms for mixtures of DML and CHOL on water, pH 5.8 at 24.5°C*

To calculate activity coefficients, γ_i, we apply the data of Figure 6 to Equation 1. Values for the two cholesterol–lecithin mixtures are presented in Table I as a function of mole fractions. The activity coefficients are greater than 1 in all instances of cholesterol mixtures.

The molecular areas on the transitional line FF′ in Figures 3–5 are presented as a function of the mole fraction in the condensed film in Figure 7; for the cholesterol–lecithin mixtures there is a decrease in area;

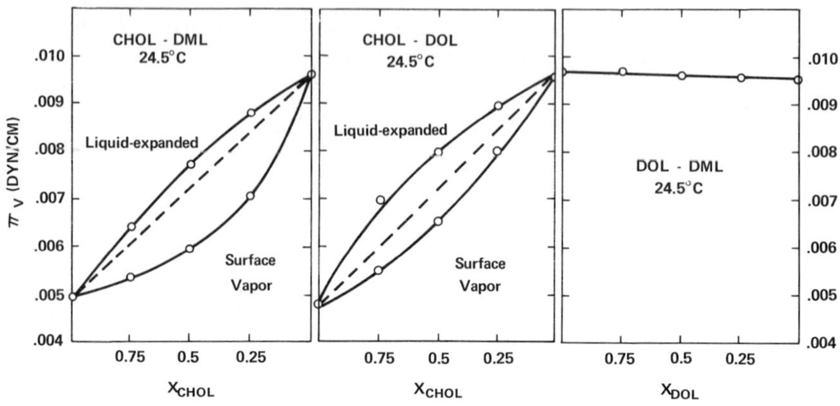

Figure 6. π_v-x surface phase diagrams for lipid mixtures on water, pH 5.8 at 24.5°C. Dotted line represents ideal mixing.

for the mixture of DOL and DML there is no change in the area of each component caused by mixing.

In earlier studies it was concluded that the surface properties of lipid mixtures are determined largely by the hydrocarbon domain of the monolayer and that regular solution theory may be applied to these systems (*8*). To test the regular solution theory, we compared the partial heats of mixing obtained from the vapor pressure study with values

Table I. Activity Coefficients of DML and DOL in Two-Component Mixed Monolayers with CHOL: Comparison with Regular Solution Heats of Mixing

x_1 (c)	x_1 (v)	π_v, $\times 10^3$ dynes/cm	γ_1	$\Delta\bar{H}_1$, cal/mole	
				γ_1	RS[a]
DOL (1) + CHOL (2)					
0.1	0.32	5.8	1.91	384	277
0.25	0.54	6.8	1.51	245	183
0.50	0.75	8.0	1.24	127	74
0.75	0.91	9.0	1.13	73	17
1.0	1.0	9.5	1.0		
DML (1) + CHOL (2)					
0.1	0.36	5.5	2.06	428	223
0.25	0.62	6.4	1.65	297	160
0.50	0.82	7.7	1.32	165	75
0.75	0.92	8.8	1.12	67	20
1.0	1.0	9.7			

[a] RS: regular solution theory $\Delta\bar{H}_1 = \phi^2_2 V_1 (\delta_1 - \delta_2)^2$, $\delta_1 = 5.8$, $\delta_2 = 4.9$ V_1: DML = 334, DOL = 442.

calculated from solution theory. For the former Equation 2 may be used, assuming that the entropy of mixing is ideal (*20*),

$$\ln \gamma_i = \frac{\Delta \overline{H}_i}{RT} \qquad (2)$$

where ΔH_i is the partial molar heat of mixing for component i. Values of ΔH_1 are in Table I. For comparison, a calculated value for ΔH_1, ob-

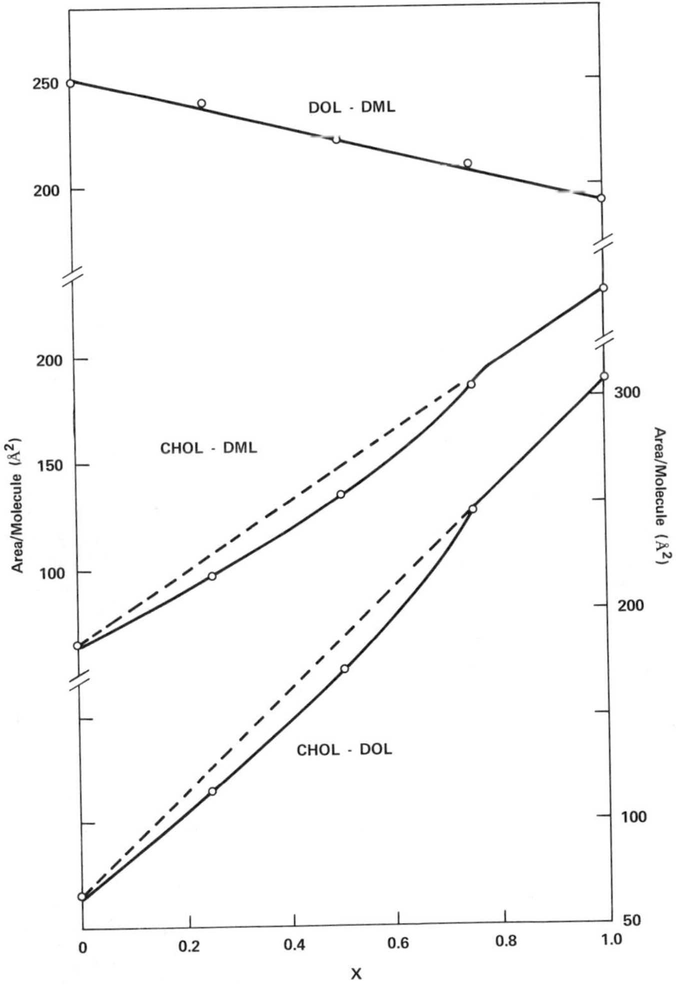

Figure 7. Average area per molecule in mixed lipid films as a function of composition. Data obtained from Figures 3–5 along F-F′, i.e., *at the low area portion of the transition region in the* π–A *isotherm.*

tained from regular solution theory, may be obtained from the following relation

$$\Delta\overline{H}_1 = \phi^2{}_2 \, V_1 \, (\delta_1 - \delta_2)^2 \tag{3}$$

where Φ_2 is the volume fraction of the component 2 (cholesterol), V_1 is the molar volume of component 1 (lecithin), and δ_1, δ_2 are the solubility parameters of each component defined for the monolayer system as

$$\delta_i = \left(\frac{\Delta H^i{}_{sv}}{V_i}\right)^{1/2} \tag{4}$$

where ΔH_{sv} is the heat of surface vaporization (*6, 8*). Considering only the hydrocarbon domains of the monolayers we estimated $\delta = 4.9$ for cholesterol and $\delta = 5.8$ for liquid-expanded films (*8*). From these values and Equation 3, ΔH_1 is calculated and compared in Table I for the lecithin–cholesterol mixtures. Mixtures of DOL and DML form ideal solutions, and $\Delta H_1 = 0$. In each case the theory of regular solutions predicts the sign and magnitude of the partial molar heat of solution. Note that the value of δ for cholesterol is subject to considerable error, and only the magnitude and sign of the calculated values given should be considered significant.

In addition, it is generally observed for mixtures of aliphatic hydrocarbons that if there is a difference in molecular size, a negative excess volume of mixing results (*21*). For the lecithin–cholesterol mixtures molecular area decreases (Figure 7), but it does not decrease with the mixture of the lecithins which forms an ideal mixture. Thus, the mixing of cholesterol and lecithin monolayers is consistent with bulk hydrocarbon mixtures with respect to both positive excess heats and negative excess volumes of mixing.

The positive heats of mixing for lecithin–cholesterol mixtures indicate that interactions between unlike molecules are smaller than the interactions between like molecules, *i.e.*, the hydrocarbon chain interactions with cholesterol are smaller than in each of the pure phases. If the excess heats of mixing become large enough, phase separation will occur. It may occur when the surface pressure is increased (*i.e.*, as the films are compressed). The point at which phase separation occurs is difficult to predict, measure, or detect; however, evidence of phase separation can be deduced from the following experiment. If excess amounts of two lipids are placed in water, the equilibrium surface pressure should reflect whether the surface film is a mixture. According to the phase rule (*11, 12, 13, 14*), if two bulk lipid phases are present, only one surface phase can be present at the air–water surface. Thus the composition of the equilibrium surface indicates whether there is mixing in the monolayer at the higher surface pressures. Earlier studies indicated that the surface pressure is a measure of the composition of the surface (*22*). Thus for cholesterol and lecithin dispersed in water at saturated concentrations, the

Table II. Equilibrium Spreading Pressures,[a] Π_e, of DML, DOL, CHOL, and Mixtures in H_2O at 29.5°C[b]

Lipid	Π_e *dynes/cm*, ±0.1
CHOL	39.3
DML	49.0
DML + CHOL (3/1)	49.0
DOL	46.3
DOL + CHOL (3/1)	46.3

[a] $\Pi_e = \gamma_0 - \gamma$ where γ_0 is the surface tension of water, and γ is the surface tension of the saturated lipid solution. Π_e values recorded by the Wilhelmy plate method (*23*), precision is ±0.1 dynes/cm.

[b] At least 0.1 g/l of each lipid was used to ensure that excess lipid was always present.

equilibrium surface pressure will be either that for pure lecithin, pure cholesterol, or neither, depending on whether only lecithin, cholesterol, or a mixture is present in the surface. In either of the first two cases, the results will indicate that phase separation has occurred at some intermediate surface pressure in which the other component or some mixture has been excluded from the surface. In the third case, mixing of the two lipids is obviously present at the elevated surface pressures. The results for DOL, DML, and CHOL are presented in Table II. For each of the mixed systems with cholesterol and lecithin, only lecithin is present in the equilibrium monolayer, as evidenced by the surface pressure equal to that of pure lecithin. Thus at some intermediate surface pressure, phase separation of cholesterol must occur. The nature of the separated phase cannot be established from this experiment without a complete analysis of the phase relations in the bulk system. It is likely that it may be a mixture of the two components in view of the complete miscibility of the components in the surface vapor pressure region of the isotherm.

Summary

Cholesterol and lecithin form completely miscible solutions in monolayers at very low surface pressures, characterized by excess positive heats and excess negative areas of mixing. At elevated surface pressures, phase separation occurs. Since these solutions conform to regular solution theory, the hydrocarbon domain of the monolayer makes the major contribution to the heats of mixing. The polar region of the monolayer may also contribute, but it may not be significant in the low pressure monolayers. Studies of the condensing effect of cholesterol at elevated surface pressures must recognize that phase separation will occur under equilibrium conditions.

Literature Cited

1. Leathes, J. B., *Lancet* (1925) **208**, 853.
2. Phillips, M. C., in *Progr. Surface Membrane Sci.* (1972) **5**, 179.
3. Gershfeld, N. L., Pagano, R. E., *J. Phys. Chem.* (1972) **76**, 1244.
4. Adam, N. K., Jessop, G., *Proc. Roy. Soc. Ser. A* (1928) **120**, 473.
5. Adam, N. K., Jessop, G., *Proc. Roy. Soc. Ser. A* (1926) **110**, 423.
6. Gershfeld, N. L., Pagano, R. E., *J. Phys. Chem.* (1972) **76**, 1231.
7. Pagano, R. E., Gershfeld, N. L., *J. Colloid Interface Sci.* (1972) **41**, 311.
8. Pagano, R. E., Gershfeld, N. L., *J. Phys. Chem.* (1972) **76**, 1238.
9. Pagano, R. E., Gershfeld, N. L., *J. Colloid Interface Sci.* (1973) **44**, 382.
10. Gershfeld, N. L., *J. Colloid Interface Sci.* (1970) **32**, 167.
11. Defay, R., Ph.D. thesis, Brussels (1932).
12. Defay, R., Prigogine, I., Bellemans, A., Everett, D. H., "Surface Tension and Adsorption," pp. 74-78, Wiley, New York, 1966.
13. Crisp, D. J., in "Surface Chemistry," pp. 17, 23, Supplement to Research, Buttersworth, London, 1949.
14. Gershfeld, N. L., in "Methods in Membrane Biology," Vol. I, E. D. Korn, Ed., p. 77, Plenum Press, 1974.
15. Prausnitz, J. M., "Molecular Thermodynamics of Fluid Phase Equilibria," p. 204, Prentice-Hall, Englewood Cliffs, N. J., 1969.
16. Gershfeld, N. L., Pagano, R. E., Friauff, W. S., Fuhrer, J., *Rev. Sci. Instrum.* (1970) **41**, 1356.
17. Colacicco, G., *Biochim. Biophys. Acta* (1971) **266**, 313.
18. Wagner, H., Horhaurmer, L., Wolff, P., *Biochim. Z.* (1961) **334**, 175.
19. Pak, C. Y. C., Gershfeld, N. L., *J. Colloid Sci.* (1964) **19**, 831.
20. Hildebrand, J. H., Scott, R. L., "The Solubility of Non-Electrolytes," 3rd edition, p. 41, Dover Publications, New York, 1950.
21. Rowlinson, J. S., "Liquids and Liquid Mixtures," pp. 142-146, Butterworths Scientific Publications, London, 1959.
22. Gershfeld, N. L., Pagano, R. E., *J. Phys. Chem.* (1972) **76**, 1244.
23. Gaines, G. L., Jr., "Insoluble Monolayers at Liquid-Gas Interfaces," p. 44, Interscience Publishers, New York, 1969.

RECEIVED October 24, 1974.

13

Thermodynamics of Penetration of Soluble Constituents into Spread Insoluble Monolayers

YOLANDE HENDRIKX and LISBETH TER-MINASSIAN-SARAGA

Laboratoire de Physico-Chimie des Surfaces et des Membranes, U.E.R. Biomédicale, Université René-Descartes, 45, rue des Saints-Pères, 75270 Paris Cedex 06, France

The penetration of soluble proteins into spread, insoluble lipid monolayers was first studied by Schulman and Rideal. Their injection technique allows measurement of the surface pressure increase of the insoluble penetrated monolayer after the soluble species is coadsorbed. The interpretation of this result cannot lead to a quantitative interpretation of the penetration unless the degree of penetration is known. This value can be determined if the soluble species is radioactivly labeled. An experimental and theoretical study of the change in monolayer surface pressure and composition during penetration by a soluble component has shown that information on the mechanism of this process can be obtained.

The interaction of a soluble constituent A with a spread monolayer of an insoluble constituent B was first studied by Schulman and Rideal (*1*). A was injected into the liquid substrate, and its penetration into B was studied by measuring the change in surface potential, ΔV, and surface pressure, $\Delta\Pi$, of the spread monolayer at constant area. Assuming that the resulting mixed $(A + B)$ monolayer was ideal, the surface density of A was inferred to be proportional to $\Delta\Pi$. A positive $\Delta\Pi$ was taken as evidence for interaction between A and B.

In our study (*2*) of the penetration of hexadecyltrimethylammonium bromide, CTAB, in egg–lecithin monolayers, we measured $\Delta\Pi$ and the surface density of the penetrating CTA^+ ions. $\Delta\Pi$ was separated into its ideal and excess parts, and only the excess component was considered relevant to the molecular interaction between the two constituents of the mixed film.

Experimental Results and Their Interpretation

The method used and the results obtained are described in Ref. 2. Two different methods of monolayer formation were investigated.

(1) Lecithin spread on CTAB solutions after equilibrium of CTA^+ ions was achieved.

(2) Lecithin spread immediately after CTAB solution surfaces were cleaned; this procedure is comparable with the injection method.

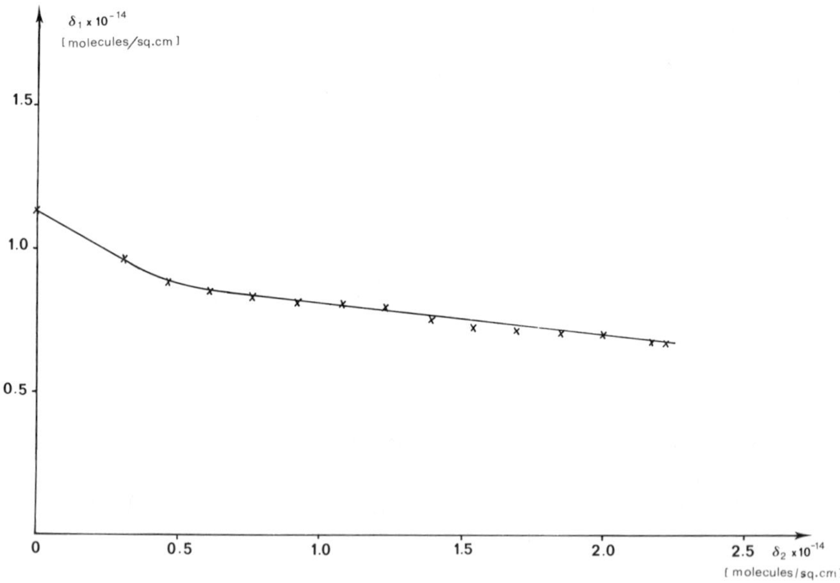

Figure 1. Decrease of surface density of the CTA^+ ions, δ_1; as a function of the lecithin surface density, δ_2. Concentration of CTAB in substrate = 2×10^{-5}M. Surface area is constant.

The second method was used since the results were much more reproducible than with the first method. We performed two types of measurements:

(1) Surface tension. CTAB solutions of various concentrations in a KH_2PO_4–NaOH buffer at pH 7 and 0.1*N* ionic strength were prepared. At their surface egg–lecithin monolayers of various surface densities were spread from a cyclohexane solution, and the surface tensions were measured at equilibrium adsorption of CTAB.

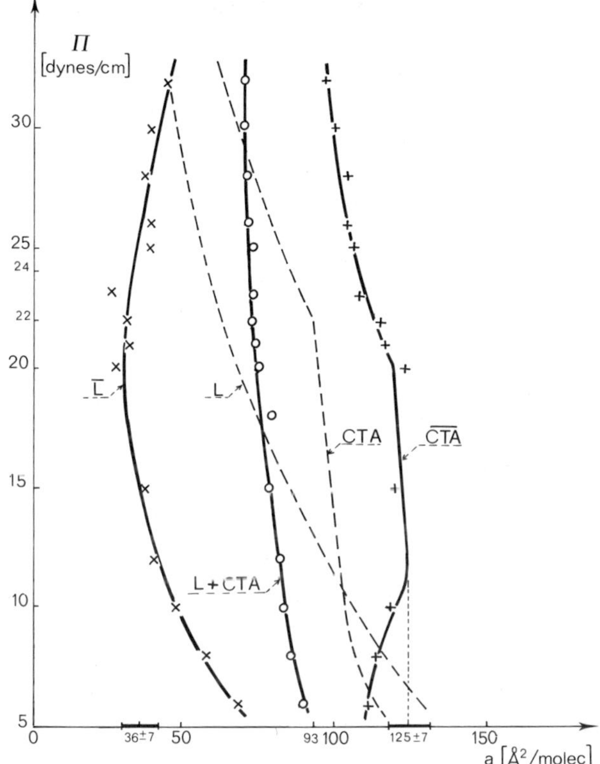

Figure 2. Surface pressure, Π, vs. *average and partial molecular areas,* a

L,CTA: pure films of (1) and (2)
$\overline{L}$,$\overline{CTA}$: partial areas of (1) and (2)
L + CTA: isotherm of the 1/1 mixed real film

(2) Surface density of adsorbed CTAB. This was measured under the conditions as described above for surface tension using ^{14}C labeled CTAB. Figure 1 represents the decrease in surface density of the CTA^+ ions, δ_1, adsorbed from a given CTAB solution with the increase of the egg–lecithin surface density δ_2. The rate of this decrease was higher when the egg–lecithin molar fraction in the mixed film was below 0.34.

The mixed films, CTAB (1) + egg–lecithin (2), were treated as two-dimensional mixtures of known composition. From the average molecular areas at a given surface pressure, we deduced the partial molecular areas $\bar{a}_1$ and $\bar{a}_2$ at the same pressure (2) using the classical Bakhuis-Rooseboom method. The pressure of the mixed films as a function of the partial molecular areas is shown in Figure 2. Also shown are the isotherms of the pure components 1 and 2 and of the 1/1 mixed real film.

We have shown (Ref. 2, Figure 6) that at surface pressures from 18 to 26 dynes/cm the partial areas of the constituents are independent of the composition of the mixed films if the molar fraction of component 2 is 0.27–0.70. In this case the error for the partial molecular areas results from the error on experimental average molecular areas equal to $\pm 10\%$. However, an additional error must be considered above 26 dynes/cm resulting from the tangent slope to the curve (Ref. 2, Figure 2) which can be estimated as $\pm 10\%$.

Furthermore, the real mixture may be less compressible than the ideal one and behaves as if the mixed films were solid (*see* Figure 2 for the mixture with molar fraction of component 2 equal to 0.5). This difference in behavior results from the difference in the compressibilities of the pure and mixed constituents 1 and 2. It shows that the interaction of the constituents in the mixed film can modify their state. We have deduced the variation of the state of CTAB (1) by mixing, applying the following approach appropriate to penetration.

Partial Free Energy of Mixing of CTAB in the Mixed Film. The chemical potential of 1 in solution is independent of the composition of the mixed film:

$$\mu_1 = \mu_{10}{}^{s} = \mu_1{}^{s} \tag{1}$$

At equilibrium:

$$\mu_1 = \zeta^{os} - \gamma_1{}^{o}\, a_1{}^{o} = \mu_1{}^{s} \tag{2}$$

The following expression for $\mu_1{}^s$ was established in Ref. 2 and is presented in the Appendix (Equation 12):

$$\mu_1{}^{s} = \zeta^{os} + RT\ln f_{1e}{}^{s} + RT\ln \frac{\delta_1\bar{a}_1}{\delta_1\bar{a}_1 + \delta_2\bar{a}_2} + RT\left(\frac{\delta_2\bar{a}_2}{\delta_1\bar{a}_1 + \delta_2\bar{a}_2}\right)\left(1 - \frac{\bar{a}_1}{\bar{a}_2}\right) - \gamma\bar{a}_1 \tag{3}$$

where f^{s}_{1e}, the activity coefficient of 1 in the mixed film, the reference state being the pure film of 1, is defined by the relation (Appendix, Equation 15):

$$RT\ln f_{1e}{}^{s} = \Delta\mu_{1e\ \mathrm{mix}} \tag{4}$$

where $\Delta\mu_{1e\,\mathrm{mix}}$, the partial free energy of mixing of component 1, is a function of the state of 1 in its interaction with 2 in the mixed film.

From Equations 2 and 3, it follows that:

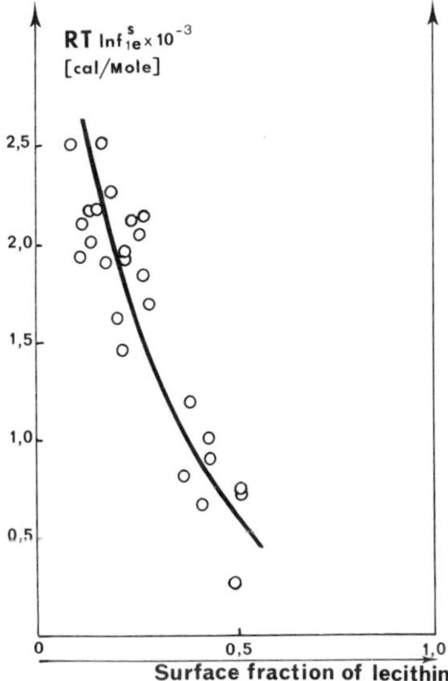

Figure 3. Partial free energy of mixing, $RT\ln f^s_{1e\,mix} = \Delta\mu_{1e\,mix}$ *of* CTA^+ *ions as a function of the surface molar fraction of lecithin*

$$\Delta\mu_{1e\ mix} = RT\ln f_{1e}{}^{s} = \gamma\bar{a}_1 - \gamma_1{}^{\circ}\, a_1{}^{\circ} - RT\ln \frac{\delta_1\bar{a}_1}{\delta_1\bar{a}_1 + \delta_2\bar{a}_2} - RT\left(\frac{\delta_2\bar{a}_2}{\delta_1\bar{a}_1 + \delta_2\bar{a}_2}\right)\left(1 - \frac{\bar{a}_1}{\bar{a}_2}\right) \quad (5)$$

Using Equation 5, $\Delta\mu_{1e\,mix}$ was calculated and is plotted in Figure 3 as a function of the surface molar fraction of 2 in the mixed films. The surface pressure ranged from 18 to 32 dynes/cm.

In the range 0–0.10 of surface molar fractions of 2, the partial free energy of mixing of CTAB increased considerably. When the molar fraction of lecithin in the mixed film increases above 0.1, the partial free

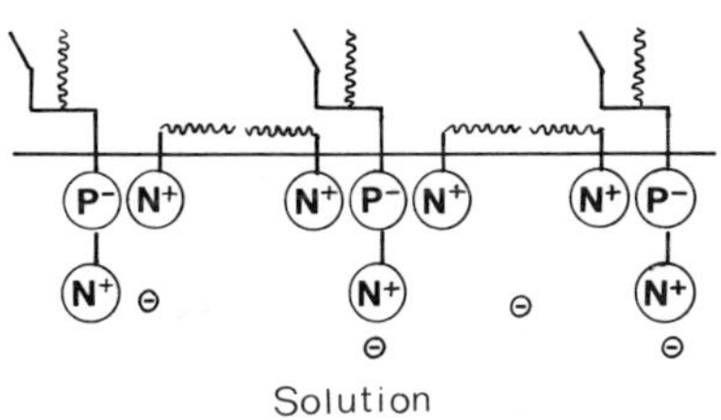

Figure 4. Model of the mixed films of CTA^+ *ions and lecithin*

N^+: trimethylammonium group
P^-: phosphate group
$P^- - N^+$ = polar moiety of the lecithin molecule

energy of mixing of CTAB decreases. Analysis (3) of the results of a mixed film where the molar ratio of lecithin is equal to 0.50 (reproduced in Figure 2) leads to the hypothesis that the mixed film may be represented by the two-dimensional lattice shown in Figure 4. In this lattice the lecithin molecules and the penetrating CTA^+ ions interact electrostatically and orient each other mutually.

The partial molecular areas of CTAB correspond to a parallel orientation of these molecules to the surface. This orientation occurs over a range of surface pressures which is larger for mixed films than for the pure CTAB films.

Even at the highest pressure studied, 32 dynes/cm (Figure 2), the partial area of CTAB is much larger than the area of CTAB in the pure monolayer. Therefore, lecithin molecules, by interacting with CTAB molecules, determine their orientation or spreading and produce the variation $\Delta\Pi$ of the surface pressure.

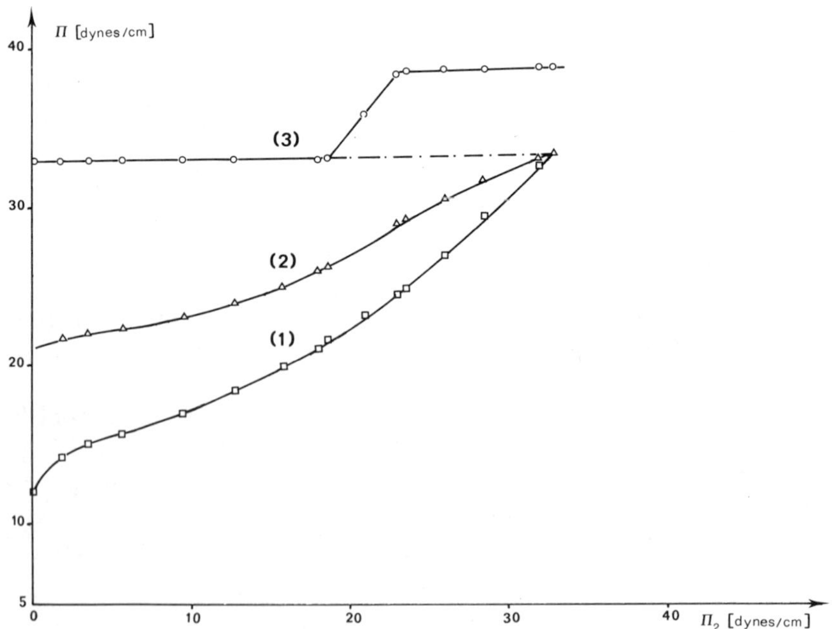

Figure 5. Surface pressure, Π, *of the mixed films,* CTA^+ *ions + lecithin* vs. *the surface pressure,* Π_2, *of the pure lecithin monolayers*

Curve 1: concentration of CTAB salt in bulk $= 5 \times 10^{-6}$M
Curve 2: concentration of CTAB salt in bulk $= 2 \times 10^{-5}$M
Curve 3: concentration of CTAB salt in bulk $= 2 \times 10^{-4}$M

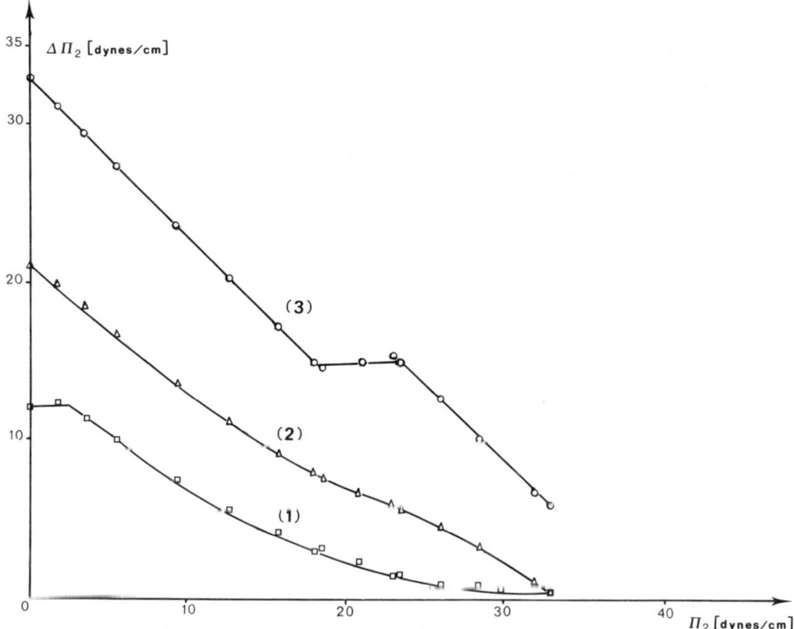

Figure 6. Effect of CTA⁺ ions on the surface pressure of the lecithin monolayers. Surface pressure increase, $\Delta\Pi_2$ vs. surface pressure, Π_2, of the pure lecithin monolayers.

Analysis of Penetration Experiments According to the Classical Approach

The experimental results for CTAB penetration in lecithin monolayers were used, and it was assumed that the mixed film was an ideal two-dimensional mixture. From the increase in surface pressure we deduced the "ideal" surface density of the penetrating CTAB molecules and compared them with the values measured by the radioactive method.

The values of the final pressures of the mixed films CTAB (1) + lecithin (2) obtained for various solutions of 1 and films of 2 are shown in Figure 5 as a function of the surface pressure of the pure 2 monolayer. Figure 6 shows a classical plot of surface pressure increase $\Delta\Pi_2$ produced by penetration of component 1 into the monolayer of pressure Π_2.

Using a Rideal-Davies type equation of state (*4*) for the mixed films as follows:

$$\Pi\,[(\delta_1 + \delta_2)\,A - \delta_1\,A_1{}^{\circ} - \delta_2\,A_2{}^{\circ}] = (\delta_1 + \delta_2)\,kT \tag{6}$$

and assuming that $\Delta\Pi$ may be considered equal to the partial pressure of 1 in the mixture, we obtain:

$$\delta_1 = \frac{\delta_2 \, (A - A_2{}^{\circ}) \, \Delta\Pi_2}{kT \; - \; (A - A_1{}^{\circ}) \, \Delta\Pi_2} \qquad (7)$$

We calculated δ_1 using Equation 7 and the following values: $A_1{}^{\circ} = 35\ A^2$/molecule, and $A_2{}^{\circ} = 56\ A^2$/molecule (*5*). The calculated and measured values of δ_1 corresponding to a CTAB (1) solution of $2 \times 10^{-5} M$ are compared in Figure 7. The absolute values δ_{10} of δ_1 at $\delta_2 = 0$ are different because Equation 6 does not contain the attractive van der Waals terms. Since we are interested in the variation of δ_1 as a function of δ_2, we plotted the relative values of δ_1—*i.e.*, (δ_1/δ_{10})—as a function of δ_2.

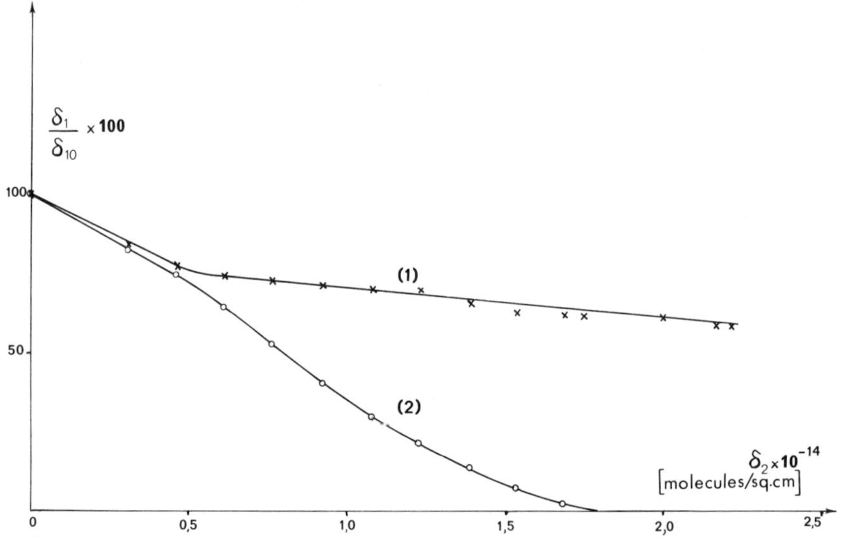

Figure 7. Relative surface density of CTA⁺ ions, $\delta_1/\delta_{10} \times 100$ vs. *lecithin surface density, δ_2. Concentration of CTAB salt in bulk* $= 2 \times 10^{-5}$M.

Curve 1: measured surface density of CTAB⁺ ions
Curve 2: calculated surface density of CTA⁺ ions according to Equation 7

It appears that at low surface densities of lecithin, to a maximum of 250 A^2/molecule, the hypothesis of perfect mixing may be realistic. At high lecithin surface densities this hypothesis leads to unrealistic conclusions for CTAB. Thus, Equation 7 would predict $\delta_1 = 0$ for surface densities of lecithin equal to 1.8×10^{14} molecules/cm² while the experimental results show that δ_1 has been decreased only up to 60%.

Because of the assumptions made in deriving Equation 7, it is unrealistic to extrapolate it this far. However, when surface densities, δ_1, of penetrating molecules are deduced from variations of the surface pressure $\Delta\Pi_2$, the assumptions made in deriving Equation 7 are made implicitly.

Thus, the results of the penetration of proteins into phospholipid monolayers (*6, 7*) are analyzed according to the assumptions we made in deriving Equation 7.

Analysis of Penetration Experiments According to Equation 5

The results in Figures 5 and 6 are shown in Figure 8 in a different way. The increase in surface pressure of the pure CTAB (1) monolayer, $\Delta\Pi_1 = \Pi - \Pi_{10}$, in the presence of the spread lecithin (2) monolayer of surface density δ_2 is plotted as a function of δ_2. The expression for $\Delta\Pi_1$ can be obtained by transforming Equation 5:

$$\Delta\Pi_1 = \gamma_1^{\circ} - \gamma = \underbrace{-\frac{\Delta\mu_{1e\,\mathrm{mix}}}{\bar{a}_1}}_{\left[\text{interaction effect}\right]} + \underbrace{\gamma_1^{\circ}\left(1 - \frac{a_1^{\circ}}{\bar{a}_1}\right)}_{\left[\text{orientation effect}\right]} + \underbrace{\frac{RT}{\bar{a}_1}\left[\ln\frac{\delta_1\bar{a}_1 + \delta_2\bar{a}_2}{\delta_1\bar{a}_1} - \left(\frac{\delta_2\bar{a}_2}{\delta_1\bar{a}_1 + \delta_2\bar{a}_2}\right)\left(1 - \frac{\bar{a}_1}{\bar{a}_2}\right)\right]}_{\left[\text{ideal mixing effect}\right]} \quad (8)$$

It follows from Equation 8 that $\Delta\Pi_1$ may be considered, in principle, as the sum of three contributions which may not be independent. The first term on the right of Equation 8 is the contribution of the change in the partial molar free energy of 1 when it is mixed with 2. One may expect $\Delta\mu_{1e\,\mathrm{mix}} > 0$ for repulsion between 1 and 2 and $\Delta\mu_{1e\,\mathrm{mix}} < 0$ for attraction between 1 and 2. Therefore, $\Delta\Pi_1$ may decrease in the first case (repulsion) and increase in the second (attraction). In one case, it was found (*2*) that $\Delta\mu_{1e\,\mathrm{mix}} > 0$ (*see* Figure 3).

The second term on the right of Equation 8 is the contribution of the reorientation—or change in conformation—of molecule 1 when it penetrates the monolayer of molecules 2. When $\bar{a}_1 > a_1^{\circ}$, this contribution to $\Pi\Delta_1$ is positive. It corresponds to an orientation of the alkyl chains of the penetrating molecules induced by the molecules of the spread compound.

For CTAB (1) and lecithin (2), when $\bar{a}_1 > a_1^{\circ}$, it seems that the repulsion between 1 and 2 impedes penetration of 1 and leads to the orientation of 1 parallel to the surface (Figure 4). Therefore, the interaction and orientation terms of Equation 8 have opposite signs. However the orientation term depends on γ_1° and hence decreases when the concentration of CTAB (1) in solution increases, as shown in Figure 8.

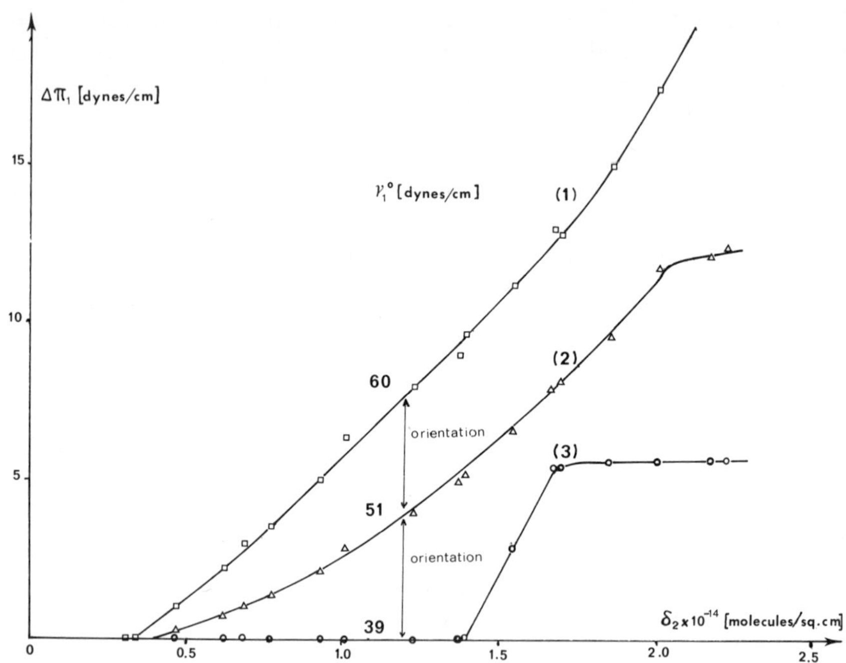

Figure 8. Surface pressure increase of pure CTAB monolayer, $\Delta\Pi_1$, vs. lecithin surface density δ_2

Curve 1: concentration of CTAB salt in bulk = 5×10^{-6}M, $\gamma_1{}^\circ = 60$ dynes/cm
Curve 2: concentration of CTAB salt in bulk = 2×10^{-5}M, $\gamma_1{}^\circ = 51$ dynes/cm
Curve 3: concentration of CTAB salt in bulk = 2×10^{-4}M, $\gamma_1{}^\circ = 39$ dynes/cm

The last term in Equation 8 is the athermal mixing of components 1 and 2 which differ in molecular size. In our case (2) (Figure 2) $\bar{a}_1 > \bar{a}_2$; thus, the contribution of this term to $\Delta\Pi_1$ is positive. However, if a reorientation of 1 occurs so that $\bar{a}_1 < \bar{a}_2$, this term may become negative.

Even in the absence of interaction or reorientation by mixing, it follows from Equation 8 that positive or negative values of $\Delta\Pi_1$ may be obtained if the components of the mixed film have different sizes. Therefore, such values of $\Delta\Pi_1$ are not necessarily an indication of orientation or interaction effects in the mixed films.

Penetration of Human Serum Albumin into Phosphatidylserine Monolayers. We have tried to analyze quantitatively the results of Kimelberg and Papahadjopoulos (7) of protein penetration in phospholipid monolayers. These authors did not measure the surface density of the penetrating protein. They studied the system phosphatidylserine monolayer—human serum albumin (HSA) at different pH values. When the pH was 7.4, the protein and the lipid repelled each other. When the pH was 4.5, the protein and lipid interacted by electrostatic forces. The authors stated that the collapse pressure of the proteins was about 15 dynes/cm.

In Figure 9 we plotted their results according to our technique: $\Delta\Pi_1$ *vs.* δ_2 where $\Pi\Delta_1$ is calculated using $\Pi_1{}^o = 13$ dynes/cm for both pH values. Although $\gamma_1{}^o$ is assumed to be the same, an increase of $\Delta\Pi_1$ is noticed. This increase may be related mainly to an attractive, negative $\Delta\mu_{1e\,mix}$ for HSA, according to our Equation 8.

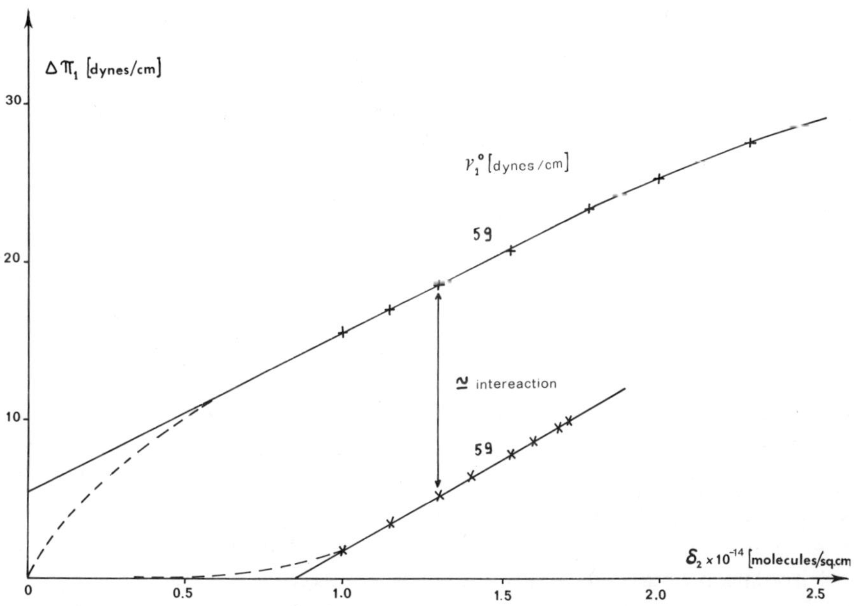

Figure 9. Plot of Kimelberg and Papahadjopoulos (6) *results according to our representation. Surface pressure increases of pure human serum albumin monolayer (HSA),* $\Delta\Pi_1$, vs. *phosphatidylserine surface density,* δ_2. *Phosphatidylserine surface densities are deduced from the surface pressure–area isotherm of Papahadjopoulos* (8).

Curve 1: pH = 7.4
Curve 2: pH = 4.5

Since $\gamma_1{}^o$ is the same at pH 4.5 and 7.4, the orientation contribution to $\Delta\Pi_1$ may be nearly constant if the molecular areas of the HSA are the same under both conditions.

Furthermore, since the protein concentration used in the substrate was very high, one can consider that the mixed monolayer has been saturated with the penetrating protein. Then the mixing terms may be the same at both pH values.

Discussion

Our results on the penetration of CTAB into lecithin monolayers and those of protein penetration into phospholipid monolayers are similar. Kimelberg and Papahadjopoulos (7) found increased "penetration" or $\Delta\Pi$ when the concentration of protein in the substrate increased. In our study, this effect was noticed when the concentration of CTAB in the bulk was increased (Figure 6) to $2 \times 10^{-4}M$. Then even for initial lecithin pressures of 33 dynes/cm a "penetration" of about 6 dynes/cm is observed. According to the discussion of Equation 8, it seems that this increased penetration may be related to the simultaneous reorientation of the alkyl chains of CTAB, followed by penetration inside the hydrophobic region of the lipid monolayer and enhanced attraction.

Kimelberg and Papahadjopoulos (7) assume that the hydrophobic areas of bound proteins can deform the phospholipid monolayer, compelling the fatty acyl liquid chains of the monolayer molecules to orientate parallel to the surface. They conclude that this effect may control the permeability of membranes, but they do not discuss how this reorientation of the acyl chains can produce positive values of $\Delta\Pi$. The reorientation term of Equation 8 explains the origin of this increase in $\Delta\Pi$ since this term is positive when $\bar{a}_1 > a_1^{\circ}$.

Unfortunately a quantitative discussion of their results is impossible since the surface densities of the penetrating protein were not measured.

When the methods of interpretation of penetration experiments—classical (Figure 6) and ours (Figure 7)— are compared for the same system, a strange result is found which is not yet understood. Extrapolation of $\Delta\Pi_2$ to a zero value of Π_2 (Figure 6) does not correspond to the extrapolation of $\Delta\Pi_1$ to a zero value of δ_2 (Figure 8). Indeed the curves of Figure 8 should have origins at $\Delta\Pi_1 = 0$, $\delta_2 = 0$. We find $\Delta\Pi_1 = 0$ when $\delta_2 \neq 0$. This extrapolated value of δ_2 varies with the concentration in solution of the penetrating constituent 1. It may also vary when the extent of interaction between the constituents is changed (Figure 9, results of Kimelberg and Papahadjopoulos).

Conclusions

An experimental and theoretical study of the variations of the surface pressure and composition of a monolayer (2) during penetration by a soluble component (1) has shown that information on the mechanism of this process can be obtained. In practice, one plots the increase of the surface pressure of the soluble constituent 1,

$$\Delta\Pi_1 = \Pi - \Pi_1^{\circ}$$

as a function of δ_2, the surface density of the spread, insoluble lipid monolayer.

Thus, below a limiting inferior value of δ_2, the spread lipid monolayer does not change the surface pressure of the penetrating substance. This remark is different from the remark concerning the limiting value or pressures Π_2 of the spread film, stopping the penetration of constituent 1. Finally, it is suggested that $\Delta\Pi_1$ is the sum of three contributions: interaction, reorientation, and mixing, which are not independent.

Appendix

Partial Free Energy of Mixing of CTAB in the Mixed Film. In the following thermodynamic development, γ, T, and p are constant. The pure film of 1 is taken as a reference. In this case:

$$\mu_1{}^{s} = \mu_{10}{}^{s} = \zeta^{os} - \gamma_1{}^{o}\, a_1{}^{o} \tag{1}$$

In an analogous way, we may write for the studied mixed film:

$$\mu_1{}^{s} = \mu_{10}{}^{s} + \Delta\mu_1{}^{s} \tag{2}$$

and:

$$\mu_2{}^{s} = \mu_{20}{}^{s} + \Delta\mu_2{}^{s} \tag{3}$$

The free enthalpies for the reference and test system are thus:

$$G^{s} = \delta_{10}\, \mu_1{}^{s} \tag{4}$$

and:

$$G^{s} = \delta_1\, \mu_1{}^{s} + \delta_2\, \mu_2{}^{s} \tag{5}$$

Using Equations 2 and 3, Equation 5 becomes:

$$G^{s} = \delta_1\, \mu_{10}{}^{s} + \delta_2\, \mu_{20}{}^{s} + \Delta G^{s}{}_{mix} \tag{6}$$

Differentiating Equation 6 with respect to δ_1, one obtains:

$$\left(\frac{\partial G}{\partial \delta_1}\right)_{\gamma, T, p} = \mu_1{}^{s} = \zeta^{os} - \gamma_1{}^{o}\, a_1{}^{o} + \left(\frac{\partial \Delta G^{s}{}_{mix}}{\partial \delta_1}\right)_{\gamma, T, p} \tag{7}$$

where:

$$\left(\frac{\partial \Delta G^{s}{}_{mix}}{\partial \delta_1}\right)_{\gamma, T, p} = \mu_1{}^{s} - \mu_{10}{}^{s} = \frac{\partial}{\partial \delta_1}(\Delta H^{s}{}_{mix})\, \gamma, T, p - \mathrm{T}\frac{\partial}{\partial \delta_1}(\Delta S^{s}{}_{mix})\, \gamma, T, p - \frac{\partial}{\partial \delta_1}[\Delta(\gamma A_s)_{mix}]\, \gamma, T, p \tag{8}$$

with

$$\Delta H^{s}{}_{\text{mix}} = \delta_1 \, RT\ln f_{1e^s} + \delta_2 \, RT\ln f_{2e^s} \tag{9}$$

$$\Delta S^{s}{}_{\text{mix}} = - R \left[\delta_1 \ln \frac{\delta_1 \bar{a}_1}{\delta_1 \bar{a}_1 + \delta_2 \bar{a}_2} + \delta_2 \ln \frac{\delta_2 \bar{a}_2}{\delta_1 \bar{a}_1 + \delta_2 \bar{a}_2} \right] \tag{10}$$

$$\begin{aligned} \Delta(\gamma A_s)_{\text{mix}} &= \gamma A_{s,\,1+2} - (\delta_1 \gamma_1{}^o \, a_1{}^o + \delta_2 \gamma_2 \, a_2{}^o) \\ &= \gamma(\delta_1 \bar{a}_1 + \delta_2 \bar{a}_2) - (\delta_1 \gamma_1{}^o \, a_2{}^o + \delta_2 \gamma_2{}^o \, a_2{}^o) \end{aligned} \tag{11}$$

Using Equations 9, 10, and 11 one obtains:

$$\mu_1{}^s = \zeta^{os} + RT\ln f_{1e^s} + RT\ln \frac{\delta_1 \bar{a}_1}{\delta_1 \bar{a}_1 + \delta_2 \bar{a}_2} + RT \left(\frac{\delta_2 \bar{a}_2}{\delta_1 \bar{a}_1 + \delta_2 \bar{a}_2} \right) \left(1 - \frac{\bar{a}_1}{\bar{a}_2} \right) - \gamma \, \bar{a}_1 \tag{12}$$

At equilibrium, in the presence of component 2,

$$\mu_1{}^s = \mu_2{}^s \tag{13}$$

and, in the absence of component 2, for the same bulk concentration of 1,

$$\mu_1{}^s = \mu_{10}{}^s = \zeta^{os} - \gamma_1{}^o a_1{}^o = \mu_1 \tag{14}$$

From Equations 12, 13, and 14 one deduces:

$$RT\ln f_{1e^s} = \Delta \mu_{1e\ \text{mix}} = \gamma \bar{a}_1 - \gamma_1 a_1{}^o - RT\ln \frac{\delta_1 \bar{a}_1}{\delta_1 \bar{a}_1 + \delta_2 \bar{a}_2} - RT \left(\frac{\delta_2 \bar{a}_2}{\delta_1 \bar{a}_1 + \delta_2 \bar{a}_2} \right) \left(1 - \frac{\bar{a}_1}{\bar{a}_2} \right) \tag{15}$$

Nomenclature

CTAB	component 1
Lecithin:	component 2
μ_1	chemical potential of 1 in solution
ζ^{os}, $\mu^s{}_{10}$	standard chemical potential and chemical potential of 1 in the pure surface film (in the absence of 2)
$\mu^s{}_{20}$	chemical potential of 2 in the pure surface film (in the absence of 1)
$\mu_1{}^s$, $\mu_2{}^s$	chemical potential of 1 and 2 in the mixed film
$\Delta\mu_1{}^s$, $\Delta\mu_2{}^s$	increment of the chemical potential of 1 and 2 in the mixed film caused by mixing
δ_{10}	surface density of 1 in the pure surface film (in the absence of 2)

δ_1, δ_2	surface density of 1 and 2 in the mixed film
a_1^o, a_2^o	molecular area of 1 and 2 in the pure film of each component
$\bar{a}_1$, $\bar{a}_2$	partial molecular area of 1 and 2 in the mixed film
A_1, A_1^o, A_2^o	mean molecular area and co-area of 1 and 2
$A_{s,\,1+2}$	total area occupied by $(\delta_1 + \delta_2)$ in the mixed film
G^s	Gibbs free enthalpy
$\Delta G^s{}_{mix}$	free enthalpy of mixing
$\Delta H^s{}_{mix}$	enthalpy of mixing
$\Delta S^s{}_{mix}$	entropy of mixing
$\Delta(\gamma A_s)_{mix}$	variation of the γA_s function of the mixing
$f^s{}_{1e}$, $f^s{}_{2e}$	activity coefficient of 1 and 2 in the mixed film
$RT \ln f^s{}_{1e} = \Delta\mu_{1e\,mix}$	partial free energy of mixing of 1
γ_1^o, γ_2^o, γ	surface tension of the solution in the absence of 2 when 1 is adsorbed; surface tension of the solution in the absence of 1 when 2 is spread; surface tension of the solution, when 1 is adsorbed and 2 is spread
Π_1^o, Π	surface pressure of the solution in the absence of 2 when 1 is adsorbed and surface pressure of the solution, when 1 is adsorbed and 2 is spread
$\Delta\Pi_1$	change in surface pressure of the pure solution of 1 when 2 is spread
Π_2, $\Delta\Pi_2$	initial pressure of 2 and change in the film pressure of 2 by penetration of 1
k, R, T, p	Boltzmann constant, gas constant, temperature and pressure

Literature Cited

1. Schulman, J. H., Rideal, E. K., *Proc. Roy. Soc.* (1937) **B 122,** 29.
2. Hendrikx, Yolande, Ter-Minassian-Saraga, Lisbeth, *J. Chim. Phys.* (1970) **67,** 1620.
3. Ter-Minassian-Saraga, Lisbeth, "Progress in Surface and Membrane Science," J. F. Danielli, M. D. Rosenberg, and D. A. Cadenhead, Eds., in press.
4. Davies, J. T., *J. Colloid Sci.* (1956) **11,** 377.
5. Watkins, J. C., *Biochim. Biophysis. Acta* (1968) **152,** 293.
6. Quinn, P. J., Dawson, R. M. C., *Biochem. J.* (1969) **113,** 791.
7. Kimelberg, H. K., Papahadjopoulos, Demetrios, *Biochim. Biophys. Acta* (1971) **233,** 805.
8. Papahadjopoulos, Demetrios, *Biochim. Biophysis. Acta* (1968) **163,** 240.

RECEIVED September 23, 1974.

14

Protein–Surfactant Interactions

I. Effects of Surfactants upon Gliadin Monolayers at the Air–Aqueous Interface

MARVIN N. YUDENFREUND and PAUL BECHER

Specialty Chemicals Research Dept., ICI U.S. Inc., Wilmington, Del. 19897

JOHN B. BROWN

Denison University, Granville, Ohio 43023

Interactions of several surface-active agents with monomolecular films of gliadin on an ammonium sulfate solution were monitored by measuring changes in surface pressure ($\Delta\pi$) and surface potential (ΔV). Minimum areas per molecule of gliadin were obtained from compression data. The presence of nonionics such as BRIJ *35 and* BRIJ *78 increased these minimum values by over 40%, indicating unfolding of the protein molecules. Calcium stearoyl-2-lactylate, on the other hand, did not increase the minimum area per molecule. Changes in $\Delta\pi$ and ΔV resulting from injection of surfactants under gliadin monolayers show that the polyoxyethylene nonionics interact with both polar and nonpolar regions of the protein molecule. A mechanism is offered to explain the effect of surfactants on the rheology of bread dough.*

In nature proteins interact with ions, lipids, and other proteins as part of the broad spectrum of necessary biological processes including membrane functionality and antigen–antibody effects. Protein functionality can be altered greatly by the interaction of proteins with surface-active agents, and the subject of protein–surfactant interaction is important in relation to food, cosmetic, and biomedical areas.

Surfactants such as mono- and diglycerides and certain types of polyoxyethylene nonionics are routinely used in yeast-raised baked goods as anti-staling agents and as dough conditioners—materials which modify the rheological properties of dough. Dough conditioning depends to a great extent on the interaction of surfactants with the proteins found in dough. Investigators attempted to elucidate the nature of this interaction through rheological studies of dough (*1, 2*) and other means (*3, 4, 5, 6*), but no clear understanding has emerged to date.

Various physical parameters—*e.g.* surface pressure, surface potential, and surface viscosity—have frequently been used to characterize monomolecular films of macromolecules at the air–aqueous interface (*7, 8*). To correlate the surface chemical properties of wheat gluten with the baking quality of the flour from which it was derived, Tschoegl and Alexander measured the surface pressure (*9*) and surface viscoelasticity (*10*) of monomolecular films of the protein under various conditions. Their results suggested that gluten forms highly coherent films in which the protein molecules associate with each other through a vast network of hydrogen bonds and salt linkages. However, they did not correlate the surface chemistry of gluten with baking properties of dough.

Monolayer techniques have been used with much success to study the interaction of proteins with surfactants (*11, 12, 13*), ions and lipids (*14*), and other proteins (*15*). This paper investigates, through well-established procedures, the surface chemistry of monolayers of a major component of heterogeneous wheat gluten protein, gliadin, and explores these interactions with various surface-active agents.

Experimental

Gluten was obtained from ProVim, a gluten-enriched flour, by a conventional dough-and-wash procedure (*16*). Gliadin was separated from the gluten by the method of Jones *et al.* (*17*). The surfactants selected for study were MYRJ 45 [polyoxyethylene (*8*) stearate], TWEEN 60 [polyoxyethylene (20) sorbitan monostearate], BRIJ 35 [polyoxyethylene (23) lauryl ether], BRIJ 76 [polyoxyethylene (10) stearyl ether], BRIJ 78 [polyoxyethylene (20) stearyl ether], and calcium stearoyl-2-lactylate. (Note that the numbers in parentheses indicate the number of monomer units of ethylene oxide.) Working solutions were made up in 70% ethanol at 2.5 mg per 100 ml solution.

The substrate was a 35% solution of ammonium sulfate (*18*). Water was twice distilled in a Corning all glass still. The ammonium sulfate solution was agitated with DARCO activated carbon overnight and was filtered for use. This solution had no surface activity.

Monolayer studies were done on a polyester resin-impregnated fiber glass Langmuir trough, 15.1 × 85 cm and 5 mm deep, coated with paraffin. This arrangement was mounted on a lathe bed equipped with a variable speed motor with a tachometer feedback control which drove the Teflon sweeps at 0.65 cm/min, ± 1%. Surface pressure measurements were made with a platinum Wilhelmy plate coupled to a Cahn RG Electro-

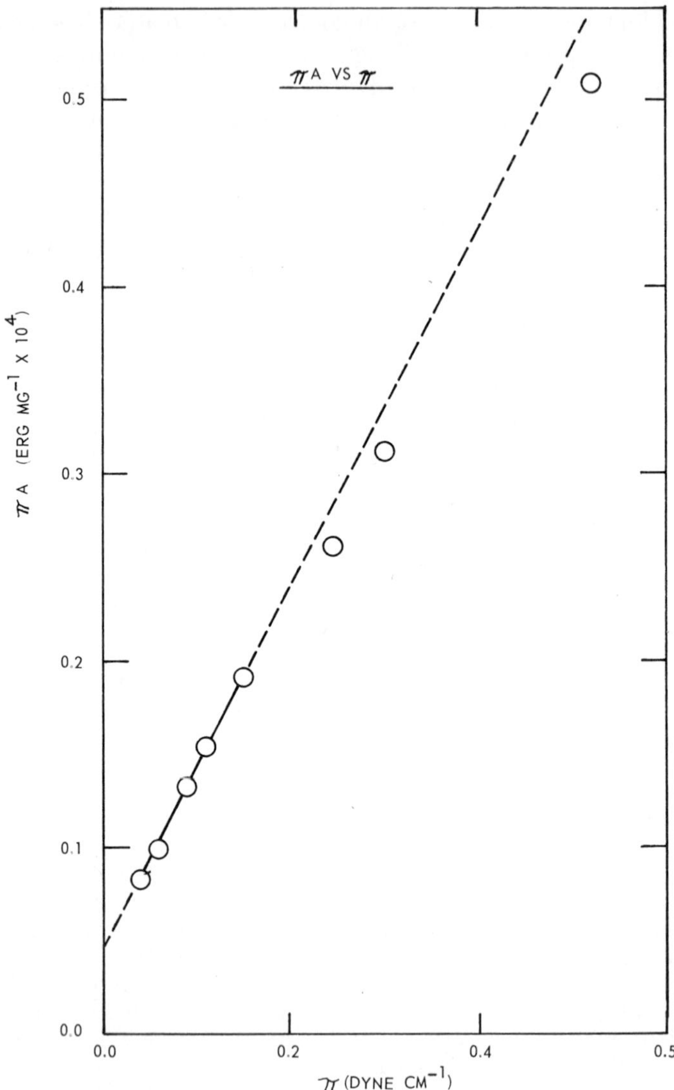

Figure 1. Linear relationship between πA and π for gliadin monolayers in the region of low surface pressure

balance with an accuracy of ± 0.05 dyne/cm. The plate was periodically electroplated with platinum black to ensure complete wettability. Surface potential measurements were obtained using an ionizing electrode of americum 241 and a platinum reference electrode in the subsolution, in conjunction with a Keithley electrometer. Data were obtained at 24° ± 1°C.

Monolayers were spread from an Agla micrometer syringe. Mixed monolayers consisted of 0.045 mg gliadin and 0.005 mg surfactant. For the penetration studies 5×10^{-4} mg quantities of surfactants were injected with a Hamilton microliter syringe under gliadin monolayers (5 ×

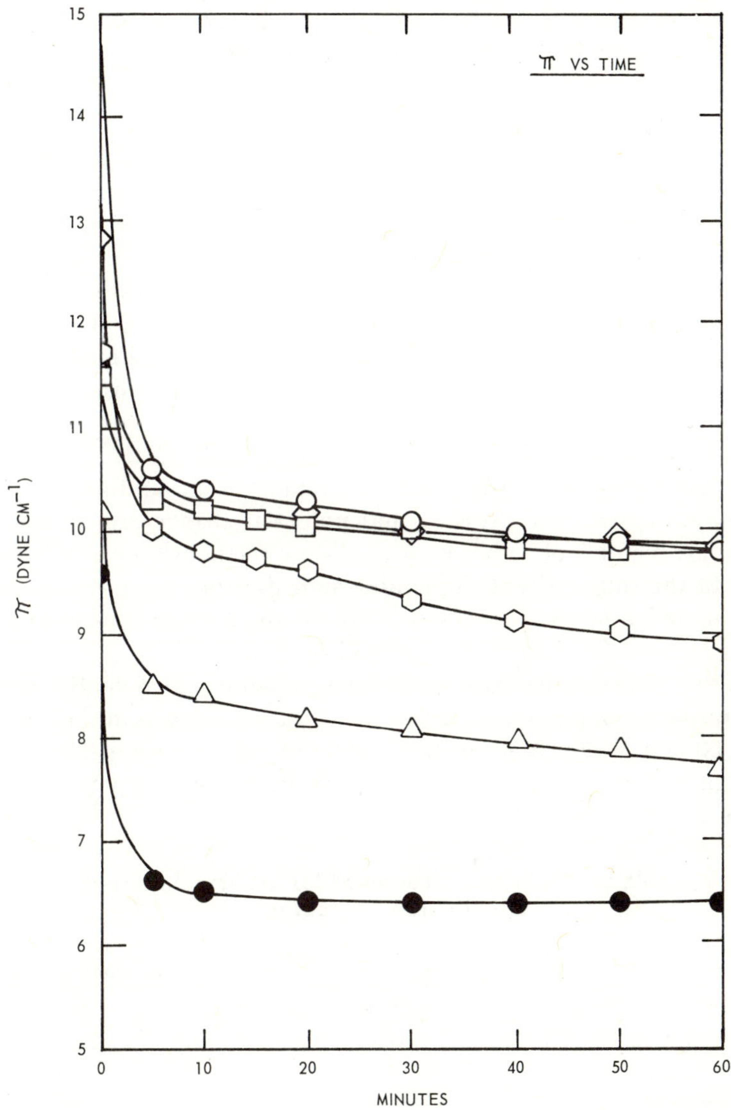

Figure 2. Change of surface pressure with time for gliadin films experiencing submonolayer injections of: calcium stearoyl-2-lactylate (●), MYRJ *45* (△), TWEEN *60* (⬡), BRIJ *35* (□), BRIJ *76* (○), BRIJ *78* (◇)

10^{-3} mg) compressed to 5.0 dynes/cm. Data presented are averages of at least three determinations. Isotherms were reproducible within ± 0.5 dyne/cm, and surface potentials varied by ± 5 mV.

To compensate for any additivity effects of the surfactants and gliadin we spread surfactant solutions, in the amounts used in the experiments, on the film balance. No significant surface pressure was observed over the entire compression range, indicating that the surfactants did not contribute directly to the surface pressure measurements.

Results

Molecular Weight. The molecular weight of gliadin was obtained from the low pressure region of the π–A isotherm. In this region the protein behaves more or less like an ideal gaseous film, and the two-dimensional equivalent of the ideal gas law can be applied (*19*): $\pi A = nRT$. A plot of πA vs. π (Figure 1) is linear at low pressure. When extrapolated to zero pressure a molecular weight of about 57,000 is obtained. This is within the range of determinations cited in the literature (*20*).

Submonolayer Injections of Surfactants. Plots of surface pressure *vs.* time are shown in Figure 2. Time zero is the time at which the injection was completed. The surface pressure, initially 5 dynes/cm, rises sharply and subsequently decreases with time to an equilibrium value. The equilibrium surface pressure and potential change are summarized in Columns 2 and 3 of Table I. A nonlinear relationship exists between the log of the difference of the equilibrium pressure, π_e, the pressure at time t, π_t, and time [*i.e.*, log (π_e–π_t) *vs.* t]; the surface pressure change is therefore not a first-order phenomenon.

The equilibrium increase in surface pressure and potential for the BRIJ emusifier compounds is the same—*viz.*, 4.8 dynes/cm and 30 mV. TWEEN 60 and MYRJ 45 are in the middle of the spectrum. The anionic surfactant, calcium stearoyl-2-lactylate, has the smallest change in surface pressure and surface potential.

Table I. Results of Studies of Mixed Monolayers of Gliadin and Surfactants

Monolayer Composition	$\Delta\pi$ *(dynes/cm)*	ΔV *(mV)*	*Minimum Molecular Area* (A^2)
Gliadin	—	—	6600
+ MYRJ 45	2.7	19	8500
+ TWEEN 60	3.9	18	8500
+ BRIJ 35	4.8	28	9500
+ BRIJ 76	4.8	31	8500
+ BRIJ 78	4.8	29	9500
+ Calcium stearoyl-2-lactylate	1.4	9	6600

Compression Isotherms. The π–A compression isotherms of gliadin and the six mixed monolayers of gliadin and surfactants are in Figures 3 and 4. The abscissa in both figures has been moved to the right by 0.2 m^2/mg for successive curves to avoid confusion. All isotherms had the same approximate shape; no collapse was observed even at surface pressures of 27 dynes/cm.

The compressibility (δ) of an insoluble monolayer can be expressed

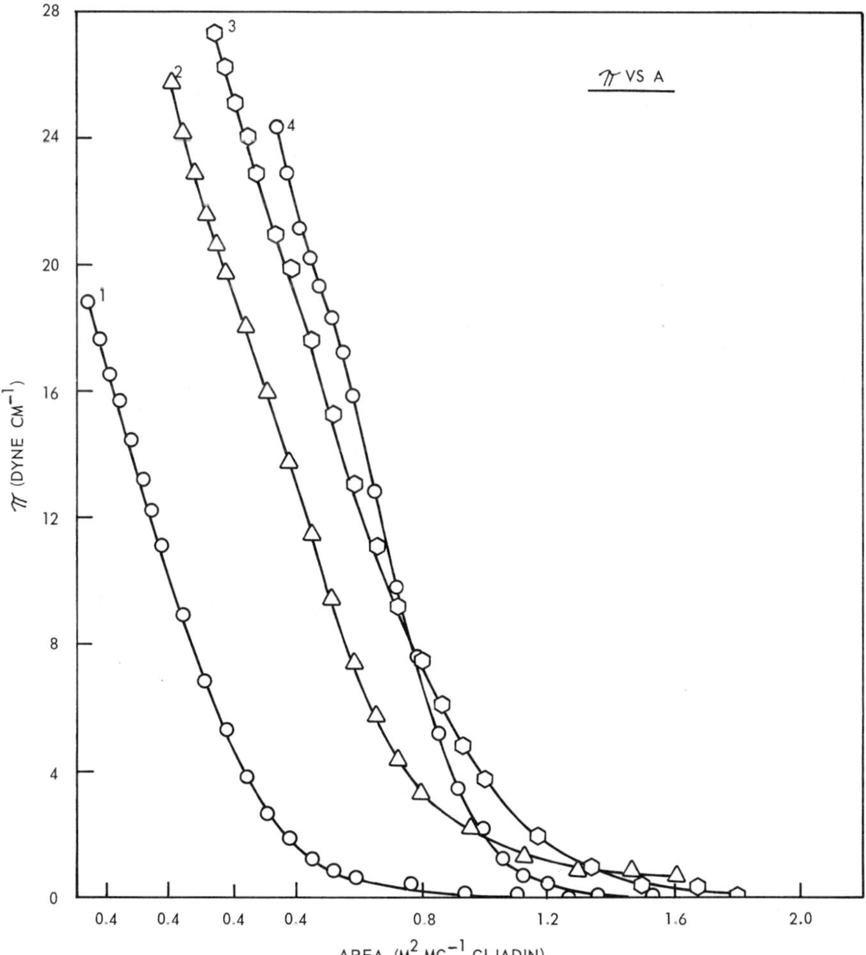

Figure 3. Compression isotherms for gliadin (Curve 1) and mixed monolayers of gliadin and the following surfactants: MYRJ *45 (Curve 2),* TWEEN *60 (Curve 3), and calcium stearoyl-2-lactylate (Curve 4). The abscissa is moved to the right by 0.2 m^2/mg for successive curves to avoid confusion.*

as:

$$\delta = (1/A)\ (\partial A/\partial \pi)_T, \tag{1}$$

where A is the molecular area of the film, and π is the surface pressure. The compressibility is derived from the slope of the isotherm. A plot of δ *vs.* A reveals a minimum which represents the minimum surface area per molecule—*i.e.*, the minimum area that a protein film can be compressed and remain uncollapsed. Minimum areas per molecule of gliadin are in Column 4 of Table I.

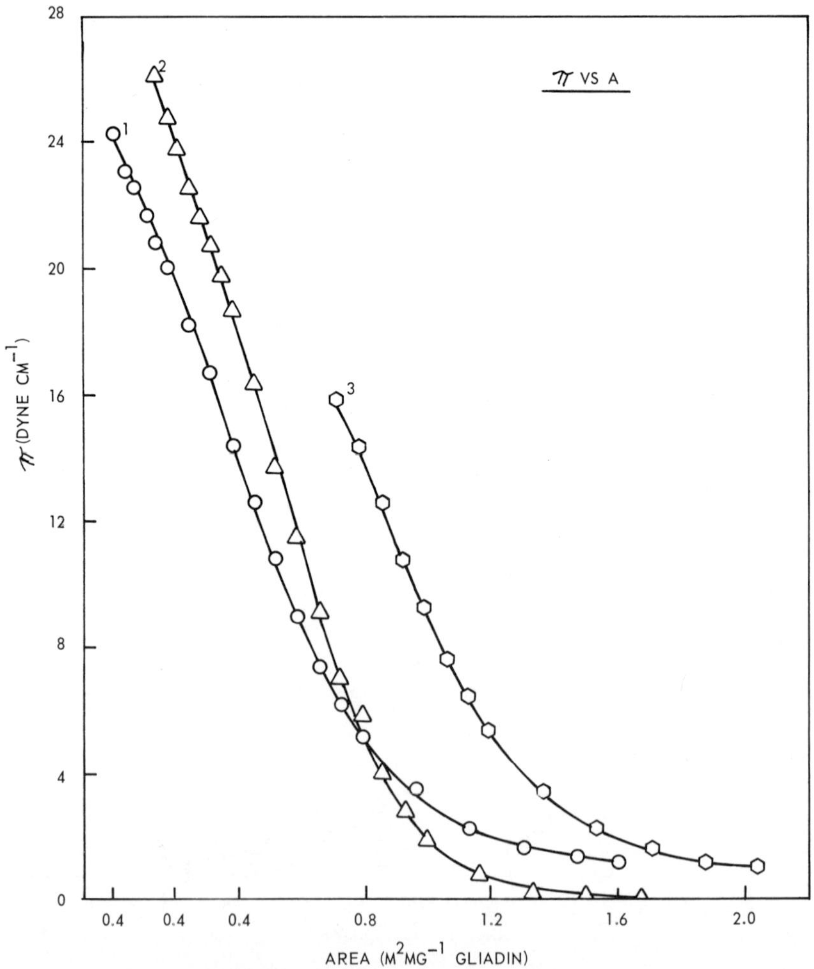

Figure 4. Compression isotherms for mixed monolayers of gliadin and the following surfactants: BRIJ *35 (Curve 1),* BRIJ *76 (Curve 2), and* BRIJ *78 (Curve 3). The abscissa is moved to the right by 0.2 m^2/mg for successive curves to avoid confusion.*

Gliadin containing BRIJ 35 and BRIJ 78 has areas per molecule over 40% greater than that of gliadin alone. Gliadin containing BRIJ 76, MYRJ 45, and TWEEN 60 exceeds pure gliadin in molecular area by 30%. The addition of calcium stearoyl-2-lactylate to gliadin appears to have no effect on the area per molecule.

Discussion

When spread at the air–aqueous interface there is some loss of tertiary structure of proteins as the molecules orient themselves with polar groups in the substrate and apolar groups above the surface. Schulman (*21*) explored the interaction of ionic surfactants with films of gliadin oriented in this manner. He concluded that if there were no association between the surfactant and the monolayer, no change in either surface pressure or surface potential would be observed. If the surfactant associated only with the polar regions of the protein, a reorientation of dipoles in the aqueous phase would produce a change in surface potential. If the surfactant not only associated with the polar groups but penetrated the protein film and was adsorbed in the apolar regions, an increase in surface pressure would occur in addition to the change in surface potential.

The data indicate that penetration of the monolayer occurs when the polyoxyethylene nonionic surfactants are injected into the substrate. Polar portions of surfactant molecules interact with their counterparts on the protein film through permanent dipole attraction and dipole-induced–dipole (van der Waals) interaction, and electrostatic attraction. The nature of this adsorption disrupts hydrogen bonding which partially stabilizes the tertiary protein structure. This makes the macromolecules more amenable to unfolding, and this has been observed with proteins in the presence of ionic surfactants (*22*) as well.

The apolar moiety of a surfactant molecule present in the aqueous substrate is exceedingly prone to adsorption in the apolar regions of the protein monolayer. Kauzmann (*23*) noted that large positive entropy changes, 20–30 eu, are observed when apolar groups in the aqueous phase are transferred to apolar environments within the protein. This occurs with very little change in enthalpy. Also Kauzmann made reference to an observation that a volume increase of 6.7 ml/mole of bovine serum albumin occurs when it is bound by sodium dodecyl sulfate. Thus the penetration of the aliphatic hydrocarbon portion of the surfactant molecule to the apolar regions of gliadin films very definitely contributes to the propensity of the protein to unfold.

The observation that the minimum molecular area of gliadin increases in the presence of small amounts of polyoxyethylene nonionic surfactants confirms the interaction mechanism alluded to in the previous paragraphs. The unfolding is extensive, with areas per molecule of gliadin treated with nonionic surfactants exceeding that of the untreated material by over 40% in some cases.

Calcium stearoyl-2-lactylate has only a slight tendency to interact with gliadin. The changes in surface pressure and surface potential are small when compared with the changes produced by the nonionic surfactants. This may be attributed to one of several possibilities: calcium stearoyl-2-lactylate may be quite soluble in the substrate and diffuse slowly to the surface, or competition for polar protein binding sites may occur between the surfactant and the substrate. Analysis of the mixed monolayer π–A isotherm shows no increase in area per molecule of gliadin, indicating that little if any unfolding of gliadin occurs in the presence of the anionic surfactant.

Calcium stearoyl-2-lactylate—a recognized bread dough conditioner (*24*)—is known to complex with certain water-soluble fractions of wheat protein (*3*), but there is little hard evidence pointing to a direct interaction between it and gluten proteins on the order of that seen between the nonionic surfactants and gliadin. Some evidence (*25*) suggests that calcium stearoyl-2-lactylate interacts with gluten proteins in the presence of starch.

MacRitchie (*26*) explored the effect of the molecular weight distribution of gluten fractions on the rheological characteristics of bread dough. He found that a flour which produced a weak, highly plastic dough could be modified by the addition of a high molecular weight protein fraction of gluten. Conversely a flour which produced a strong, highly viscous dough could be balanced by the addition of low molecular weight gluten fractions. He attributes the rheological properties of dough to the phenomenon of entanglement–coupling of side chains of neighboring protein molecules. Entanglement–coupling essentially links a large mass of protein molecules together.

We demonstrated that certain nonionic surfactants cause gliadin to unfold by a mechanism involving the disruption of tertiary molecular structure. Gliadin comprises 40–50% of wheat gluten; the remainder is the grossly heterogeneous fraction—glutenin. This consists of gliadin-like subunits linked by disulfide bonds (*27*), and one may therefore expect glutenin to interact with surfactants in a manner similar to that of gliadin —*i.e.*, to unfold. The unfolding of wheat gluten by surfactants significantly increases the number of residues available for entanglement–coupling.

Acknowledgment

We express our gratitude to the late Leland F. Gleysteen for providing us with guidance, encouragement, and inspiration in this investigation.

Literature Cited

1. Knightly, W. F., *Bakers Digest* (1973) **47,** 64.
2. Langhans, R. K., Thalheimer, W. G., *Cereal Chem.* (1971) **48,** 283.
3. Fullington, J. G., *Cereal Chem.* (1974) **51,** 250.
4. DelVecchio, A. J., Stutz, R. L., Tenney, R. J., *Meetg. Amer. Chem. Soc., New York, 164th,* August 29, 1972.
5. Grosskreutz, J. C., *Cereal Chem.* (1961) **38,** 336.
6. Birnbaum, H., *Bakers Digest* (1971) **45,** 22.
7. Malcolm, B. R., *J. Poly. Sci. Part C* (1971) **34,** 87.
8. Suzuki, A., *Kolloid-Z. Z. Poly.* (1972) **250,** 365.
9. Tschoegl, N. W., Alexander, A. E., *J. Colloid Sci.* (1960) **15,** 155.
10. Tschoegl, N. W., Alexander, A. E., *J. Colloid Sci.* (1960) **15,** 168.
11. Bull, H. B., *J. Amer. Chem. Soc.* (1945) **67,** 10.
12. Vilallonga, F. A., Garrett, E. R., *J. Pharm. Sci.* (1972) **61,** 1720.
13. Vilallonga, F. A., Garrett, E. R., *J. Pharm. Sci.* (1973) **72,** 1605.
14. MacRitchie, F., *J. Macromol. Sci.-Chem.* (1970) **A4(5),** 1169.
15. Arnold, J. D., Pak, C. Y., *J. Colloid Sci.* (1962) **17,** 348.
16. Dill, D. B., Alsburg, C. L., *Cereal Chem.* (1924) **1,** 222.
17. Jones, R. W., Taylor, N. W., Senti, F. R., *Arch. Biochem. Biophys.* (1959) **84,** 363.
18. Bull, H. B., *J. Amer. Chem. Soc.* (1945) **67,** 4.
19. Gaines, Jr., G. L., "Insoluble Monolayers at Liquid-Gas Interfaces," p. 270, Interscience, New York, 1966.
20. Pence, J. W., Nimmo, C. C., Hepburn, F. N., "Wheat Chemistry and Technology," I. Hlynka, Ed., p. 245, American Association of Cereal Chemists, Inc., St. Paul, Minn., 1964.
21. Schulman, J. H., *Trans. Faraday Soc.* (1937) **33,** 1116.
22. Fraser, M. J., *J. Pharm. Pharmacol.* (1957) **9,** 497.
23. Kauzmann, W., *Adv. Protein Chem.* (1959) **14,** 1.
24. Finney, K. F., Shogren, M. D., *Bakers Digest* (1971) **45,** 40.
25. Chung, O. K., Tsen, C. C., *Meetg. Amer. Ass. Cereal Chem., St. Louis, Nov. 1973.*
26. MacRitchie, F., *J. Sci. Food Agr.* (1973) **24,** 1325.
27. Bietz, J. A., Wall, J. S., *Cereal Chem.* (1973) **50,** 537.

RECEIVED October 14, 1974.

15

Some Effects of Hydrocarbon Solvent Structure on the Phase Behavior of Distearoyl Lecithin Monolayers at the Hydrocarbon/Water Interface

CRAIG M. JACKSON and BEATRICE Y. J. YUE

Department of Biological Chemistry, Division of Biology and Biomedical Sciences, Washington University, St. Louis, Mo. 63110

Monolayers of distearoyl lecithin at hydrocarbon/water interfaces undergo temperature and fatty acid chain length dependent phase separation. In addition to these variables, it is shown here that the area and surface pressure at which phase separation begins also depend upon the structure of the hydrocarbon solvent of the hydrocarbon oil/aqueous solution interfacial system. Although the two-dimensional heats of transition for these phase separations depend little on the structure of the hydrocarbon solvent, the work of compression required to bring the monomolecular film to the state at which phase separation begins depends markedly upon the hydrocarbon solvent. Clearly any model for the behavior of phospholipid monolayers at hydrocarbon/water interfaces must account not only for the structure of the phospholipid but also for the influence of the medium in which the phospholipid hydrocarbon chains are immersed.

During the past five or 10 years interest in elucidating the relationship between molecular structure of biological membrane components and their function has increased immensely. As a result of this interest classical and modern physicochemical techniques have been used in studies specifically designed to elucidate how the structure of phospholipids determines the liquid crystal properties of aqueous dispersions of this class of membrane components. Our own interest in surface catalysis in blood coagulation which is caused by phospholipid "membrane" participation in this process has led us to study the surface properties of phospholipid bilayer vesicles and phospholipid monolayers.

The development of a convenient apparatus for investigating monomolecular films at hydrocarbon oil/water interfaces by Brooks and Pethica (*1*) specifically led us to investigate the behavior of phospholipid monolayers at hydrocarbon/water interfaces. During these studies, which were done in collaboration with the Unilever Research group at Port Sunlight, we observed that fatty acyl chain length dependent phase separation occurs in phospholipid monolayers at the hydrocarbon/water interface (*2*). Although detailed analysis of the compression isotherms is only now underway, the behavior of distearoyl lecithin films at the heptane/aqueous NaCl interface as a function of temperature, NaCl concentration, and pH has been described (*3*). Here we report our observation that the phase behavior of distearoyl lecithin monolayers depends upon the structure of the hydrocarbon solvent in addition to the previously investigated parameters.

Materials and Experimental Techniques

All hydrocarbon solvents were purchased from Phillips Petroleum Co., Bartlesville, Okla., and were of 99 mole % or greater purity. Prior to use the hydrocarbon solvents were filtered through a 5×40 cm column of neutral alumina, Brockmann activity grade I (Sigma Chemical Co., St. Louis). Water was prepared by filtration through activated charcoal and mixed bed ion exchange cartridges (Continental Water Conditioning Corp., El Paso, Tex.), distilled once from alkaline $KMnO_4$ and finally redistilled in a borosilicate glass still. Surface tension of the water was within 0.1 dyne/cm of the accepted literature values (*4*). Sodium chloride, ExP grade was purchased from Heico, Inc., Delaware Water Gap, Pa. and was roasted at 750°C prior to preparation of the aqueous subphase solution. Surface balance temperatures were maintained within ± 0.2°C. All experiments were carried out in a positive pressure clean room in which the input air supply was filtered through a microporous paper filter (Farr Company, Los Angeles, Calif.).

Distearoyl lecithin (1,2-dioctadecanoyl-*sn*-glycero-3-phosphorylcholine) was synthesized by acylation of *sn*-glycero-3-phosphorylcholine with stearic anhydride/potassium stearate (*5*). Stearic acid (Sigma Chemical Co., St. Louis) was greater than 99 mole % pure by gas liquid chromatography of the fatty acid methyl esters. The distearoyl lecithin gave a single spot by silica gel thin layer chromatography and had a single, sharp thermal transition at 56°C by differential scanning calorimetry in the presence of excess water (*6*). The same lot of lecithin was used for all isotherm determinations. Monolayers were spread from solutions of the appropriate hydrocarbon solvent and absolute ethanol, 10:1 (v/v) or greater as described by Mingins and Taylor (*7*).

As a result of possible incomplete spreading and the uncertainty in the concentration of the phospholipid in the spreading solutions, some

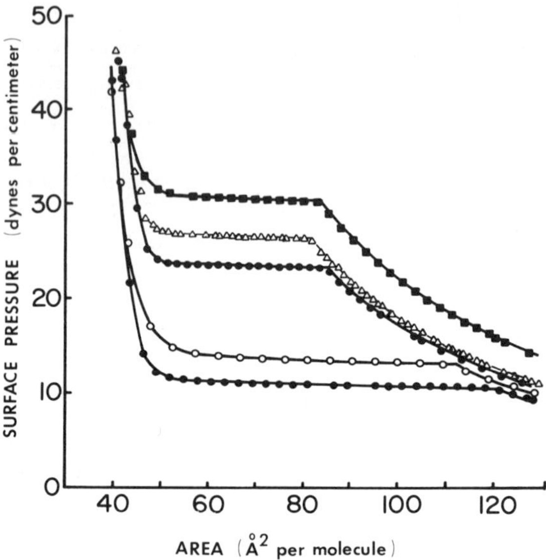

*Figure 1. Surface pressure–area per molecule isotherms for distearoyllecithin at various hydrocarbon/aqueous NaCl interfaces. Cyclohexane/0.1*M *NaCl, 19.7°C* (■); n-*heptane/0.1*M *NaCl, 20.0°C* (△); *isooctane/0.1*M *NaCl, 20.3°C, upper curve, and 3.1°C, lower curve* (●); n-*nonane/0.1*M *NaCl, 3.1°C* (○).

determinate error in the value of the area per molecule always exists in surface pressure–area per molecule isotherms such as these. To exclude artifactual difference in the area at which phase separation begins for any particular pair of isotherms, all isotherms were "normalized" to the *n*-heptane/aqueous NaCl isotherms (*13*) by multiplying the experimentally determined areas per molecule for each film by a constant. This normalization constant was determined by fitting each isotherm both in the low area close packed region and at areas per molecule greater than 350 A^2. Since no solvent dependence can be observed at high areas per molecule and in the close packed region where the films can be expected to be practically hydrocarbon solvent free, these two regions can legitimately be used in such comparisons. The constants used for this fitting varied from 0.984 to 1.016 relative to the *n*-heptane film which was taken as reference 1.0. Because of the large differences in π transition, and the small percentage change that such normalization yields, this exercise only validates the comparisons made here. Any difference in temperature between the films being compared was compensated by using the values of $d\pi/dT$ and dA/dT obtained from the data of Ref. 3.

Results and Discussion

Representative surface pressure/area per molecule isotherms from monolayers of distearoyl lecithin at the interface between 0.1*M* NaCl and cyclohexane, *n*-heptane, and isooctane at 20°C and *n*-nonane and isooctane at 3°C are shown in Figure 1. Two completely independent isotherms which were actually determined some months apart for the *n*-heptane/0.1*M* NaCl interface are plotted to illustrate the precision and reproducibility of the method and the data. Quite clearly the area and surface pressure at which phase separation begins depend on the hydrocarbon component of the oil/water interfacial system. The areas and surface pressures at which phase separation occurs for these and the other solvents which have been investigated are summarized in Table I.

Table I. Area and Surface Pressure at which Phase Separation Begins in Distearoyl Lecithin Films at Various Hydrocarbon/Aqueous NaCl Interfaces

	Temperature Group					
	3°C			*20°C*		
Solvent	A_t	π_t	*T(°C)*	A_t	π_t	T(°C)
n-Pentane	116	11.2	3.1	86	23.7	17.9
n-Hexane	110	13.4	3.1	81	25.8	18.8
Cyclohexane	[a]	—	—	83	30.4	19.7
n-Heptane	106	13.5	4.0	82	26.5	20.0
Isooctane	119	10.6	3.1	84	23.3	20.3
n-Nonane	114	12.4	3.1	80	24.9	20.9
n-Undecane	128	9.3	3.1	85	21.6	40.9

[a] Cyclohexane is a solid at this temperature.

At 20°C the phase transition areas vary over 6 A^2 per molecule. Although this range is outside the experimental error for the normalized isotherms, such small variation must be considered to be on the borderline of significance. More striking however is the approximately 10 dyne/cm range of variation in surface pressure that results from the differences in hydrocarbon solvent effects on the work of compression.

Extensive investigation of the monolayer isotherm behavior at the *n*-heptane and isooctane/aqueous NaCl interfaces (*8*) has shown that $d\pi(\text{trans})dT$ and ΔA's for each film the heats of transition can be calculated from the two-dimensional Clapeyron equation (*3*). Based upon the variation observed in the heats calculated in a previous study these heats are not significantly different, even at 3°C. In contrast, however (*see* both isotherms of Figure 1 and data of Table I), the work of compression from high areas to the area at which phase separation begins depends upon the hydrocarbon solvent. Although not shown in Figure 1, all isotherms become coincident at areas per molecule in excess of approximately 400 A^2. However, sufficient high area data do not yet exist for all these solvents, and integration of the isotherms to determine the work of compression has not been completed.

Attempts to fit these and other compression isotherms between 400 A^2 and the area at which phase transition (*9, 10, 11, 12*) begins with a variety of two-dimensional equations of state indicates that a constant value for the co-area or the partial molecular area term of these equations cannot suffice in describing the behavior of these monolayers. Such unsuccessful attempts were actually the motivating factor for these investigations of the influence of the hydrocarbon solvent on the monolayer compression isotherms.

Acknowledgement

The authors thank J. Mingins and B. A. Pethica of Unilever Research Laboratory, Port Sunlight, Cheshrire, England, for many stimulating and helpful discussions.

Literature Cited

1. Brooks, J. H., Pethica, B. A., *Trans. Faraday Soc.* (1964) **60,** 208-215.
2. Taylor, J. A. G., Mingins, J., Pethica, B. A., Tan, B. Y. J., Jackson, C. M., *Biochim. Biophys. Acta* (1973) **323,** 157-160.
3. Yue, B. Y. J., Jackson, C. M., Taylor, J. A. G., Mingins, J., Pethica, B. A., *J. Chem. Soc., Faraday Trans.*, in press.
4. "Handbook of Chemistry and Physics," 54th ed., Chemical Rubber Co., Cleveland, Ohio, 1972.
5. Cubero-Robbs, E., van den Berg, D., *Biochim. Biophys. Acta* (1969) **187,** 520.
6. Phillips, M. C., Williams, R. M., Chapman, D., *Chem. Phys. Lipids* (1969) **3,** 234.
7. Mingins, J., Taylor, J. A. G., "A Manual for the Measurement of Interfacial Tension, Pressure and Potential at Air or Non-polar Oil/Water Interface," Unilever Res. Ltd., Port Sunlight, Cheshire, England, 1970.
8. Yue, B. Y. J., Ph.D. Dissertation, Washington University, St. Louis, Mo., 1974.
9. Defay, R., Prigogine, I., Bellemans, Everett, D. H., "Surface Tension and Adsorption," pp. 208-216, Longmans, Green & Co., Ltd., London, 1966.
10. Fowkes, F. M., *J. Phys. Chem.* (1961) **65,** 355-359.
11. Gershfeld, N. L., Pagano, R. E., *J. Phys. Chem.* (1972) **76,** 1231-1237.
12. Langmuir, I., *J. Chem. Phys.* (1933) **1,** 756-776.

RECEIVED October 18, 1974. Work sponsored by grant HL-14147 from the Heart and Lung Institute of the National Institutes of Health.

16

Enzymic Degradation of Ethylene Glycol Adipate Oligomers at the Air–Water Interface

W. HECQ, C. BERLINER, J. M. RUYSSCHAERT, and J. JAFFÉ

Laboratoire de Chimie—Physique des Macromolécules aux Interfaces, Faculté des Science, Université Libre de Bruxelles, Belgique

The degradation of ethylene glycol adipate oligomers by pancreatic lipase was investigated at the air–water interface. The enzyme was dissolved in the subphase, and the tritiated substrate was spread at the interface. Surface radioactivity measurements allow us to determine the concentration of substrate at the air–water interface and to describe the evolution of the interfacial reaction. The kinetics were studied as a function of pH, temperature, surface concentration, and length of the oligomers. Labeling (tritium) of the enzyme and the substrate gave quantitative information about two simultaneous processes: adsorption of the enzyme at the interface and the disappearance of the substrate from the interface. All the results demonstrate clearly a well defined activity of pancreatic lipase on this new substrate.

Interfacial phenomena play a fundamental role in biological systems. It is important to know if surface energy and anisotropy affect the conformation of biological macromolecules. Well defined physicochemical models might simplify this problem (*1–8*); spread monolayers at the air–water interface exemplify this kind of model. For polypeptides which are introduced as simple models of proteins, no "surface denaturation" of the spread macromolecules occurred (*9, 10, 11*). Protein structures are too complex to yield direct information about eventual changes of conformation, but one can detect the presence or the disappearance of biological activity—*e.g.*, enzymic activity. The enzyme would be denatured if the conformation were modified by the anisotropy of the interface.

Table I. Homodispersity of Samples

Fraction	*Molecular Weight* *Viscosity*	*Osmotic Vapor Pressure*	*End Groups Titration*	*Number of Monomers*	*Specific Activity μC/mg*
C_1	800	780	770	4	—
C_2	1100	1080	945	6	321.33
C_3	1200	1250	1327	7	64.57
C_4	1450	1440	1532	8	150.51
C_5	4200	4180	4221	23	418.95

A surface enzymic reaction presumes two simultaneous processes: adsorption of the enzyme at the interface and the disappearance of the substrate from the interface (*2*). Therefore, it is essential to know at every moment the surface concentration of the substrate and of the enzyme.

The present work investigates the action of pancreatic lipase on spread monolayers of ethylene glycol adipate oligomers. The tritiated hydrolysis products of the uniformly labeled substrate escape into the subphase. Surface radioactivity measurements allow us to determine the substrate concentration at the interface (*2, 3*) and the extent of the enzymic reaction. The kinetics were investigated as a function of subphase pH and temperature and substrate length and concentration. By

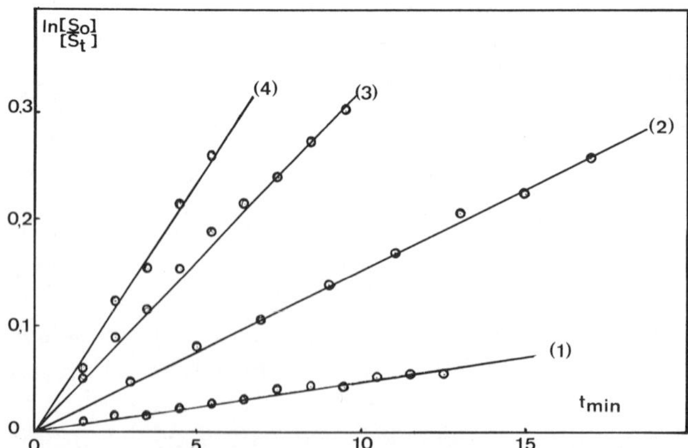

Figure 1. Evolution of the kinetics at the air–water interface for several concentrations of enzyme in the subphase: (1) 2 mg/l, (2) 9 mg/l, (3) 18 mg/l, (4) 27 mg/l. Substrate concentration: 0.5 mg/m², pH$_{subphase}$ = 8.2, T$_{subphase}$ = 20°C (fraction C_5).

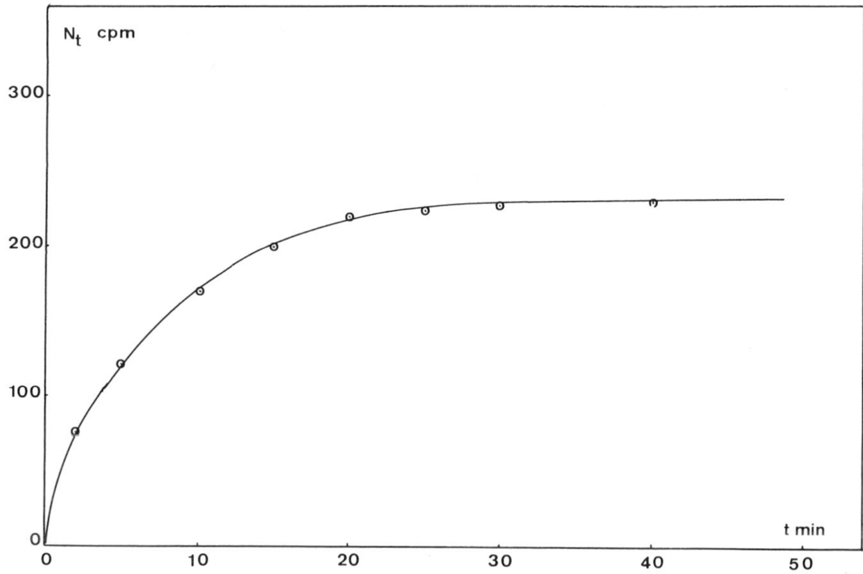

Figure 2. Adsorption of tritiated pancreatic lipase (2 mg/l) in the presence of the substrate (0.5 mg/m²). Surface radioactivity: $T_{subphase}$ *= 20°C,* $pH_{subphase}$ *= 8.2 (fraction* C_5*).*

labeling the enzyme, its adsorption at the air–water interface can be followed.

Materials and Methods

Materials. Ethylene glycol adipate oligomers were synthesized using the method of Youngson and Melville (*12*). Condensation of ethylene glycol with adipic acid produced a polymer with a molecular weight nearly proportional to the reaction time.

The samples were fractionated by gel chromatography (Sephadex LH 20) in 1,4-dioxan. The fractions were characterized as follows:

(1.) by viscosity measurements in chloroform at 25°C with a Couette viscosimeter (*13*). Molecular weights were determined from the Mark–Houwink equation (*14*).

(2.) by using a vapor pressure osmometer. Measurements were made in tetrahydrofuran with a Mechrolab vapor pressure osmometer (type 302).

(3.) by titration of the end groups (*15*).

Table I indicates the homodispersity of the samples.

All the oligomers were tritiated in pure tritium gas. To increase the rate at which tritium atoms exchanged with hydrogen, the system received a silent electric discharge for 45 min (*16*). The tritiated samples were dissolved in 1,4-dioxan and were precipitated with petroleum ether. The

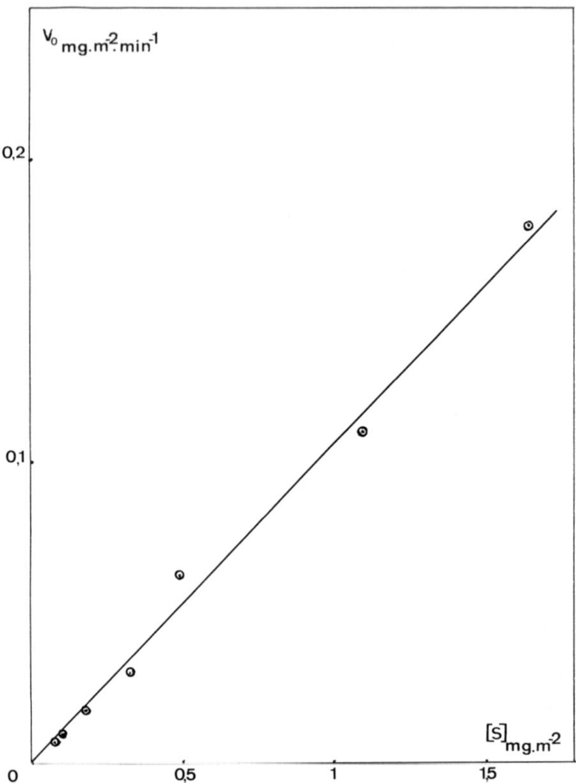

Figure 3. Relation between initial velocity v_0 *and substrate concentration:* $T_{subphase} = 20°C$, $pH_{subphase} = 8.2$ *(fraction* C_5*)*

molecular weights of the fractions were checked, and no degradation caused by the tritiation was detected. Moreover, at equal surface concentrations, the surface pressures of the tritiated and non-tritiated oligomers were identical. Specific activities are given in Table I.

The pancreatic lipase (Worthington) was tritiated by acetylation and purified by I. R. Miller using a method already described (*17*). The enzymic activity is 116 U/mg; the specific activity is 6.4 μC/mg.

Methods. Surface pressure was measured on a Cahn RG electrobalance using the Wilhelmy plate method. The surface radioactivity was measured with a gas flow counter (*2, 3*). Oligomers were spread with an Agla syringe from a benzene solution. Buffered solutions (tris HCl $10^{-1}M$) were used to prepare the subphase, and the enzyme was dissolved in the support. Adsorption was studied without stirring the bulk phase. The Ca^{2+} ($CaCl_2$) concentration in the subphase, essential for a correct enzymic activity (*18*), was fixed at 5 m*M*.

Results and Discussion

Influence of Enzyme Concentration. If the kinetics are represented by a Michaelis–Menten process, then,

$$E_a + S \underset{k_{-1}}{\overset{k_1}{\rightleftharpoons}} ES \xrightarrow{k_2} E + P$$

and

$$v = -\frac{dS}{dt} = \frac{k_2}{K_m} E_a S = K_{ap} E_a S \quad (1)$$

where K_m is the Michaelis equilibrium constant, k_2 the complex decomposition rate constant, S the surface concentration of the substrate, and E_a the enzyme concentration in the layer adjacent to the surface. Integration of Equation 1 gives

$$\ln S_o/S_t = K_{ap} E_a t \quad (2)$$

where S_0 is the surface concentration at time 0, S_t the surface concentration at time t.

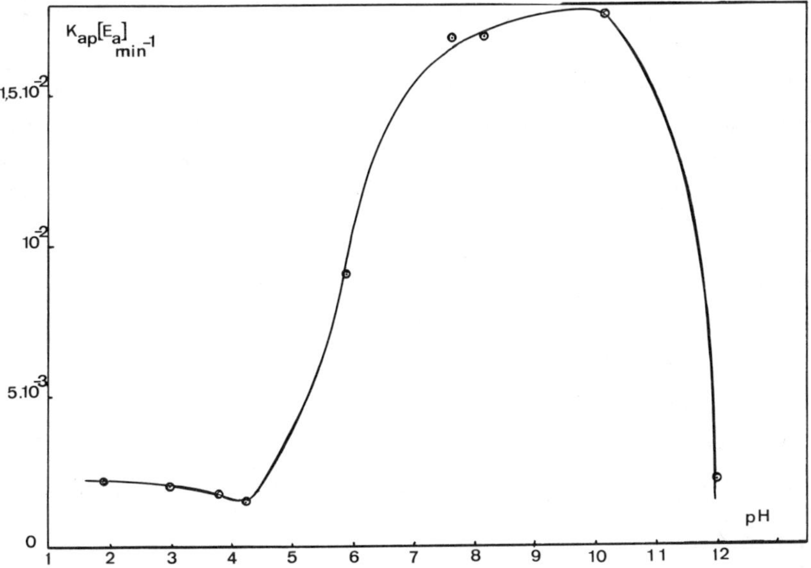

Figure 4. Evolution of $K_{ap}E_a$ as a function of subphase pH. Substrate concentration: 0.5 mg/m². Enzyme concentration in the subphase: 9 mg/l, $T_{subphase}$ = 20°C (fraction C_5).

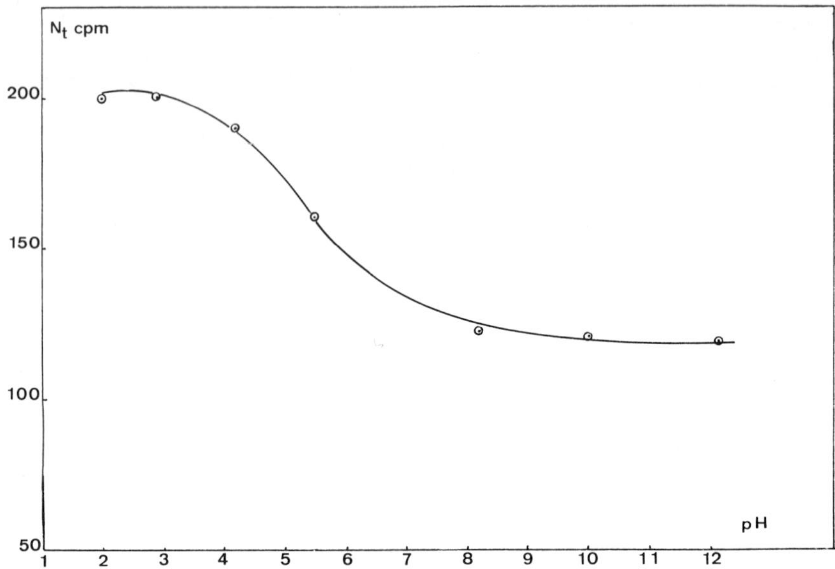

Figure 5. Evolution of surface radioactivity from tritiated pancreatic lipase (2 mg/l) after 5 min adsorption. Influence of subphase pH. Substrate concentration: 0.5 mg/m², $T_{subphase}$ = 20°C. Surface radioactivity is in cpm (counts/minute) (fraction C_5).

Figure 1 shows clearly that the total process can be represented by Equation 2 for several bulk enzyme concentrations. However, the evolution of the surface radioactivity of the tritiated enzyme under the oligomer film (Figure 2) indicates that enzyme activity and adsorption of the enzyme at the interface are not interrelated—namely, the enzyme kinetics at 1 min are identical at 10 min, when most of the enzyme adsorption takes place.

To reconcile the results of Figures 1 and 2, assume that the number of molecules that participated in the reaction are rapidly obtained, or suppose that the quantities of enzyme that carry out the hydrolysis are insufficient compared with the quantities of adsorbed enzymes (E_a).

Influence of Substrate Surface Concentration. The influence of the substrate concentration on the kinetics was investigated by several authors (*2, 4, 7*). Generally, the rate of enzymic cleavage is greatly reduced at high and low surface concentrations. At high concentration, the diminution probably arises from steric hindrance of the enzyme–substrate interaction (*2*).

Figure 3 shows a linear relation between the initial rate of the enzymic reaction and the surface concentration of the substrate at the interface. These results demonstrate that if the substrate has a well defined orientation, no reduction of activity is observed in this system. The possibility of surface inactivation of the enzyme can thus be excluded.

Influence of Subphase pH. The evolution of the kinetics as a function of pH is shown in Figure 4. Maximum activity occurs between pH 8 and 10. This agrees with the range of Desnuelle for another substrate (*18*).

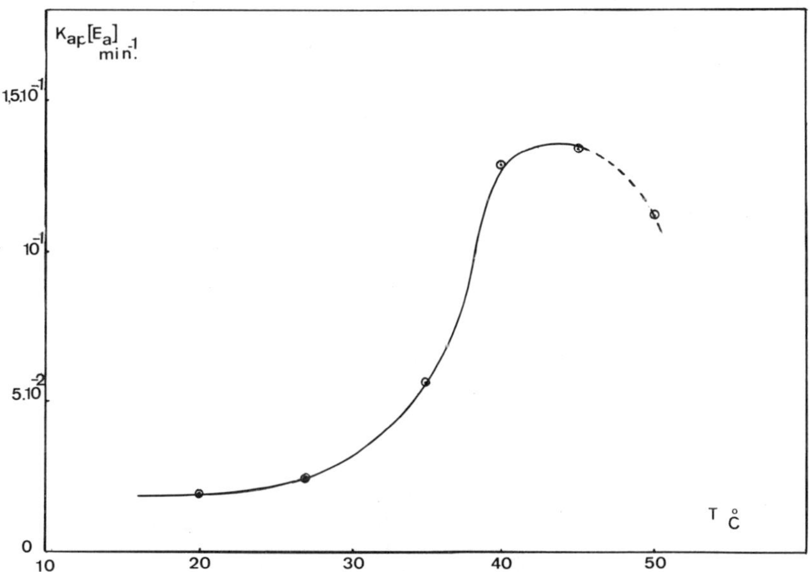

Figure 6. Evolution of $K_{ap}E_a$ *as a function of subphase temperature. Substrate concentration: 0.5 mg/m*^{2}*. Enzyme concentration in the subphase: 9 mg/l,* $pH_{subphase} = 8.2$ *(fraction* C_5*).*

The evolution of the enzyme adsorption was studied with tritiated molecules at pH 2–11 (Figure 5). Even for a low adsorption of enzyme molecules, maximum activity can be obtained. Thus, either the molecules initially adsorbed contribute to the kinetics, or the enzyme required for activity is not permanently adsorbed. If it were, the small quantities would be immeasurable with our techniques and indistinguishable from the ordinary adsorption or film penetration.

Influence of Subphase Temperature. Enzymic activity at the interface increases clearly with subphase temperature (Figure 6). When comparing enzymic adsorption with enzymic activity (Figure 7), the rate of enzymic cleavage increases when the number of macromolecules of the enzyme that reaches the surface increases. The two processes are parallel but are not immediately related.

Influence of the Molecular Weight of the Substrate. The choice of oligomers permits a study of the rate of enzymic cleavage as a function of substrate length.

Table II shows that substrate length is important up to a number of eight monomers. Ester bonds are probably less accessible for short than for long oligomers. This hypothesis agrees with the surface pressure isotherms obtained for the oligomers (Figure 8). Thus, at equal

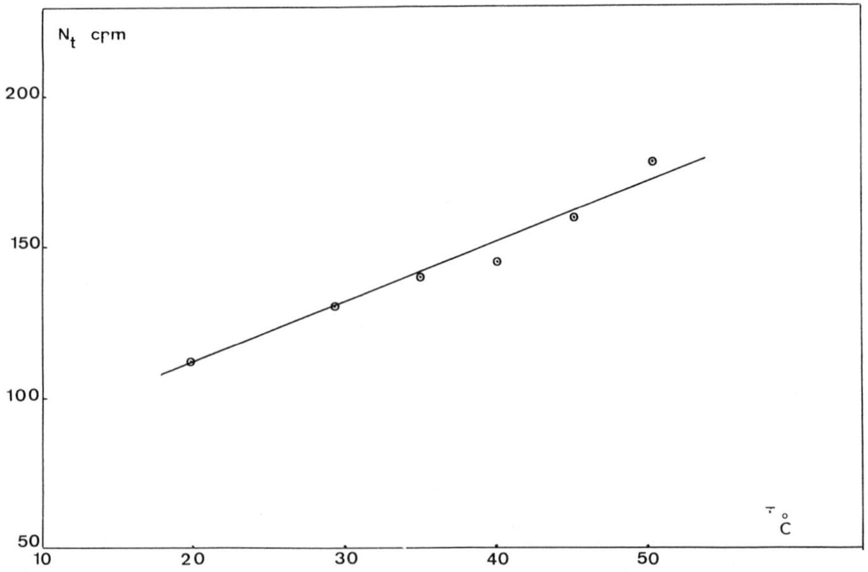

Figure 7. Evolution of surface radioactivity from tritiated pancreatic lipase (2 mg/l) after 2.5 min of adsorption. Influence of subphase temperature. Substrate concentration: 0.5 mg/m², $pH_{subphase} = 8.2$. Surface radioactivity is in cpm (counts/minute) (fraction C_5).

surface concentrations the degree of spreading is best for longer chains. Enzymic hydrolysis depends however on the possibility of forming the enzyme substrate complex. The basic hydrolysis permits a simplification of the problem because, in this case, the notion of orientation appears to be less significant. The rate of cleavage is the most significant for short oligomers (Table II). A possible explanation is that the OH^- ions hydrolyze the ester bonds in the immersed part of the monolayer and the ester bonds in the plane of the interface. In the enzymic reaction, only the hydrolysis of ester bonds maintained in the plane of the interface is possible.

Table II. Influence of Substrate Length on the Kinetics

Fraction	$K_{ap}\ E_a$ *(min^{-1})*	K_{ap} *OH$^-$ (pH = 12.5) (min^{-1})*
C_2	5×10^{-3}	0.43×10^{-3}
C_3	10^{-2}	0.31×10^{-3}
C_4	1.5×10^{-2}	0.076×10^{-3}
C_5	1.6×10^{-2}	0.06×10^{-3}

Conclusion

One problem in studying enzymic reactions at the air–water interface is following simultaneously the adsorption of the enzyme at the interface and the disappearance of the substrate from the interface. In this study, enzyme (pancreatic lipase) and substrate (ethylene glycol adipate oligomers) were tritiated, and a method of measuring the radioactivity at the interface permitted us to follow the two processes. For the well defined system of pancreatic lipase–ethylene glycol adipate oligomers no simple relation between the rate of enzymic cleavage and the adsorption process was obtained, probably because shortly after the reaction begins, the enzyme concentration necessary to establish the kinetics at the interface is reached, and the adsorption of other enzyme

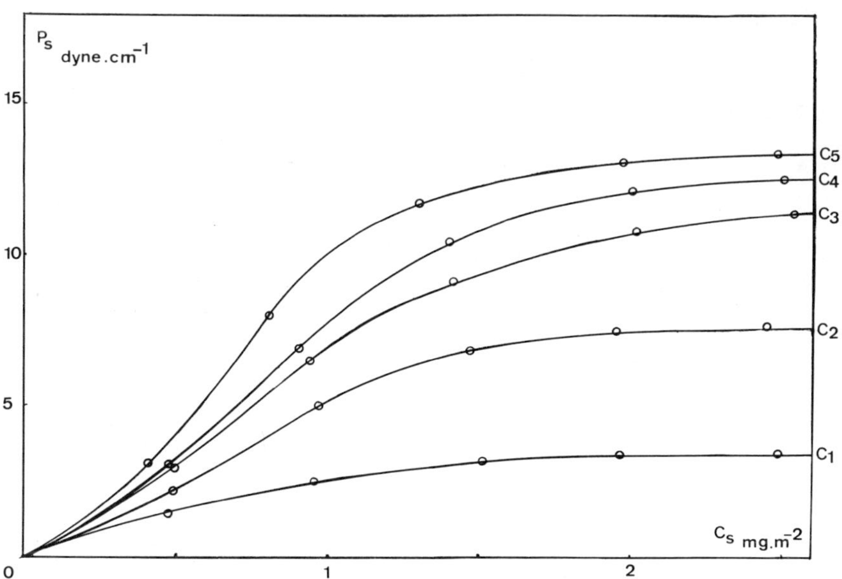

Figure 8. Surface pressure vs. *surface concentration. Influence of oligomer length.* $T_{subphase} = 20°C$, $pH_{subphase} = 8.2$ *(fraction C_5).*

molecules does not modify the total process. Thus, an eventual "surface inactivation" of the enzyme at low substrate concentration can be excluded.

It appears also that for our system the rate of enzymic cleavage is not modified by the surface concentration of the substrate but depends on oligomer length. Thus the influence of the substrate concentration is completely different from the behavior obtained with lipid substrates of pancreatic lipase (*19*).

Literature Cited

1. Hughes, A., *Biochem. J.* (1935) **29,** 437.
2. Miller, I. R., Ruysschaert, J. M., *J. Colloid Interface Sci.* (1971) **35,** 340.
3. Kummer, J., Ruysschaert, J. M., Jaffé, J., "Chemie, Physikalische Chemie und Anwendungstechnik der Grenzflächenaktive Stoffe," Vol. II, pp. 1, 284, Carl Hanser Verlag, München, 1973.
4. Dervichian, D. G., *Biochimie* (1971) **53,** 25.
5. Olive, J., Dervichian, D. G., *Biochimie* (1971) **53,** 207.
6. Zografi, G., Verger, R., de Haas, G. H., *Chem. Phys. Lipids* (1971) **7,** 185.
7. Verger, R., de Haas, G. H., *Chem. Phys. Lipids* (1973) **10,** 127.
8. Esposito, S., Semeriva, M., Desnuelle, P., *Biochim. Biophys. Acta* (1973) **302,** No. 2, 243.
9. Loeb, G. I., Baier, R. E., *J. Colloid Interface Sci.* (1968) **38,** 27.
10. Malcolm, B. R., *Proc. Roy. Soc. Ser. A* (1968) **363,** 305.
11. Jaffé, J., Ruysschaert, J. M., Hecq, W., *Biochim. Biophys. Acta* (1970) **207,** 11.
12. Youngson, J. W., Melville, H. W., *J. Chem. Soc. London* (1950) Pt. II, 1613.
13. Jaffé, J., Ruysschaert, J. M., *J. Sci. Instr.* (1965) **42,** 628.
14. Zavaglia, E. A., Billmeyer, F. N., *Proc. Paint Res. Inst.* (1964) **36,** 222.
15. Müller, E., "Coll. Huben-Weyl," Vol. XIV/2, pp. 1-47.
16. Dorfman, L. M., Wilzbach, K. E., *J. Phys. Chem.* (1959) **799,** 63.
17. Miller, I. R., Great, H., *Biopolymer* (1972) **11,** 2537.
18. Desnuelle, P., *Biochimie* (1971) **53,** 841.
19. Desnuelle, P., *Ber. Intern. Kongr. Grenz. 6th* (1973) **I,** 21.

RECEIVED September 23, 1974.

17

Interaction of Proteins with Phospholipid Monolayers

M. C. PHILLIPS, M. T. A. EVANS, and H. HAUSER

Unilever Research Laboratory Colworth/Welwyn, The Frythe, Welwyn, Herts., England

By using a surface radioactivity technique, the penetration of the hydrophobic and flexible 1-^{14}C-acetyl-β-casein and the rigid and globular 1-^{14}C-acetyl-lysozyme molecules into phospholipid monolayers in different physical states was monitored. The adsorption of β-casein to lecithin monolayers is described by a model in which it is assumed that the protein condenses the lecithin molecules so that the degree of penetration is a function of the lateral compressibility of the phospholipid monolayer. The interaction of β-casein with phospholipid monolayers is dominated by the hydrophobicity of the macromolecule, but lysozyme tends to accumulate mostly beneath phospholipid monolayers; in this situation, electrostatic interactions between the lipid and protein are important.

Much of our understanding of the phase behavior of insoluble monolayers of lipids at the air–water interface is derived from Adam's studies of fatty acid monolayers (*1*). It is now clear that the phase behavior of phospholipid monolayers (*2*) parallels that of the fatty acids; we make use of these structure variations in our study of the interactions of phosphatidylcholine (lecithin) monolayers with proteins. Because of the biological significance of the interfacial behavior of lipids and proteins, there is a long history of studies on such systems. When Adam was studying lipid monolayers, other noted contemporary surface chemists were studying protein monolayers (*3*) and the interactions of proteins with lipid monolayers (*4*). The latter interaction has been studied by many so-called "penetration" experiments where the protein is injected into the substrate below insoluble lipid monolayers that are spread on the air–water interface. These studies initially concerned highly water-soluble, globular proteins and more recently very surface-active, hydrophobic proteins and peptides. The latter proteins were studied by Colacicco and coworkers; he has reviewed the subject and discussed possible mechanisms of protein penetration (*5, 6, 7*).

There are two principal problems with penetration experiments: the adsorption characteristics of the protein have to be understood, and the amount of protein that adsorbs to the interface when lipid is present has to be determined. Previously, most researchers used the change in film pressure ($\Delta\pi$) as a measure of the amount of protein that interacted with the lipid monolayer. However, this approach implicitly assumes that the adsorption of protein can be described by Gibbs' adsorption equation, but as pointed out by Colacicco (*6*), this is invalid for proteins which adsorb irreversibly. Because the surface concentration of protein is unknown, radiolabeled proteins have been used (*8, 9, 10*). This work has been concerned exclusively with highly water-soluble proteins whose prime mode of interaction with monolayers (and bilayers) is electrostatic. In these cases a simple description of the packing in the mixed lipid–protein films was impossible (*6*).

Our objective here is to investigate the interaction of a radiolabeled hydrophobic protein with monolayers of lecithin which are neutral and which can be either condensed or expanded. The role of the compressibility of the lecithin monolayers is considered explicitly. The well defined β-casein molecule (*11*), which is very hydrophobic, is used for this study, and where comparative data for a globular protein are required, lysozyme is utilized.

Experimental

Materials. PROTEINS. The preparation of 1-^{14}C-acetyl derivatives of bovine β-casein *A* and hen egg white lysozyme has been described (*12, 13*). Modification of β-casein involved reaction of the protein with 1-^{14}C-acetic anhydride so that two lysine residues along the disordered, single polypeptide chain of 209 amino acids (*11*) were acetylated. A similar procedure was done with lysozyme so that one or two lysine residues on the exterior of the globular protein (*14*) were acetylated. These modifications ensured that the derivatives which were isolated in the lyophilized state had suitable specific activities (typically 1.5 μC/mg dry protein). The modification slightly increased the surface activity of lysozyme but not that of β-casein (*12, 13*); in this paper the 1-^{14}C-acetyl derivatives are referred to as β-casein and lysozyme.

LIPIDS. The chromatographically pure 1,2-dibehenoyl lecithin (C_{22}) was synthesized in this laboratory and described before (*15*). The 1,2-dipalmitoyl lecithin (C_{16}) was purchased from Fluka AG Chemical Co., Buchs, Switzerland, and further purified by passage through a silicic acid column. Chromatographically pure hen egg lecithin and phosphatidyl-

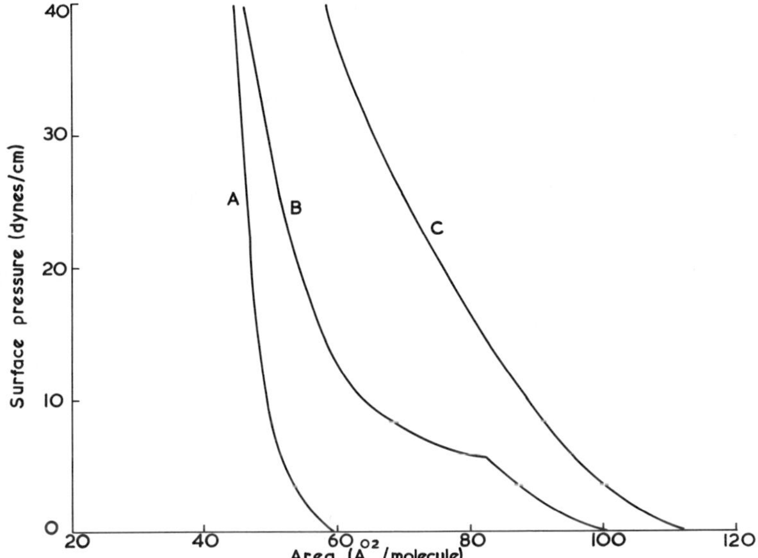

Figure 1. Surface pressure (π)–molecular area (A) *isotherms for dibehenoyl lecithin (A), dipalmitoyl lecithin (B), and egg yolk lecithin (C) on phosphate buffer (pH 7,* I = *0.1) at room temperature*

ethanolamine, and the monosodium salt of ox brain phosphtidylserine were purchased from Lipid Products, South Nutfield, U.K. Sodium dicetyl phosphate and arachidic acid were obtained from Albright & Wilson, Ltd., Birmingham, England, and Applied Science Labs, State College, Pa. The materials used in the preparation of all solutions, buffers, etc. are described elsewhere (*15, 16*).

Methods. The apparatus and procedure for determining the surface pressure (π)—molecular area, A, isotherms for phospholipid monolayers spread at the air–water interface have been described before (*15, 17*). Surface concentrations, Γ, of β-casein and lysozyme on adsorption were determined by measuring the surface radioactivity (ΔR) during adsorption (*12, 13*). The contribution of radioactivity from the substrate was estimated by measuring the count rate of solutions of $Na_2{}^{14}CO_3$. Calibration curves for converting ΔR to Γ were obtained by spreading 1-^{14}C-β-casein as an insoluble monolayer, compressing it in the normal way (*16, 18*), and recording ΔR and Γ. To determine changes in surface pressure ($\Delta\pi$) and surface concentration of protein on adsorption of protein to phospholipid monolayers, the lipid was spread from hexane–ethanol (9:1 v/v) solutions and adjusted to the appropriate initial film pressure (π_i) by the movement of a Teflon barrier. A Teflon trough (20.8 cm × 5.1 cm × 0.8 cm) contained 100 ml of phosphate buffer substrate (pH 7, I = 0.1) so that the surface area to volume ratio of the substrate was 1 cm^{-1}. Small volumes of a phosphate buffer solution of protein were injected into the substrate from behind the Teflon barrier.

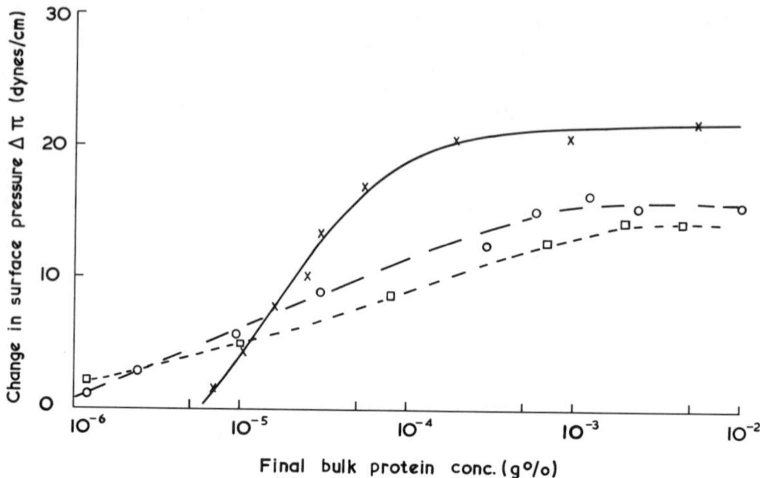

Figure 2. Steady-state surface pressure (π) as a function of the final substrate concentrations of 1-^{14}C-acetyl-β-casein for adsorption with stirring to the air–water interface (×), an egg lecithin monolayer spread to an initial film pressure of 10 dynes/cm (○), and a dibehenoyl lecithin monolayer spread to an initial film pressure of 10 dynes/cm (□). Substrate: phosphate buffer (pH 7, I = 0.1) at 22° ± 2°C.

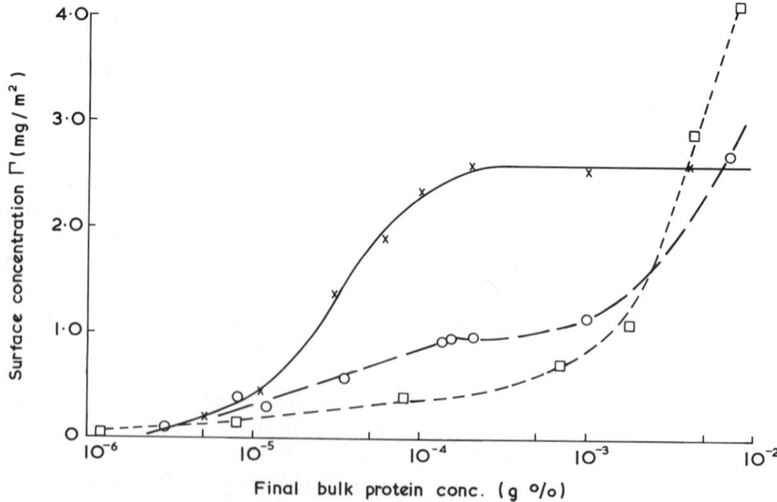

Figure 3. Steady-state surface concentrations (Γ) as a function of the final substrate concentrations of 1-^{14}C-acetyl-β-casein for adsorption with stirring to the air–water interface, egg lecithin and dibehenoyl lecithin monolayers. Symbols and conditions are the same as for Figure 2.

The contents of the trough were then stirred at a constant rate with a magnetic stirrer, and $\Delta\pi$ was measured with a roughened mica Wilhelmy plate (*13*). Adsorption proceeded with the lipid film held at constant area ($\sim$ 70 cm^2) until steady-state values of $\Delta\pi$ and ΔR were attained. $\Delta\pi$ and ΔR could be reproduced to ± 1 dynes/cm and ± 1 counts/sec, *i.e.*, Γ ± 0.1 mg/m^2. All experiments were carried out at 22° ± 2°C.

Results

The π–A curves at room temperature for dibehenoyl, dipalmitoyl, and egg yolk lecithins are shown in Figure 1. The condensed isotherm for dibehenoyl lecithin and the curve for dipalmitoyl lecithin which undergoes a two-dimensional condensation at 5.6 dynes/cm at 20.5°C are consistent with earlier investigations (*15, 17*). The fully expanded isotherm for egg yolk lecithin was expected in view of the high level of unsaturation of this lipid.

The steady-state values of $\Delta\pi$ and Γ were attained at the air–water interface after stirring with different concentrations of β-casein in the trough (Figures 2, 3). When the final substrate concentration of protein (C_p) was less than 5×10^{-6} g %, no changes in π were detected although Γ had a finite value. At this point the film is essentially gaseous and exerts a very small π. As C_p increases to 10^{-4} %, adsorption proceeds with a concomitant increase in $\Delta\pi$ to about 22 dynes/cm. Monolayer coverage ($\Gamma \sim 2.5$ mg/m^2; 7.7 A^2/residue) is complete when C_p is equivalent to 10^{-4} %; there is no further adsorption when C_p is increased to 10^{-2} %. The adsorption of β-casein is irreversible over the region where π changes with C_p because removal of the protein that remains in the substrate after adsorption does not lead to any desorption of protein. The maximum amount of irreversibly adsorbed β-casein is 2.6 ± 0.1 mg/m^2; this material can only be removed from the interface by compression with a barrier. When $C_p = 10^{-5}$ %, approximately half of the total protein finishes up at the interface; when C_p equals 10^{-3} %, only a few percent of the total β-casein is adsorbed ($C_p = 10^{-3}$ g % $\equiv 4 \times 10^{-7} M$).

The effects of lecithin monolayers on $\Delta\pi$ and Γ at different C_p's are shown in Figures 2 and 3. The lecithin monolayers were spread to initial film pressures (π_i) of 10 dynes/cm so that the molecular areas for the dibehenoyl and egg yolk lecithins were about 50 and 88 A^2/molecule. When C_p is less than 10^{-5} %, lecithin monolayers give larger values of $\Delta\pi$ than are observed at the clean air–water interface. In this region Γ is probably reduced, but the adsorption of protein is so limited that it approaches the limits of experimental error. However, when $10^{-5} < C_p < 4 \times 10^{-3}$ %, the lecithin monolayers cause large reductions in Γ and $\Delta\pi$, and the condensed dibehenoyl lecithin film has the greater effect. The reduction in Γ is real and not an artifact caused by the lipid monolayer

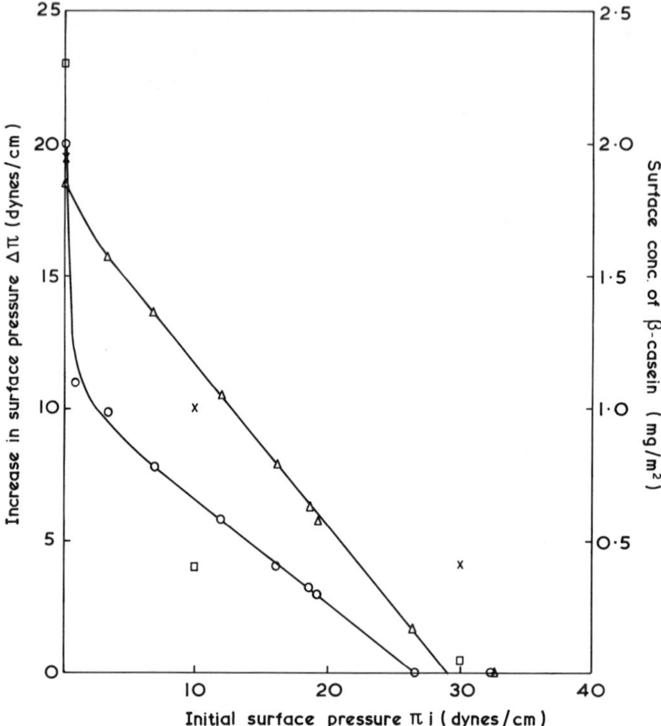

Figure 4. Dependence of change in surface pressure ($\Delta\pi$) and surface concentration (Γ) of 1-^{14}C-acetyl-β-casein on the initial film pressure (π_i) of egg lecithin and dibehenoyl lecithin monolayers. The $\Delta\pi$ ($\triangle$) and Γ ($\bigcirc$) data for egg lecithin monolayers were obtained with an initial substrate concentration ($C_p{}^i$) of β-casein of 5.4 × 10^{-5} %. The $\Delta\pi$ ($\times$) and Γ ($\square$) data for dibehenoyl lecithin monolayers were obtained with $C_p{}^i = 10^{-4}$ %. The other conditions are described in Figure 2.

that affects the detection of radiation by the gas-flow counter; spreading an insoluble lipid monolayer on $Na_2{}^{14}CO_3$ solutions does not change the count rate. The maximum film pressure ($\pi_i + \Delta\pi$) in these cases is 24–26 dynes/cm; this is similar to the maximum pressure exerted by β-casein alone. With both types of lecithin monolayers, Γ increases significantly when $C_p \gtrsim 10^{-3}$ %; when $C_p > 4 \times 10^{-3}$ %, more protein is adsorbed to the lecithin monolayer than to the clean air–water interface. These increases in Γ are not accompanied by any significant changes in $\Delta\pi$, and the contributing β-casein molecules can exchange with others in the substrate. This exchange was demonstrated by allowing adsorption to proceed to a steady-state with $C_p = 7 \times 10^{-3}$ % and then increasing C_p

by a factor of 2 with unlabelled β-casein; after a 15-min stir, a gradual decrease in ΔR was observed although $\Delta\pi$ remained constant. Thus, it seems that the surface concentration of irreversibly adsorbed β-casein is decreased by an insoluble lecithin monolayer.

Measurements were made when π_i varied and C_p was constant. Figure 4 shows that at a given initial C_p^i, $\Delta\pi$ with egg lecithin monolayers decreases more or less linearly with increasing π_i. The few points for dibehenoyl lecithin monolayers show similar trends. The linear decrease in $\Delta\pi$ with increasing π_i has been observed for several systems (*8, 19, 20*). β-Casein does not penetrate egg lecithin monolayers when $\pi_i = 29$ dynes/cm; generally, protein penetration ceases when π_i is equivalent to 30 dynes/cm (*6, 8*). Figure 4 clearly shows that, compared with the clean air–water interface, a lecithin monolayer spread to a π_i as low as

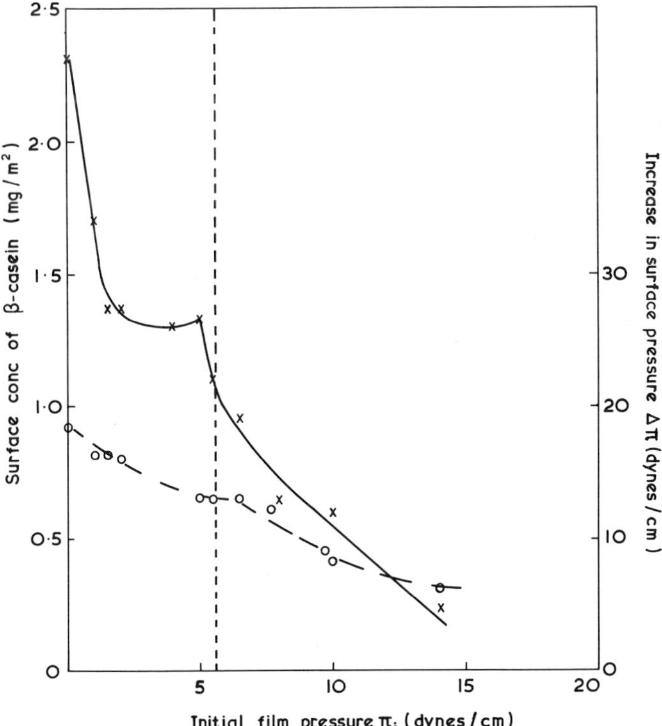

Figure 5. Dependence of $\Delta\pi$ (○) and Γ (×) of 1-^{14}C-acetyl-β-casein ($C_p^i = 10^{-4}$ %) on the π_i of dipalmitoyl lecithin monolayers at 20.5°C. The film pressure at which the two-dimensional condensation starts is shown by the vertical dashed line. The other conditions were as described in Figure 2.

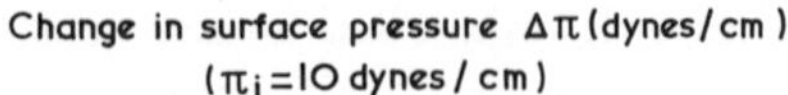

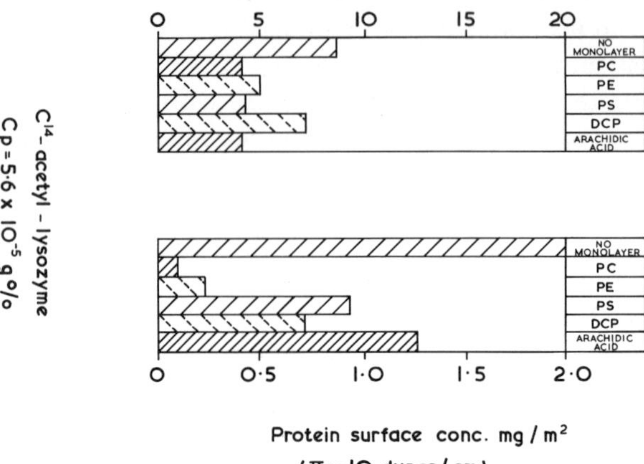

Figure 6. Variation of $\Delta\pi$ and Γ for 1-^{14}C-acetyl-lysozyme ($C_p{}^i = 5.6 \times 10^{-5}$ %) adsorbing to various lipid monolayers at $\pi_i = 10$ dynes/cm. Substrate was phosphate buffer (pH 7, I = 0.1), and the monolayers were egg lecithin (PC), egg phosphatidylethanolamine (PE), ox brain phosphatidylserine (PS), dicetyl phosphate (DCP), and arachidic acid.

1 dyne/cm leads to a dramatic decrease in Γ. The subsequent linear decrease in Γ with increasing π_i seems consistent with observations on other systems (*8*). From the variation in $\Delta\pi$ and Γ with π_i for dipalmitoyl lecithin monolayers (Figure 5), the effect of a phase transition in the lipid monolayer is seen. When $\pi_i > 5.6$ dynes/cm, two-dimensional condensation does not occur during protein adsorption and $\Delta\pi$ and Γ decrease smoothly with increasing π_i. However, when $\pi_i < 5.6$ dynes/cm, the phase transition occurs and leads to a local maximum in the plot of Γ *vs.* π_i (Figure 5). Again there is an initial rapid decrease in Γ at low π_i.

An indication of the role of electrostatic effects follows. The net charge on a β-casein molecule at pH 7 is (−)13 (*11*), but $\Delta\pi$ and Γ are only slightly affected when, under given conditions, egg lecithin is replaced by lipids (*e.g.* phosphatidylserine, dicetyl phosphate, and arachidic acid) with a net negative charge. Electrostatic effects play a minor role in the interaction of β-casein with insoluble lipid monolayers. A comparative series of experiments with lysozyme show some major effects. The results are summarized in Figure 6 where $\Delta\pi$ and Γ for lysozyme adsorbing to a variety of lipid monolayers at $C_p{}^i = 5.6 \times 10^{-5}$ % and

$\pi_i = 10$ dynes/cm are shown. The presence of a lipid monolayer always reduces $\Delta\pi$ and Γ below the values for the clean air–water interface. The neutral phosphatidylcholine (lecithin) and phosphatidylethanolamine monolayers dramatically decrease Γ; the negatively charged monolayers at pH 7 increase Γ (Figure 6). This enhanced adsorption presumably arises from the electrostatic attraction between the lipid and protein because lysozyme is a basic protein (isoelectric point p$I = 10.5$). The importance of electrostatic attraction for adsorption of lysozyme was confirmed by repeating the experiments with phosphatidylserine either at pH 10.5 or in the presence of 2*M* sodium chloride. In both cases Γ was reduced.

Discussion

Adsorption at Air–Water Interface. The data for the adsorption of β-casein at the clean air–water interface (Figures 2 and 3) can be combined to give surface pressure–molecular area (A) curves for the adsorbed protein films. For β-casein the resultant π A curve is similar to that obtained by spreading the protein (*12, 16, 18*); therefore, this flexible protein has essentially the same conformation when spread or adsorbed. At low π, $A = 2$ m^2/mg or 38 A^2/residue, so there is sufficient space for the peptide backbone to lie in the plane of the interface with no loops or tails of residues protruding into either bulk phase. A conformational change occurs at $\pi = 7$ dynes/cm where loop-tail formation commences (*12*). Monolayer coverage ($A \sim 0.4$ m^2/mg $= 1610$ A^2/molecule) is complete when $C_p \sim 10^{-4}$ %. At this point the protein is in a close-packed,

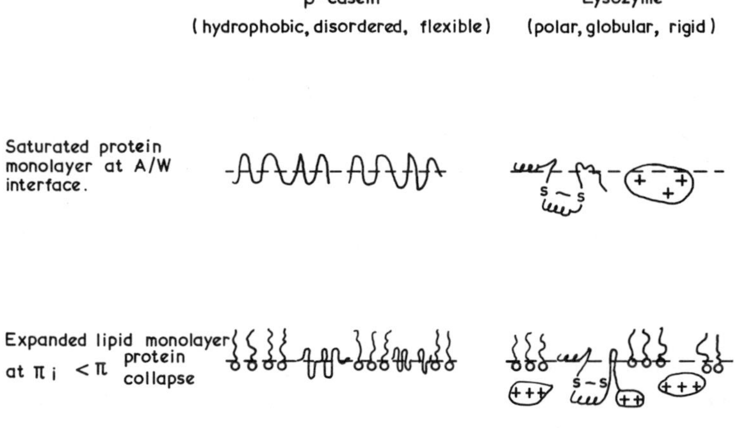

Figure 7. Schematic of the structures of films of β-casein and lysozyme and mixed lipid–protein monolayers

looped configuration (Figure 7) and, by ellipsometry, is approximately 60 A thick (to be published). These loops give rise to steric–electrostatic repulsion between the film and protein molecules in the substrate, and further increases in Γ do not occur when $C_p \leq 10^{-2}$ %. Multilayer formation occurs when $C_p > 10^{-2}$ % at 22°C and films more than 100 A thick are formed (to be published).

Unpublished studies show that when the same analysis is carried out with equivalent lysozyme data, the films are more ordered than those of β-casein; at a given Γ, π for lysozyme is lower than that of β-casein. This strong cohesion in the lysozyme films presumably arises from the structured regions that contain α-helices, disulfide bridges and perhaps native molecules (Figure 7). A large amount of residual structure is expected because the lysozyme molecule is relatively difficult to denature.

Effect of Lipid Monolayers on Protein Adsorption. Originally, lipid–protein interaction at the interface was based on results obtained at low C_p's ($< 5 \times 10^{-6}$ % for β-casein) where $\Delta\pi$ is zero at the clean air–water interface but positive when an insoluble lipid monolayer is present. However, the data in Figures 2 and 3 for a large range of C_p's indicate that under most conditions the presence of a lecithin monolayer inhibits protein adsorption rather than promoting it by the lipid–protein interaction. Because protein cannot occupy the same areas of the interface as the lipid molecules, the area of interface available to the protein is decreased (*12*). When this occurs at low Γ ($C_p < 5 \times 10^{-6}$ %), the resultant compression of the protein, rather than enhanced protein adsorption arising from lipid–protein interaction, leads to positive values of $\Delta\pi$.

It is simple to describe the stoichiometry of a mixed lipid–protein film after protein adsorption is complete if the penetrating protein molecules simply act as a mechanical barrier and compress the lipid molecules in the manner used to obtain π–A curves. The mixed film is described by a geometric model in which the area per lipid molecules decreases from π_i to $(\pi_i + \Delta\pi)$ consistent with the π–A isotherm of the lipid. The adsorption of the hydrophobic β-casein molecules to lecithin monolayers (*12*) can be described by this model if additional condensation of the lipid and/or protein is included. The degree of condensation is a function of C_p; the lipid–protein mole ratios in the mixed films (data in Figures 2 and 3) range from *ca.* 500:1 to 50:1.

Effect of Lipid Compressibility on Protein Penetration. With this simple model it is possible to estimate the maximum amount of protein that accumulates within a lipid monolayer and to elucidate the role of the compressibility of the lipid monolayer in protein penetration. A protein molecule will penetrate a lipid monolayer at π_i and initial area, X, when some characteristic area, ΔX, is cleared. The energy, E, required to create this space is the same as that needed to compress the monolayer from an area X to $(X - \Delta X)$:

$$E = \int_{X}^{X - \Delta X} \pi_i \, dX = -\pi_i \, \Delta X \tag{1}$$

The increasing π_i increases E so that there is a rising activation energy barrier which eventually prevents penetration. The pressure (π_f) at which protein penetration ceases is about 30 dynes/cm (*cf.* Figure 4). The area made available per lipid molecule after compressing π_i to π_f is

$$[A_l^i - A_l^f] = (\pi_i - \pi_f) \frac{dA}{d\pi} \tag{2}$$

where A_l^i and A_l^f are the lipid molecular areas at π_i and π_f. The total area made available in a constant surface area, X, that contains X/A_l^i lipid molecules is

$$(\pi_i - \pi_f) \frac{X}{A_l^i} \cdot \frac{dA}{d\pi} = (\pi_f - \pi_i) \, XC \tag{3}$$

where C is the compressibility $[= -1/A(dA/d\pi)_T]$ of the lipid monolayer; C is taken as a constant over the range $\pi_i \rightarrow \pi_f$ which is reasonable for monolayers without a phase transition. In the simple geometric model (*i.e.*, ideal mixing of lipid and protein at π_f), the surface concentration of protein is determined by the number of molecules that are packed into this available space. The number of protein molecules, N, that ultimately occupy the space will be

$$N = \frac{X(\pi_f - \pi_i) \, C}{A_p} \tag{4}$$

where A_p equals average area per protein molecule. The surface concentration, Γ, of protein in the mixed film is N/X so that

$$\Gamma = \frac{(\pi_f - \pi_i) \, C}{A_p} = \frac{\Delta\pi \, C}{A_p} \tag{5}$$

The maximum possible available area arises by compressing the lipid monolayer to its collapse pressure ($\sim$ 45 dynes/cm for egg lecithin) during protein adsorption. Assuming that $A_l^f = 50$ A^2/molecule and $A_p = 0.4$ m^2/mg, with the above equations, the maximum amount of β-casein which can be accommodated in an egg lecithin monolayer is equivalent to $\Gamma \sim 1.1$ mg/m^2. This is close to the inflection point in the plot of Γ

$-C_p$ (Figure 3); this suggests that, at $C_p \sim 10^{-3}$ %, β-casein condenses the lipid by the maximum possible amount, despite the fact that π_f is only about 26 dynes/cm. The higher values of Γ observed at $C_p > 10^{-3}$ % arise from the onset of multilayer formation. This is supported by the observation that $\Delta\pi$ attains its maximum at this point (Figure 2), and, under these conditions, there are exchangeable protein molecules associated with the interface. It is not obvious why the presence of a lecithin monolayer leads to multilayer formation at a lower C_p than at the clean air–water interface (*i.e.*, 10^{-3} rather than 10^{-2} %). It may be caused by some partially penetrated β-casein molecules that act as nucleation sites. Similar calculations for dibehenoyl lecithin, assuming that the minimum area to which the molecules can be compressed is 44 A^2/molecule, indicate that the maximum amount of β-casein which can penetrate is equivalent to $\Gamma \sim 0.3$ mg/m^2. Comparison with the data in Figure 3 suggests that the protein molecules tend to penetrate the condensed lecithin monolayer partially because $\Gamma > 0.3$ mg/m^2 when C_p is equivalent to or greater than 5×10^{-5} %. Since π does not attain its maximum value until $C_p \sim 2 \times 10^{-3}$ % (Figure 2), partial penetration must occur over this C_p range. Again, multilayer formation occurs when $C_p > 2 \times 10^{-3}$ % (Figure 3).

These calculations suggest that β-casein and egg yolk lecithin do not mix ideally, but the interaction between them promotes a condensation or non-additivity of molecular areas. Also, π_f in the egg lecithin–β-casein system can be as high as 29 dynes/cm (Figure 4) whereas the maximum pressure exerted by the protein alone is about 22 dynes/cm (Figure 2). Of course, the assumption that the interfacial conformation of the protein is unchanged by the presence of lipid is valid only when π is less than the collapse pressure of the pure protein film. Between this point and 29 dynes/cm the protein has probably only partially penetrated. Further information about the nature of the lipid–protein interaction can be gained from the data in Figure 4.

Any deviation of the molecular area of the mixed film from the weighted average of the areas of the components can arise from the condensation of lipids and/or proteins. However, if it is assumed that at π_f the protein occupies a molecular area consistent with the data in Figures 2 and 3 so that only the lipid is condensed, then the data in Figure 4 show that when $C_p^i = 5 \times 10^{-5}$ %, β-casein compresses egg lecithin molecules to an average area of about 68 A^2/molecule. Since protein molecules embedded in phospholipid bilayers are surrounded by a boundary layer of restricted lipid molecules (*21*), it is interesting to determine whether the above condensation in β-casein–egg lecithin monolayers is consistent with this concept. If each β-casein molecule occupies a circular area of 1600 A^2 and each egg lecithin molecule has a diameter of 8 A, a layer of lecithin, one molecule thick, around the circumference of a protein molecule will contain about 20 lipid molecules. The experimentally observed condensation is accounted for if these molecules are restricted to a molecular area of 50 A^2/molecule while the remainder of the lipid molecules occupy the area at π_f expected from the π–A curve. The idea that layers of lipid molecules solvate hydrophobic protein molecules in mixed monolayers requires corroboration. However, the interaction readily explains the greater stability against collapse of proteins in this situation as compared with the lower stability of pure protein films at the air–water interface.

Stoichiometry and packing are quite different when the lipid molecules are initially in a quasi-crystalline array (*e.g.*, dibehenoyl lecithin monlayers). We do not have sufficient data to test the boundary layer hypothesis in this case, but since β-casein seems to penetrate dibehenoyl lecithin monolayers only partially, it is possible that the protein is also condensed. It is clear from Equation 5 that the Γ's for a protein that penetrates into two different lipid monolayers should be in the ratio of their compressibilities as long as $\Delta\pi$ and A_p are the same in both cases. For example, the compressibilities for dibehenoyl and egg lecithin monolayers at $\pi = 10$ dynes/cm are in the ratio 1:2, and the Γ's for β-casein are in the ratio 1:1.7 (Figure 4). The data at $\pi_i = 10$ dynes/cm for dipalmitoyl lecithin (Figure 5) cannot be correlated in this way since C, contrary to the situation with fully expanded or condensed monolayers, changes markedly between π_i and π_f (Figure 1). However, the form of the Γ–π_i plot for dipalmitoyl lecithin monolayers (Figure 5) confirms the importance of the compressibility of the lipid in penetration; $dA/d\pi$ is very large at the phase transition (Figure 1), and the sudden increase in C leads to the maximum in Γ at $\pi_i \sim 5$ dynes/cm. Additives, such as cholesterol, which remove the phase transition, condense lipid monolayers (*2*), and decrease their compressibility should reduce the degree of penetration of hydrophobic proteins such as β-casein.

Summary

The interaction of β-casein molecules with lecithin monolayers is governed by the surface activity (hydrophobicity) of the macromolecule. Hydrophobicity does not guarantee interaction of a protein with lecithin in dispersion (*22*), but it favors penetration of proteins into monolayers at the air–water interface. The whole molecule seems to penetrate in the case of β-casein (Figure 7); this leads to the compression of the lipid molecules and perhaps the formation of a layer of relatively restricted lecithin molecules around the periphery of the protein molecules. The situation is quite different for a very polar protein such as lysozyme where complete penetration does not occur, and the adsorption is markedly affected by the electrostatic interaction between lipid and protein (Figure 7).

Acknowledgment

We are grateful to P. Lankester for valuable technical assistance.

Literature Cited

1. Adam, N. K., "The Physics and Chemistry of Surfaces," Dover, New York, 1968.
2. Phillips, M. C., *Progr. Surface Membrane Sci.* (1972) **5,** 139.
3. Langmuir, I., *Cold Spring Harbor Symp. Quant. Biol., 1938* No. **6,** 171.
4. Schulman, J. H., Rideal, E. K., *Proc. Roy. Soc. Ser. B* (1937) **122,** 46.
5. Colacicco, G., *J. Colloid Interface Sci.* (1969) **29,** 345.
6. Colacicco, G., *Lipids* (1970) **5,** 636.
7. Colacicco, G., *Ann. N.Y. Acad. Sci.* (1972) **195,** 224.
8. Dawson, R. M. C., Quinn, P. J., *Advan. Exp. Med. and Biol.* (1971) **14,** 1.
9. Khaiat, A., Miller, I. R., *Biochim. Biophys. Acta* (1969) **183,** 309.
10. Miller, I. R., Bach, D., "Surface and Colloid Science," Vol. 6, p. 185, E. Matijevic, Ed., Wiley, New York, 1973.
11. Ribadeau Dumas, B., Brignan, G., Grosclaude, F., Mercier, J. C., *Eur. J. Biochem.* (1972) **25,** 505.
12. Phillips, M. C., Evans, M. T. A., Hauser, H., *Proc. Int. Congr. Surface Activity, 6th, Munich, 1973,* **2,** 381.
13. Adams, D. J., Evans, M. T. A., Mitchell, J. R., Phillips, M. C., Rees, P. M., *J. Polymer Sci.* (1971) *Pt. C* **34,** 167.
14. Blake, C. C. F., Koenig, D. F., Mair, G. A., North, A. C. T., Phillips, D. C., Sarma, V. R., *Nature* (1965) **206,** 757.
15. Phillips, M. C., Chapman, D., *Biochim. Biophys. Acta* (1968) **163,** 301.
16. Mitchell, J., Irons, L., Palmer, G. J., *Biochim. Biophys. Acta* (1970) **200,** 138.
17. Paltauf, F., Hauser, H., Phillips, M. C., *Biochim. Biophys. Acta* (1971) **249,** 539.
18. Evans, M. T. A., Mitchell, J., Mussellwhite, P. R., Irons, L., "Surface Chemistry of Biological Systems," M. Blank, Ed., p. 1, Plenum Press, New York, 1970.
19. Camejo, G., Colacicco, G., Rapport, M. M., *J. Lipid Res.* (1968) **9,** 562.
20. Quinn, P. J., Dawson, R. M. C., *Biochem. J.* (1970) **119,** 21.
21. Jost, P. C., Griffith, O. H., Capaldi, R. A., Vanderkooi, G., *Proc. Nat. Acad. Sci. U.S.* (1973) **70,** 480.
22. Davis, M. A. F., Hauser, H., Leslie, R. B., Phillips, M. C., *Biochim. Biophys. Acta* (1973) **317,** 214.

RECEIVED September 24, 1974.

18

Effects of Cations on Biologically Active Surfaces

Divalent Cation Selectivity of the Membrane Na–K ATPase

MARTIN BLANK and JOHN S. BRITTEN

Department of Physiology, College of Physicians and Surgeons, Columbia University, New York, N.Y. 10032

The Na–K ATPase is a unique enzyme present in biological membranes that causes the transport of Na^+ and K^+ ions across the membrane when ATP is hydrolyzed. There are a number of sites on the catalytic surfaces of the enzyme, and various ions or other solutes, interacting at these sites, can activate or inhibit ATP hydrolysis. We found that the divalent cations, Co^{2+}, Ni^{2+}, and Zn^{2+}, activate the Na–K ATPase in the presence of excess ATP. The characteristics of the activation are similar to those seen with the normal activation by Mg^{2+} or Mn^{2+}. By comparing divalent cations that activate the enzyme with those that do not, the activators are distinguished by a common range of ionic radii and by a tendency to form relatively fluid networks in protein monolayers. These results suggest that mechanical coupling between the two surfaces of the enzyme may be part of the mechanism linking ATP hydrolysis to Na^+ and K^+ ion transport.

In 1922, N. K. Adam (*1*) found that the area of a condensed monolayer depended on the length of time the subphase was present in the brass tray that held the supporting aqueous phase. It was subsequently shown that trace amounts of divalent cations derived from the brass caused the large changes in surface area (and surface potential) if sufficient time elapsed for the ions to diffuse to the surface film. It is now known that the specific adsorption of small numbers of ions can cause large changes in the surface properties of many systems, particularly biological systems, where one can detect dramatic alterations in such properties of proteins as enzyme activity. Recently, we measured the toxic effects caused by the adsorption of various cations on the Na–K ATPase (*2*) and attempted to characterize the specific binding sites on the enzyme. This enzyme, which is widely distributed in the membranes of animal cells, splits ATP in the presence of Mg^{2+}, Na^{+}, and K^{+} and in so doing causes the active transport of Na^{+} and K^{+} ions across cell membranes (*3*).

In our earlier study (*2*) we measured the concentration of ions required to inactivate the enzyme and tried to determine whether the inactivation could be changed by competition with the normal ions, Mg^{2+}, Na^{+}, and K^{+}. From these studies we assigned the inactivation effects of some cations to actions at specific sites. The cations that could not be associated by the demonstration of competitive inhibition with Mg^{2+}, Na^{+}, or K^{+} sites were classed as "non-specific" inhibitors. These cations act at relatively low concentrations, and the concentrations of ions giving 50% inhibition of enzyme activity are correlated with the oxidation potential of the ion and with the binding constants to ethylenediamine, histidine, and imidazole. These results suggest that the non-specific ion inhibitors may react at one site—a "histidine-like" residue near the active center of the enzyme.

In this and other studies we noted that many divalent cations, such as Ca^{2+}, Cd^{2+}, Cu^{2+}, Fe^{2+}, Hg^{2+}, Pb^{2+}, and Zn^{2+}, inhibit enzyme activity (*3, 4, 5*). While studying the mechanism of the inhibition by these divalent cations, we observed that some of the known enzyme inhibitors became enzyme activators if ATP were present in excess. Stimulated by this observation, we surveyed the activation effects of the divalent cations in some detail.

A Zn^{2+}-dependent, ouabain-inhibited ATPase was found in both rabbit kidney and calf brain preparations, thus confirming the observations of Atkinson and Lowe (*6*). A Co^{2+}-dependent, ouabain-inhibited ATPase was present in the same preparations, thus confirming the observations of Rendi and Uhr (*7*). Also, a Ni^{2+}-dependent, ouabain-inhibited ATPase was found. When assayed in the presence of the divalent cations, Co^{2+}, Ni^{2+}, and Zn^{2+}, the Na-K ATPase activity in these relatively crude membrane preparations was explored, in some detail, and compared with the activity found with Mg^{2+} or Mn^{2+}.

In this paper we focus on the activation of the enzyme by divalent cations and attempt to combine this more recent work with our knowledge about the inactivation caused by these and the other cations studied. We also present some evidence concerning the interactions between the different ion binding sites on this complicated catalytic surface.

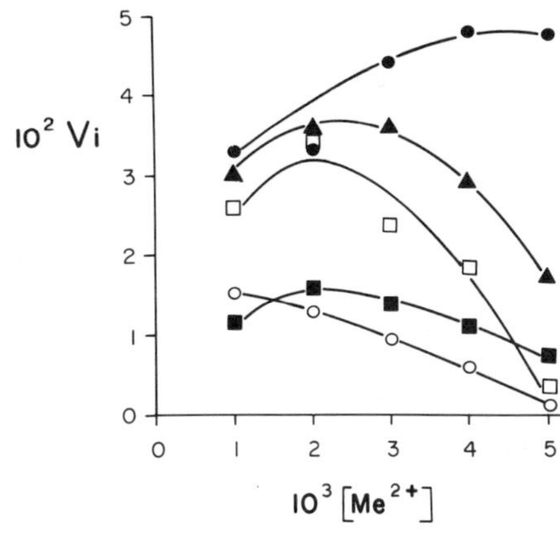

Figure 1. Variation of initial velocity, V_i in μmoles PO_4/min/mg protein, with divalent cation concentration, Me^{2+}, over the range 1–5 mM in the presence of 5 mM ATP. The 1-ml reaction mixture contained 9 μmoles KAc, 89 μmoles NaAc, and 20 μmoles tris acetate, pH 6.8, incubated at 37°C for 22 minutes. $CoCl_2$ (□), $MgCl_2$ (●), $MnCl_2$ (▲), $NiCl_2$ (■), $ZnCl_2$ (○).

Methods

Membrane fragments containing the Na–K ATPase were obtained from frozen rabbit kidneys by a modification of the method of Kinsolving *et al.* (*8*). Mg^{2+} and ethylenediaminetetraacetic acid (EDTA) were omitted from all wash solutions, and the enzyme was stored in 0.02*M* tris–acetate buffer of pH 6.8. A second preparation was obtained from calf brain by the method of Skou and Hilberg (*9*). Assays were conducted by adding the enzyme preparation to solutions containing Na^+, K^+, heavy metal ions, and ATP. The difference in the rate of release of phosphate ions, estimated by the method of Taussky and Shorr (*10*), in the absence and the presence of ouabain was taken as a measure of Na–K ATPase activity.

A Perkin-Elmer model 290 atomic absorption spectrometer was used to determine the Mg content. The Metrohm Polarecord model E 261, with a device for rapid polarography, was used to analyze the bound divalent cations, Zn^{2+}, Co^{2+}, and Ni^{2+}.

Results

The comparative activating effect of Co^{2+}, Mg^{2+}, Mn^{2+}, Ni^{2+}, and Zn^{2+} on Na–K ATPase activity is demonstrated in Figure 1. In the studies of the rabbit kidney enzyme shown, ATP was at a constant concentration of 5 m*M* while the divalent cation concentration [Me^{2+}] was varied from 1 to 5 m*M*. The importance of excess ATP is apparent in the eventual decrease in rate with increasing [Me^{2+}]. At a lower concentration range (0.1–1.0 m*M*) the activating effect of the cations was quite regular. Double reciprocal plots of divalent cation concentration *vs.* initial reaction velocity were linear for all cations, except Zn^{2+}, over a ninefold range of concentrations (Figure 2). Since ATP was present in at least a fivefold excess over this range, it appears that [Me^{2+}] was negligibly small and that the true variable was [Me^{2+}–ATP]. (*See* Perrin and Sharma (*11*)

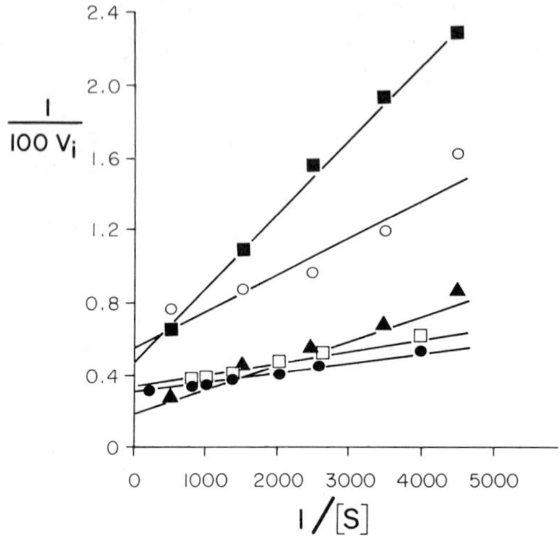

Figure 2. Double reciprocal plots of $1/V_i$ vs. $1/Me^{2+}$. *Symbols and conditions are identical to those in Figure 1 except for the lower range of* $[Me^{2+}]$.

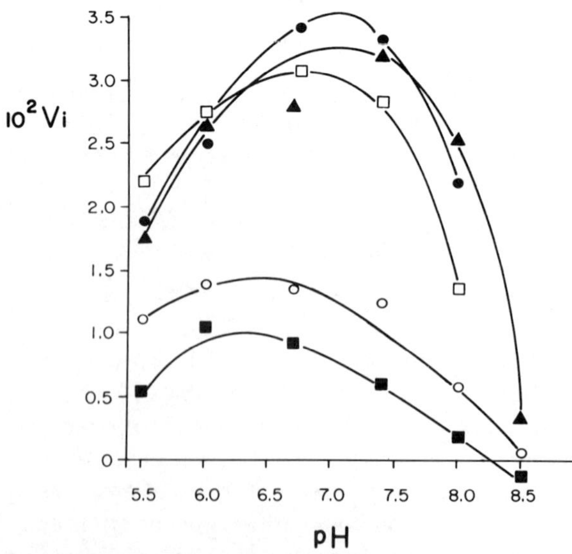

Figure 3. Variation of initial velocity, V_i *with pH. Symbols and conditions are given in Figure 1 except that the tris acetate buffer concentration was 0.1*M *at the pH's indicated.*

for binding constants of Me–ATP complexes.) Note that some Me^{2+} may have been bound to membrane-charged groups so that the expected concentration of Me^{2+}–ATP, calculated from the amount of Me^{2+} added, may be erroneously high.

When analyzed by the method of least squares no significant differences were found in the $1/V_i$ intercepts of the lines for Co^{2+}, Mg^{2+}, and Mn^{2+}. However, the slopes of these lines differed by more than two standard deviations. The line for Ni^{2+} differed significantly in intercept and slope from those for Co^{2+}, Mg^{2+}, and Mn^{2+}. The large error in the Zn^{2+} curve prevented detailed comparisons with the other cations.

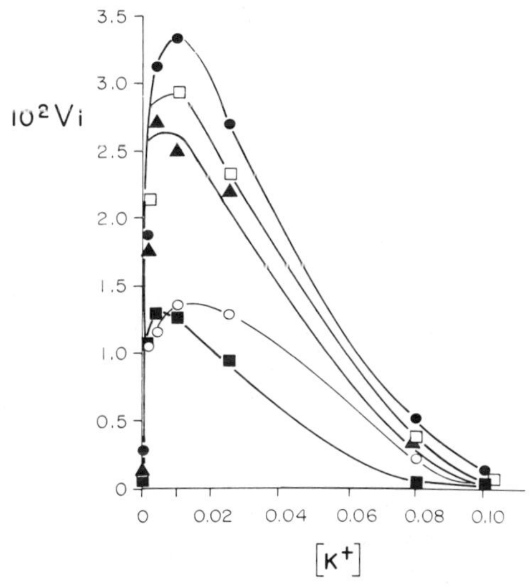

Figure 4. Variation in initial velocity, V_i, *with* K^+ *in M/L. Condition and symbols are those given in Figure 1 except that KAc was varied as indicated while the NaAc + KAc total was maintained constant at 0.1*M.

The variation in Na–K ATPase activity with pH and with Na:K ratio is shown in Figures 3 and 4. Although some differences are apparent in the curves obtained with the various divalent cations, the qualitative behavior is the same for all. However, the relative activating effect is pH dependent. The patterns of activity variation in Figures 3 and 4 could be duplicated using controls that contained choline acetate instead of Na^+–K^+ acetate and ouabain, and so they do not reflect variations in degree of ouabain inhibition. No activating effect under the conditions studied could be found for Ba^{2+}, Ca^{2+}, Cd^{2+}, Cu^{2+}, Fe^{2+}, Pb^{2+}, or Sr^{2+}.

To test the tissue specificity of these observations the effect of the activating ions was surveyed using Na–K ATPase prepared from calf brain

Table I. Ion Binding by Calf Brain ATPase[a]

Determination	μmoles/mg Protein		
	Zn^{2+}	Ni^{2+}	Co^{2+}
(1)	2.2	1.1	1.1
(2)	1.9	0.84	0.26
(3)	1.1	0.60	
(4)	0.79		
Average	1.5	0.85	0.68

[a] Values were determined polarographically with calf brain enzyme (2-5 mg prot./ml) in 0.135*M* NaCl, 0.05*M* KCl—pH 7.1.

according to Skou and Hilberg (9). Again, Co^{2+}, Ni^{2+}, and Zn^{2+} activated the enzyme in the absence of Mg^{2+}, using either ouabain or choline controls.

To exclude the possibility that the observed activation by Co^{2+}, Ni^{2+}, and Zn^{2+} was secondary to the release of bound Mg^{2+} and subsequent activation by Mg^{2+}, the binding of these cations to the calf brain enzyme preparation was determined. The polarographic diffusion current caused by the activating ion was measured in the presence of added aliquots of the ion. The difference between the extrapolated current at zero addition of the ion and the current actually observed gave a value for the bound ion (*see* Table I). Approximately 10^{-6} mole/mg protein of enzyme preparation were bound for each of the three ions studied. In the usual enzymatic assay system, which contains about 0.5 mg of protein/ml, approximately 0.5 μmole of Co^{2+}, Ni^{2+}, or Zn^{2+} might be bound. If an equivalent amount of Mg^{2+} were released as a result of ion exchange, the Mg^{2+} could account for the observed enzymatic activity. This possibility was eliminated, however, by direct analysis of the Mg^{2+} content of the enzyme preparation. One ml of calf brain enzyme preparation containing 5.8 mg of protein was digested in concentrated nitric acid and analyzed by atomic absorption spectrometry; it contained only 0.11 μmole of Mg^{2+}/mg protein. Thus the amount of Mg^{2+} potentially available for release by Co^{2+}, Ni^{2+}, or Zn^{2+} in the enzyme assay system was only 0.006 μmole. Since assays conducted in the presence of this amount of added Mg^{2+} had no Na^+–K^+ stimulated ATPase activity, it appears unlikely that the activation observed with Co^{2+}, Ni^{2+}, or Zn^{2+} is caused by the release of Mg^{2+}.

Discussion

The Divalent Cation Activators. While definitive proof must await isolation of the enzyme, it seems likely that the ATP complexes of Co^{2+}, Mg^{2+}, Mn^{2+}, Ni^{2+}, and Zn^{2+} are substrates of the Na–K ATPase. Mg^{2+} appears to be unique among these cations in that appreciable concentrations of free Mg^{2+} are compatible with enzyme activity. [It has been estimated that the optimal ratio of Mg^{2+}:ATP is roughly 2:1 (*12*), although considerable variation is possible as the components of the assay system are changed (*13*). The other cations in the presence of approximately equimolar concentrations of Mg^{2+} and ATP inhibit Na–K ATPase activity (*5*).] On comparing these divalent cations that activate with those that do not activate, differences in ionic radii appear to separate the two groups. Activating, divalent cations have ionic radii in the range 0.65–0.80 A, with Mg^{2+} at the lower limit; the non-activators have radii significantly above the upper limit (*14*). The exception to this separation is the Fe^{2+} ion, which has an ionic radius of 0.75 A. Perhaps the facile redox changes in the Fe^{2+} to Fe^{3+} system interfere with the expected activating effect. (Rendi and Uhr (*7*) found that Fe^{2+} has an activating effect; we were unable to confirm this.)

The present findings emphasize that the interaction with the Na^+ ion is the unique feature in the action of this enzyme. Many K^+ and Mg^{2+} substitutes are known (*9, 15, 16*). Even ATP may be replaced easily with CTP, ITP, or GTP (*4, 13*) in the partial reactions of the enzyme. However, the presence of Na^+ appears to be an absolute requirement.

Mechanisms of Activation and Inactivation. The Na–K ATPase, which is part of the cell membrane and accessible to both intra- and extracellular solutions, is active when the cations normally present adsorb at the proper sites on the enzyme surface. The binding of MgATP is promoted by Na^+, the two sites being present on the inner surface of the membrane. ATP splitting is then promoted by K^+ ion binding to the outer surface of the membrane. In the absence of these ions the enzyme is not active except when the substitute cations are available at the proper sites.

From our results it appears possible that this inactivation is caused by changes in the mechanical properties of the enzyme molecule. Since the result of the sequence of binding and splitting reactions is the transport of Na^+ and K^+ ions, it is probable that the enzyme is fluid and able to change its conformation during the sequence. In studying the activation of the enzyme by adding phospholipids, Kimelberg and Papahadjopoulos (*17*) achieved a maximal activity only when the fatty acyl chains were fluid. Protein fluidity appears to be as necessary for the activity of the enzyme as fluidity of the lipid portion of the membrane.

In a recent study (*18*) on the effects of cations on the apparent yield stress of ovalbumin monolayers, we found that Zn^{2+} and Co^{2+} ions resulted in the formation of a two-dimensional network that had a lower apparent yield than that formed by Mg^{2+}, Ca^{2+}, Ba^{2+}, Cu^{2+}, Pb^{2+}, and La^{3+}, which are strong inactivators, formed networks that were less fluid. If it is possible to generalize from these measurements on a protein monolayer to the ATPase protein, it appears that the inhibitory cations reduce the fluidity more than do the activating cations.

The balance between the binding of the activating cations to the two sides of the enzyme in the membrane and the activity of the enzyme is probably the basis for the activity of an ion pump that reflects the ion concentrations in the intra- and extracellular fluids. The linkage between the enzyme activity and the ion translocation is not known, but our results, together with those on lipid activation, suggest that mechanical coupling between the two surfaces may be critical to the proper functioning of the enzyme.

Literature Cited

1. Adam, N. K., "The Physics and Chemistry of Surfaces," p. 97, Dover, New York, 1968.
2. Britten, J. S., Blank, M., *J. Colloid Interface Sci.* (1973) **43**, 564.
3. Skou, J. C., *Physiol. Rev.* (1965) **45**, 596.
4. Post, R. L., Sen, A. K., in "Methods in Enzymology," R. W. Estabrook and M. E. Pullman, Eds., pp. 762-768, Academic Press, New York, 1967.
5. Bader, H., Wilkes, A. B., Jean, D. H., *Biochim. Biophys. Acta* (1970) **198,** 583.
6. Atkinson, A., Lowe, A. G., *Biochim. Biophys. Acta* (1972) **266,** 103.
7. Rendi, R., Uhr, M. L., *Biochim. Biophys. Acta* (1964) **89**, 520.
8. Kinsolving, C. R., Post, R. L., Beaver, D. L., *J. Cell. Comp. Physiol.* (1963) **62**, 85.
9. Skou, J. C., Hilberg, C., *Biochim. Biophys. Acta* (1969) **185,** 198.
10. Taussky, H. H., Shorr, E., *J. Biol. Chem.* (1953) **202,** 675.
11. Perrin, D. D., Sharma, V. S., *Biochim. Biophys. Acta* (1966) **127,** 35.
12. Skou, J. C., *Biochim. Biophys. Acta* (1957) **23,** 394.
13. Skou, J. C., *Biochim. Biophys. Acta* (1974) **339,** 246.
14. Wells, A. F., "Structural Inorganic Chemistry," p. 70, Oxford, 1950.
15. Fahn, S., Koval, G. J., Albers, R. W., *J. Biol. Chem.* (1966) **241,** 1882.
16. Britten, J. S., Blank, M., *Biochim. Biophys. Acta* (1968) **159,** 160.
17. Kimelberg, H. K., Papahadjopoulos, D., *Biochim. Biophys. Acta* (1972) **282,** 27.
18. Blank, M., Lee, B. B., Britten, J. S., *J. Colloid Interface Sci.* (1973) **43,** 539.

RECEIVED September 23, 1974. Supported in part by research grant HD06908 from the U.S. Public Health Service.

19

Surface Properties of Membrane Systems: Influence of Chemistries on Surface Viscosity

GIUSEPPE COLACICCO, MUKUL K. BASU, JEFFREY LITTMAN, and EMILE M. SCARPELLI

Departments of Pediatrics and Pathology, Albert Einstein College of Medicine, Bronx, N.Y. 10461

We used a torsion oscillation method to estimate surface viscosity by measuring the oscillation period (t) *and the difference* Δt *above the period on the aqueous subphase as a function of the chemical nature of lipid and protein, type of electrolyte, temperature, film pressure, lipid–lipid, and lipid–protein interactions. With a series* (G_s) *of glycosphingolipids* (N-*palmitoyl sphingosyl glycosides) that have different sugar moieties,* Δt *on NaCl or* $CaCl_2$ *decreased as the number of monosaccharide units (S) increased: ceramide (S = 0) > galactoside (S = 1) > lactoside (S = 2) > hematoside (S = 3) > ganglioside (S = 5). At high film pressure surface viscosity was very high with bovine serum albumin (BSA) and with dipalmitoyl lecithin (DPL); it was very low for ribonuclease (RNase) > sphingomyelin > cholesterol and for DPL-cholesterol and DPL-albumin mixed films. We conclude that surface viscosity depends on the chemical nature and configuration of the polar groups of lipid and proteins and on the formation of a copolymerization lattice mediated by water and not on the MW of surfactant molecules or on their drag of subphase water.*

N. K. Adam suggested that although surface tension, γ, represents the mathematical line of the surface free energy of water, its molecular meaning must still be determined (*1*). Indeed most existing treatments of the parameters of water surfaces escape our grasp because of the lack of a model. In applying monolayer techniques to biology, the molecular model has paramount importance. In the light of this model we set out to establish significant correlations between molecular structure and surface parameters such as surface tension, surface pressure, surface potential, and surface viscosity (*2*). These efforts illustrate the architecture and function of biological surfaces such as found in myelin membranes (*3, 4*), mitochondrial membranes (*4, 5*), lipoprotein particles (*4, 5*), and the air–water interfaces of the respiratory system (*6*). Our present efforts explore the alveolar surfaces of the mammalian lung.

Alveoli of the mammalian lung present two surfaces of interest. The first is the interface between the plasma membrane of the epithelial lining cell and the acellular alveolar lining layer (ALL) which covers it and presumably has the characteristics of a lipid–water interface. The second, apparently an air–water interface, marks the point of contact between ALL and alveolar air (*6, 7, 8, 9*). Whereas the ALL generally appears as a homogeneous medium by electron microscopy, both lamellar and particulate structures have been observed (*10, 11*). Although molecular organization and interfacial forces at the alveolar surface are primary determinants of alveolar function and stability (*6, 7*), relatively little has been reported concerning alveolar surface properties other than surface tension at the air–water interface; there is virtually no information about molecular organization *in situ*. Since the ALL and other membrane systems contain both lipid and protein constituents, this study probes the architecture and function through an investigation of surface viscosity of selected model systems.

An important feature of the mosaic or subunit model of the molecular organization of lipid and lipid–protein films and membranes (*4, 5, 12*) is the cohesion between subunits and between molecules within each subunit (*2, 13*). The latter can be measured as microviscosity by polarization of fluorescence (*14*) of the surface film; the former is measured as surface viscosity by various techniques (*15, 16*). The relevance of this parameter to the exploration of biological surfaces cannot be overemphasized. Surface viscosity, in fact, has also been invoked as a force stabilizing pulmonary alveoli (*6, 7, 13, 16*).

Both simple and complex species (*e.g.*, water, lipids, proteins, lipoproteins, antibodies, enzymes, viruses, and small and large electrolyte ions) may be adsorbed on or transported across the interfaces of biological membranes, membrane subunits, and particles. These translocations are associated with membrane or particle formation, fusion, exchange, and lysis. The processes may be relevant to the movement, exchange, and clearance of extracellular material at the alveolar lining layer of the lung (*7*). Within this framework, the role of surface and hypophase viscosity has not been investigated.

Information about fluidity and viscosity of bilayers of artificial and natural membranes has been obtained from electron spin resonance studies in which the mobility of the spin-labelled species along the surface plane of the membrane is determined (17). However, the monolayer of either lipid, protein, or lipid–protein systems at the air–water interface, makes an ideal model because several parameters can be measured simultaneously. Surface tension, surface pressure, surface potential, surface viscosity, surface fluorescence and microviscosities, surface radioactivity, and spectroscopy may be determined on the same film. Moreover, the films can be picked up on grids from which they may be observed by electron microscopy, studied further for composition, and analyzed for structure by x-ray diffraction and spectroscopy. This approach can provide a clear understanding of the function and morphology of the lipid and lipid–protein surfaces of experimental membranes. However, the first objective is to obtain molecular correlations of surface tension, pressure, potential, and viscosity.

This manuscript correlates the chemical structure of selected lipids and proteins and their surface viscosity. Our studies probed certain concepts (Figure 1): how and if surface viscosity depends on MW, the

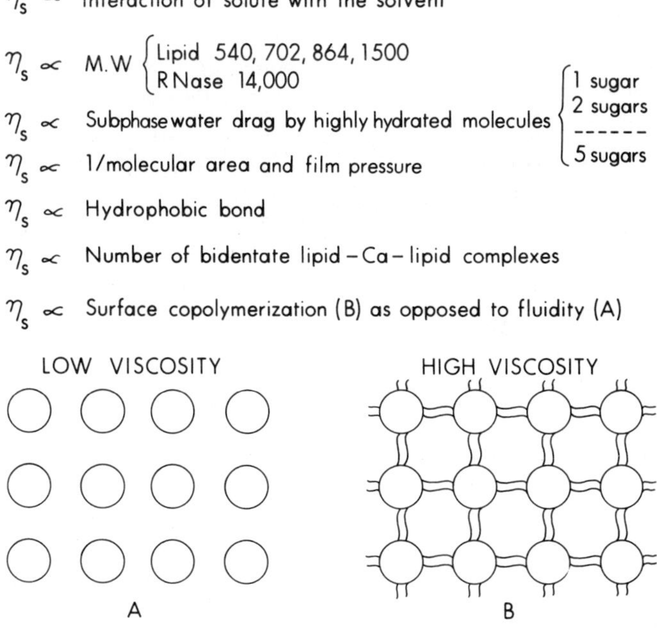

Figure 1. The existing concepts of surface viscosity. This paper tests the validity of such concepts.

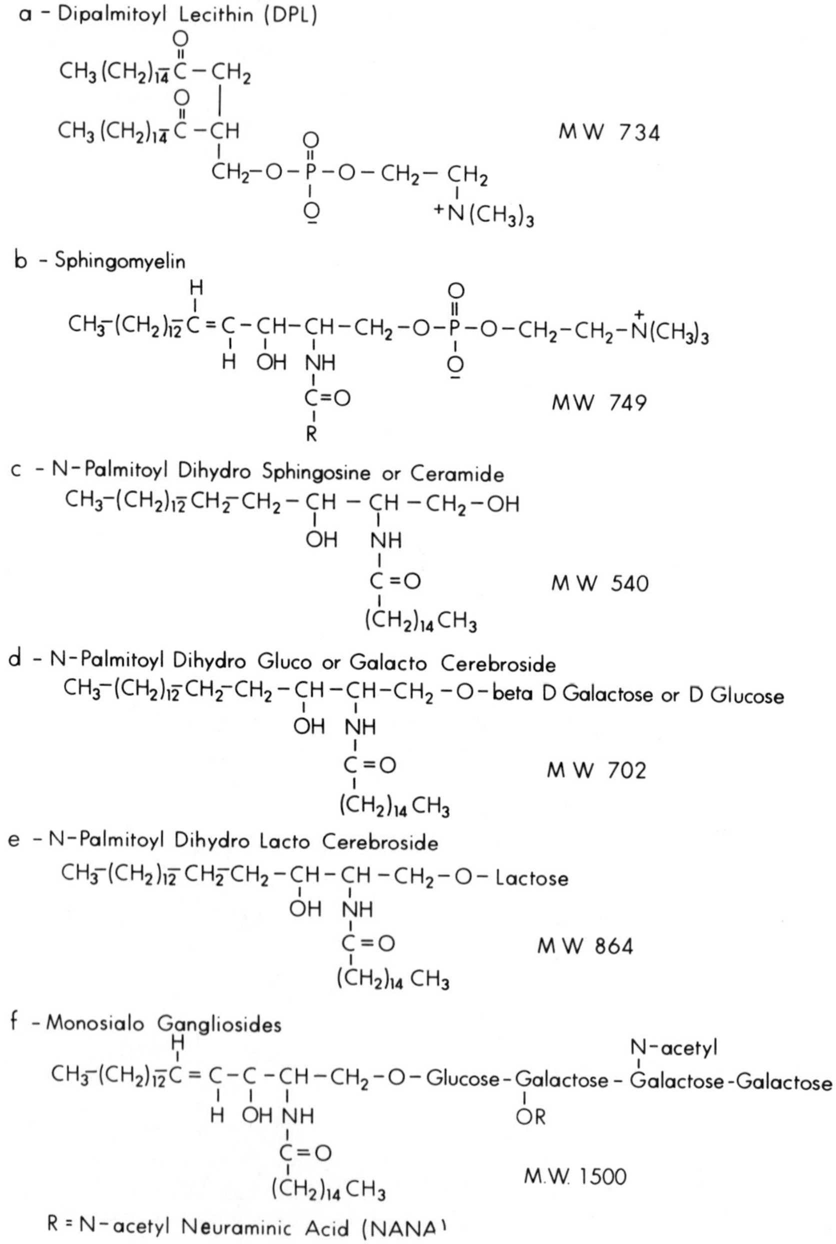

a - Dipalmitoyl Lecithin (DPL)

$$CH_3(CH_2)_{14}\overset{O}{\overset{\|}{C}}-CH_2$$
$$CH_3(CH_2)_{14}\overset{O}{\overset{\|}{C}}-CH$$
$$CH_2-O-\overset{O}{\overset{\|}{P}}(O^-)-O-CH_2-CH_2-{}^{+}N(CH_3)_3$$

MW 734

b - Sphingomyelin

$$CH_3-(CH_2)_{12}-CH=CH-CH(OH)-CH(NH-C(=O)-R)-CH_2-O-\overset{O}{\overset{\|}{P}}(O^-)-O-CH_2-CH_2-\overset{+}{N}(CH_3)_3$$

MW 749

c - N-Palmitoyl Dihydro Sphingosine or Ceramide

$$CH_3-(CH_2)_{12}-CH_2-CH_2-CH(OH)-CH(NH-C(=O)-(CH_2)_{14}CH_3)-CH_2-OH$$

MW 540

d - N-Palmitoyl Dihydro Gluco or Galacto Cerebroside

$CH_3-(CH_2)_{12}-CH_2-CH_2-CH(OH)-CH(NH-C(=O)-(CH_2)_{14}CH_3)-CH_2-O-$ beta D Galactose or D Glucose

MW 702

e - N-Palmitoyl Dihydro Lacto Cerebroside

$CH_3-(CH_2)_{12}-CH_2-CH_2-CH(OH)-CH(NH-C(=O)-(CH_2)_{14}CH_3)-CH_2-O-$ Lactose

MW 864

f - Monosialo Gangliosides

$CH_3-(CH_2)_{12}-CH=CH-C(OH)-CH(NH-C(=O)-(CH_2)_{14}CH_3)-CH_2-O-$ Glucose - Galactose(OR) - N-acetyl Galactose - Galactose

M.W. 1500

R = N-acetyl Neuraminic Acid (NANA)

Figure 2. Molecular structure of lipids used

distance between molecules, film pressure, the quantity of water dragged by the film in shearing, and the nature of the polar groups of lipids, proteins, lipid mixtures, and lipid–protein systems. We use a damping oscillation method to measure the surface viscosity of the films of typical lipids, lipid mixtures, protein, and lipid–protein mixtures. Although membrane proteins (*3, 4, 5*) are available, we chose two non-membrane proteins—RNase and BSA—because of their long use in monolayer studies and surface viscosity measurements (*2, 6, 16*). Our choice was also influenced by the following considerations: (a) serum albumin is the most abundant protein in pulmonary washings (*18*); (b) dipalmitoyl lecithin is the prevalent phospholipid of the surfactant system of the lung (*7, 18*); (c) both lecithin and sphingomyelin are important membrane constituents and are determinats of the functional maturity of the lung before birth (*19*); and (d) cholesterol, notorious for its effects on membranes and lipid and protein films (*2*), is also present in pulmonary extracts (*6, 7, 8, 9*).

We examined a series of synthetic glycosphingolipids in which the dihydroceramide part (*N*-palmitoyl dihydrosphingosyl) was kept constant while the number of sugar units was increased from zero in dihydroceramide to two in dihydroceramide lactoside. In addition, two natural glycosphingolipids were used—a hematoside containing three sugar units and a terminal *N*-acetyl neuraminic acid (NANA) and a monosialoganglioside containing four sugar units and one NANA or sialic acid residue (Figure 2). There are two reasons for using this series of glycolipids. First, their surface behavior probably relates to their location in the surface topography of membranes and particles and to the function of their sugar moieties in immunochemical reactions; also the same sugar moieties are found on the surface of glycoproteins (*2*). Second, the series is useful in studying the influence of sugar, molecular weight, molecular area, water of hydration, and water drag on surface viscosity. Finally the structural differences between these lipids and the lecithins and sphingomyelins above are instructive.

Materials and Methods

Lipids and Proteins. Cholesterol, beef brain sphingomyelin, monosialoganglioside (that has the sugar sequence glucose-, galactose-(NANA), *N*-acetyl galactosamine, galactose) and synthetic dipalmitoyl lecithin (DPL) were purchased from Supelco, Bellefonte, Pa. *N*-palmitoyl dihydrosphingosine (dihydroceramide), *N*-palmitoyl dihydrosphingosyl glucocerebroside, and galactocerebroside (dihydroceramide glucoside or galactoside), *N*-palmitoyl dihydrosphingosyl lactocerebroside (dihydroceramide lactoside) were obtained from Miles Laboratories, Bloomington, Ind. A natural hematoside from bovine adrenal medulla (*N*-acyl-sphingosyl-glucosyl-galactosyl-NANA) was a gift of R. Ledeen from the Department of Neurology, Albert Einstein College of Medicine. All lipids were homogeneous as determined by thin layer chromatography when 100 μg of material was applied; they were not purified further.

BSA was a product of Pentex, Kankakee, Ill., and salt-free RNase was obtained from Sigma, St. Louis, Mo.

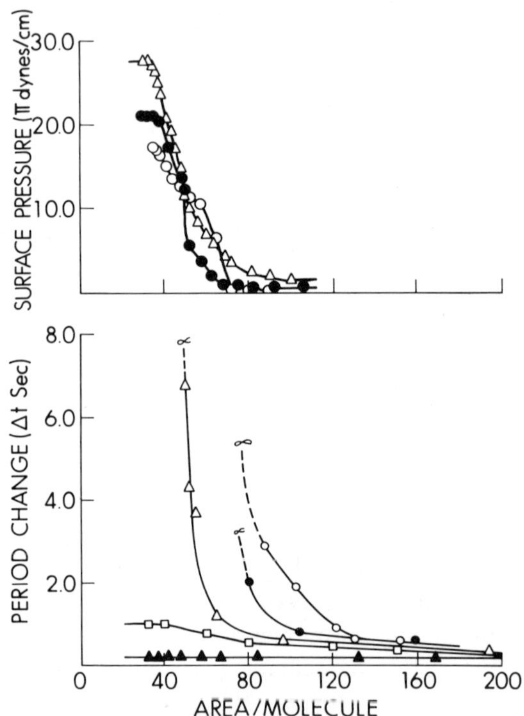

Figure 3. Surface pressure–area and surface viscosity (Δt)–area curves of glycosphingolipids on 0.15M NaCl at 25°C. (○) ceramide; (●) glucocerebroside; (△) lactocerebroside; (□) haematoside; (▲) monosialoganglioside.

Monolayer Techniques. Surface tension and surface potential were measured by a torsion balance with the Wilhelmy plate and a radioactive (^{226}Ra) air electrode (*5, 20*).

Surface viscosity was determined by a torsion oscillation method in which the period of oscillation of a paraffin-coated mica ring was measured on a clean water surface and lipid-covered surface (*6, 13*). Since the period of oscillation is proportional to the surface viscosity, the difference in period (Δt) between lipid or protein film and a clean aqueous surface measures viscosity and relates to either molecular area, film pressure, or time of interaction (t). The period values were reproducible to ±0.1 sec, and an average of five measurements was taken in each case.

The troughs contained 500 ml of subphase and consisted of borosilicate glass dishes with a jacket for temperature control. Experiments were carried out at 25°C, which conforms to most of the literature data, and at 37°C, which is body temperature. The subphase consisted of 0.15*M* NaCl or 0.075*M* $CaCl_2$. The electrolyte solutions were prepared with reagent grade salts and with water that was redistilled from glass over alkaline permanganate.

Results

Influence of the Sugar Moiety on Surface Viscosity of Glycosphingolipids. The series: ceramide > gluco- or galactocerebroside > lactocerebroside > hematoside > ganglioside (Figure 3) indicates a decrease in surface viscosity with an increase in monosaccharide units.

Influence of Electrolyte and Temperature. As expected, the surface potential of the ganglioside was higher on $CaCl_2$ than on NaCl; this suggested the formation of a bidentate G_5–Ca–G_5 complex *via* the car-

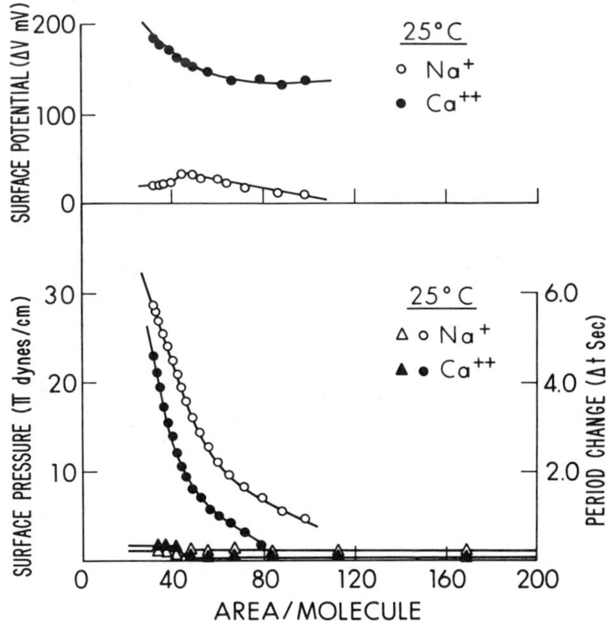

Figure 4. Surface pressure–area (○,●, low panel, left hand side ordinate), surface potential–area (○,●, upper panel), surface viscosity–area (△,▲, low panel right hand side ordinate) curves of monosialoganglioside at 25°C on 0.15M NaCl (○,△) and 0.075M $CaCl_2$ (●,▲)

boxylate group of the NANA moiety. The surface viscosity however was affected little by the electrolyte (Figure 4).

In spite of the absence of ionic groups, a specific electrolyte effect occurred in the non-sialogylcosphingolipids; with gluco- and galacto-cerebroside Na^+ induced viscosity better than Ca^{2+} (Figure 5) at 25°C and 37°C. In contrast, with ceramide, Ca^{2+} was more effective than Na^+ (Figure 6).

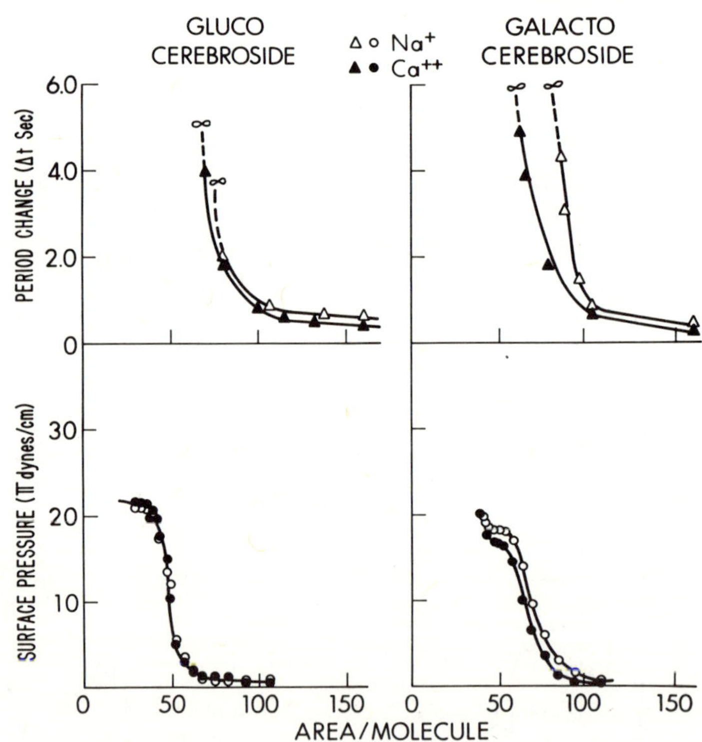

Figure 5. Surface pressure–area (○,●) and surface viscosity–area (△,▲) curves of glucocerebroside (left hand panel) and galactocerebroside (right hand panel) on 0.15M NaCl (○,△) and 0.075M $CaCl_2$ (●,▲) at 25°C. The galactocerebroside is more viscous on NaCl than on $CaCl_2$.

With lactoside, the temperature influenced the effect of the electrolyte such that at large areas (> 65 A^2/molecule) Δt on Na^+ was greater than Δt on Ca^{2+} at 37°C; the opposite was true at 25°C, *i.e.*, an inversion, $Ca^{2+} > Na^+$ to $Na^+ > Ca^{2+}$, occurred (Figure 7) for areas smaller than 65 A^2/molecule. In general higher temperature caused a decrease in viscosity; Figures 6 and 7 show that the viscosity (Δt) *vs.* area curves are shifted towards smaller areas.

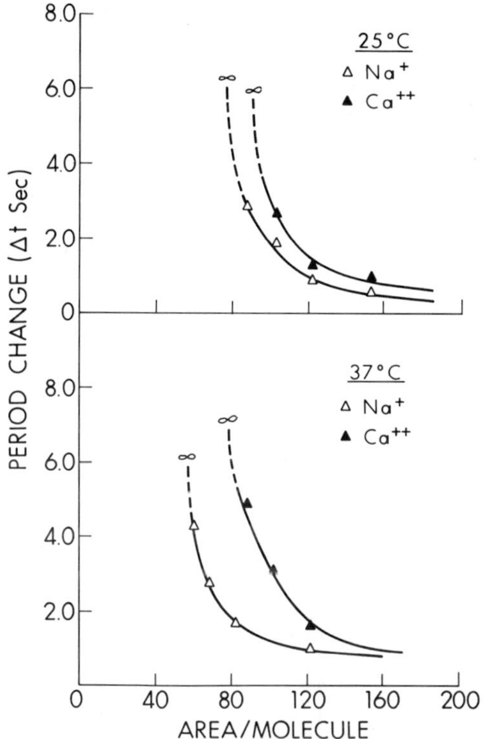

Figure 6. Surface viscosity (Δt)*–area curves of ceramide (compound c in Figure 2) at 25° and 37°C, on 0.15*M *NaCl* (△) *and on 0.075*M *$CaCl_2$* (▲)

Influence of Chemical Groupings. Ceramide (not shown) and dihydroceramide have similar high surface viscosities which indicates that the unsaturation in the polar region of the sphingosine moiety has no significant effect on surface viscosity or surface potential (*21*). Similarly, when the phosphoryl choline group is attached to ceramide or dihydroceramide, the resulting sphingomyelin and dihydrosphingomyelin have the same surface viscosity, which, however, is much lower than the viscosity of the corresponding ceramides and glycolipids. Apparently, viscosity is restored when the phosphoryl choline group is replaced by either galactose (glucose) or lactose as in galactocerebroside (Figure 8) and lactoside (Figure 7). In contrast, when the phosphoryl choline group is attached to the dipalmitoyl glyceryl residue in DPL, the viscosity is very large and increases with film pressure. Note the low surface potentials of dihydrosphingomyelin (*21*) and sphingomyelin and the difference in surface potential between sphingomyelin on one side and *N*-palmitoyl dihydrosphingosyl galactocerebroside and DPL (*13*) on the other side.

DPL-Cholesterol System. The increase of film pressure with the quantitative addition of a DPL solution in chloroform–methanol (85:15) on the surface of 0.15*M* NaCl shows a typical surface saturation curve (Figure 9). Surface viscosity increased with each addition and with film

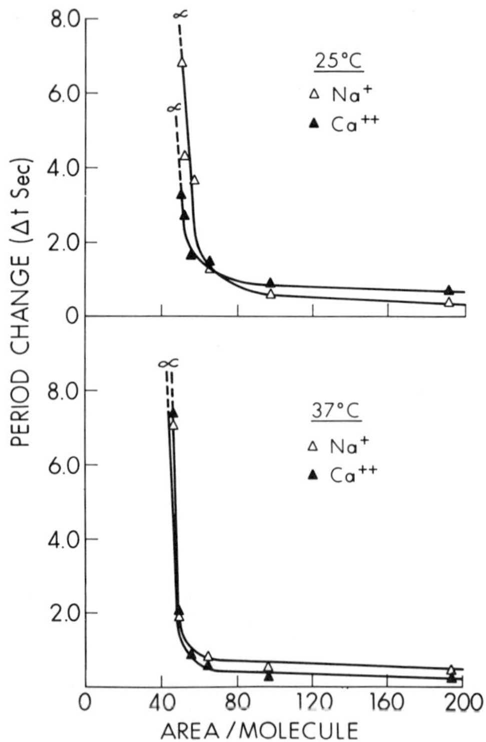

Figure 7. Surface viscosity (Δt)*–area curves of lactocerebroside at 25° and 37°C, on 0.15*M *NaCl* (△) *and on 0.075*M *$CaCl_2$* (▲)

pressure, and it was not measurable before collapse pressure, where the film became rigid. When certain quantities of cholesterol in chloroform–methanol (85:15) were added to the saturated DPL film, surface pressure remained unchanged, but surface viscosity decreased rapidly when the amount of cholesterol increased to 1:1 mole DPL: cholesterol (about 30 μl of cholesterol in Figure 9); the difference in oscillation period between water and film was only 1/2 sec at that ratio. Cholesterol alone had no surface viscosity (Figure 9), and the solvent alone [chloroform–methanol (85:15)] did not change the surface pressure or surface viscosity of the DPL film.

At 37°C the DPL film still had some viscosity ($\Delta t = 1$ sec) which was abolished by cholesterol. However, the surface viscosity of DPL vanished at 40°, 41°, 42°, and 45°C, and cholesterol had no further effect.

Protein and Lipid-Protein Systems. The high surface viscosity of BSA and the low surface viscosity of RNase had been observed by a torsion rotational method that used an extremely large torque (*2, 6*). In the present experiments, the protein is dispersed in the aqueous subphase to a final concentration of 10 μg/ml. Both film pressure and surface viscosity are measured as a function of time (to 40 min). The film pressure of BSA was 21 dynes/cm at mixing and remained constant for 40 min. Also, surface viscosity reached a high value instantaneously (although the first measurement was made at 5 min) and remained constant for 40 min. In contrast, RNase built up pressure slowly, from 3 dynes/cm

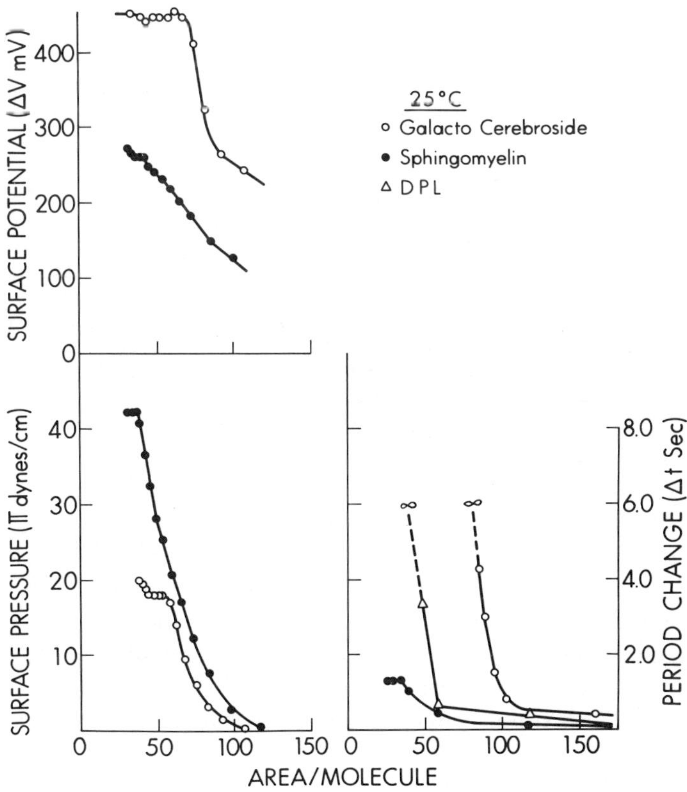

Figure 8. Surface pressure–area, surface potential–area, and surface viscosity (Δt)*–area curves of sphingomyelin* (●)*, galactocerebroside* (○)*, and dipalmitoyl lecithin* (△) *on 0.15*M *NaCl at 25°C*

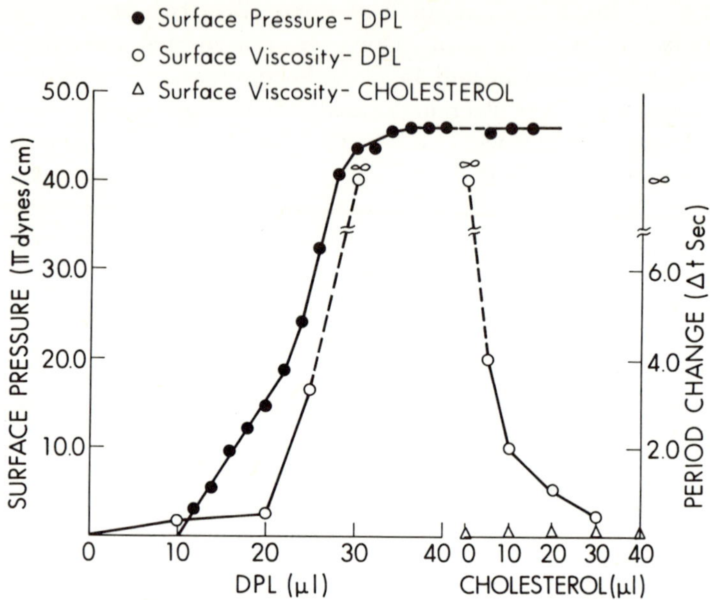

*Figure 9. Influence of cholesterol on surface viscosity (Δt) of a saturated film of DPL on 0.15*M *NaCl at 25°C. A film of DPL was formed by applying microliter quantities of DPL at a concentration of 1.5 μg/μl in chloroform–methanol (85:15) onto a fixed area. Surface pressure (●) and surface viscosity (Δt) (○) were measured. After surface saturation at about 46 dynes/cm pressure, cholesterol 1 μg/μl in chloroform–methanol (85:15) was applied onto the saturated DPL film. Surface pressure (●) was not affected; surface viscosity fell dramatically. A film of cholesterol alone (△) has no surface viscosity, and the solvent alone does not affect either surface pressure or viscosity.*

at mixing to 10 dynes/cm at 40 min; also the surface viscosity rose slowly to modest values ($\Delta t = 1$ sec).

At 40 min either DPL or cholesterol in chloroform–methanol (85:15) was added quantitatively to the protein film. With both proteins, film pressure increased to 45 dynes/cm after DPL addition and to 40 dynes/cm after cholesterol addition. In both cases, relatively small quantities of lipid obliterated the viscosity that had been established by the protein; with BSA, DPL was more effective than cholesterol.

Discussion

We can analyze the relationship of surface viscosity to various physicochemical properties of surfactants at the air–water interface in the light of current concepts (Figure 1). First however, we should analyze the meaning of the measurements with respect to the proposed model of the mosaic structure of lipid and lipid–protein surfaces (Figure 1 and Refs. *2, 4, 5*).

Validity and Significance of the Methods of Measurements. In our experimental setup, a Δt of 1 sec corresponds to about 1 mpoise of surface viscosity above that of the clean water surface. With this apparatus we can measure only up to several millipoises. The viscosity values in the literature vary from a few millipoises for fatty acids (*22*) and phospholipid monolayers (*6, 13*), to a few hundred centipoises for the microviscosity of fluorescent probes in erythrocyte membranes (*14*), and to 2 poises in the case of the rotational diffusion of the rhodopsin molecule in the retinal rod membrane disc (*23*).

Our probe, a float nearly 10 cm in diameter, is coarse; with it, we probe the relatively weak forces that exist between the subunits and the solvent (Figure 1A) or between subunit and subunit (Figure 1B), or both. Since a systematic study is not available, we do not know which intermolecular structures are probed by the different methods; surface viscosity values differ among the methods (6; Ref. *24*, p. 104).

Since our objective is to point out large differences in viscosity due to variations in chemical structure of the lipid or lipid–protein system, we do not distinguish between Newtonian and non–Newtonian viscosity. Most of our measurements are made at relatively low film pressures where viscosities are usually Newtonian (*15*). Also, in doing the oscillation, we used a constant initial velocity since the pendulum was released after torsion to the same angle (180°) from the rest position. We did not calculate surface viscosity because its theoretical meaning was not available (*15, 22, 24*). Some experiments (Figures 3, 8, 9) were also performed with a hydrophilic float of the same dimensions (10 cm external diameter). Although the periods of oscillation of the hydrophilic float were appreciably larger ($\Delta t > 2\times$) than those observed with the hydrophobic float, the trends were the same in both cases. Therefore, effects, such as the absence of viscosity in the cholesterol films (Figure 9, Refs. *25, 26*), could not be because a hydrophobic float slipped over a rigid film.

Molecular Weight and Organization of Surfactants. The present data suggest that surface viscosity is not determined by the mass or MW of the monolayer's kinetic unit, whether the unit is a simple surfactant molecule, a hydrated molecule, a polymeric complex, or a large molecule such as RNase (MW 14.000). At a film pressure of 10 dynes/cm, RNase has a very low surface viscosity as compared with the glycosphingolipids at the same film pressure (Figures 10 and 5).

In addition, despite comparable molecular areas (Figure 3), viscosity decreased in the order: ceramide (MW 540) > galactocerebroside

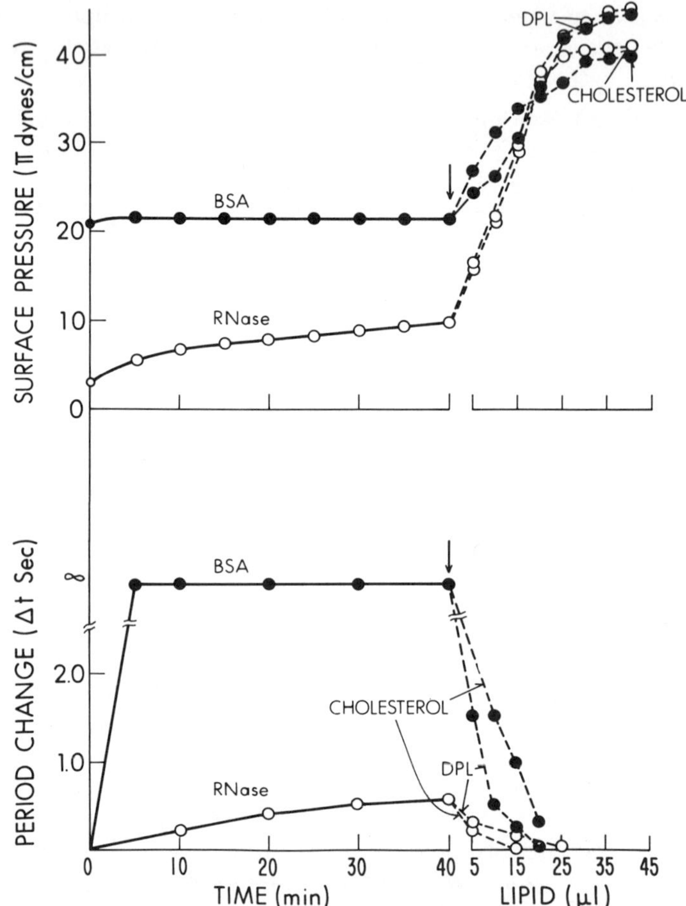

Figure 10. Influence of cholesterol on the surface viscosity of protein films adsorbed from 0.15M NaCl at 25°C. Protein: BSA (●) 10 μg/ml, RNase (○) 10 μg/ml. At 40 min, microliter quantities of lipid–DPL or cholesterol [1 μg/μl in chloroform–methanol (85:15)] were added on the protein film. The solvent alone did not affect either surface pressure or viscosity.

(MW 702) > lactoside (MW 859). The high surface viscosity of BSA as compared with RNase is probably the result of structural characteristics of the protein rather than size.

The surface viscosity of gangliosides may be discussed in relation to their surface potentials. The higher surface potential of gangliosides on $CaCl_2$ compared with that on NaCl indicates that the carboxyl group of the NANA moiety of the ganglioside probably binds Ca^{2+}; in this case a bidentate lipid–Ca–lipid complex of MW 3,000 (about six times ceramide) is expected. Nonetheless their surface viscosities are similar. Also note the higher surface viscosity of DPL on $CaCl_2$ (not shown) (*13*) as compared with NaCl (Figure 9). If the higher surface viscosity on $CaCl_2$ can be explained by a bidentate DPL–Ca–DPL complex, how can one explain the high surface viscosity of DPL on NaCl (Figure 9) since Na^+ cannot form a bidentate complex with that lipid? However, if Na^+ and Ca^{2+} can open the choline–phosphate salt linkage by ion exchange, the resultant fixed negative and positive charges from the phosphate and choline groups could produce an extensive lattice by electrostatic interactions. Additional explanations for surface viscosity and the selectivity of some uncharged glycolipids (Figures 5, 6) for Na^+ and Ca^{2+} suggest an influence from the electrolyte ion on the structure of the organized water around the polar groups of the lipid (*13*).

Finally, a larger molecular weight is expected from the hydration of either the surfactant molecules or the polymeric membrane subunits. Such a phenomenon would increase the drag of the underlying viscous liquids (*15, 22, 24*). If the subphase water drag were an important component of the observed surface viscosity, the ganglioside would have a greater Δt than ceramide because of its five highly hydrated sugar units, but the situation is the opposite.

Molecular Area and Film Pressure. A smaller area/molecule and a greater film pressure caused an increase of surface viscosity (Figures 3–8). However, at the same area/molecule, ceramide was more viscous than ganglioside. A more striking contrast is seen in Figures 9 and 10. When modest quantities of DPL or cholesterol are added to saturated films of BSA and RNase, the pressure increases markedly (*e.g.*, from 21 to 45 dynes/cm with BSA) whereas the viscosity dropped from large, immeasurable values to practically zero.

Hydrophobic Interactions. It has been suggested that surface viscosity arises from the interaction between the hydrocarbon chains of the lipid (*25*); the interaction would be favored by the smaller molecular areas and high film pressures. Cholesterol and sphingomyelin at all pressures and DPL at low pressures have no appreciable surface viscosity above that of the clean water surface. At 85 A^2/molecule, dihydrosphingomyelin (not shown), DPL and sphingomyelin (Figure 8), and *N*-palmitoyl galactocerebroside, which have comparable MW and hydrophobic moieties, have greatly different surface viscosities, *i.e.*, Δt is practically zero for sphingomyelins and DPL and very large for the glycolipid.

Polar Groups and Surface Copolymerization. Thus the "hydrophobic bond," which does not exist between the hydrophobic chains (*4*), is not responsible for surface viscosity, but polar groups and molecular organization may be. Surface viscosity can be explained in terms of the surface lattice that is produced by a type of copolymerization *via* the lateral interactions between the hydrated polar groups of the oriented surfactant molecules (*2, 6, 13*). This surface copolymerization has some similarities with that proposed by Deamer and Cornwell (*27*). They visualized a copolymerization unit of four fatty acid molecules with Ca^{2+} at the basis of surface viscosity; the loss of surface viscosity on Ca^{2+}, with respect to Na^{+}, under DPL films at low pressure confirms our previous finding (*13*) and supports our suggestion that the lipid–Ca–lipid bidentate complex may not be the cause of surface viscosity. The high pressure of biological membrane systems (*28*) favors molecular proximity and such interactions. In the Langmuir trough, the pressure is maintained by the barrier; *in vivo* and in bulk, the geometry and the curvature of the particles contain the monolayer and bilayer systems. The various curvatures must relate to the chemical nature of the membrane components, to their interactions with water, and to the attending surface tension values.

In our approach, the surface lattice is produced by saturating the lateral valences of the polar groups of the surfactant molecules. Such interactions are favored by molecular proximity and by the length and saturation of the hydrophobic chains; they are mediated by water (*2, 4*) and result in the formation of a gel of a certain rigidity (Figure 1B). Nevertheless, surface viscosity bears the classical definition of viscosity—*i.e.*, the surface subunits in Figure 1A would experience shearing of their water of hydration against the solvent in the membrane regions without a need for the intersubunit binding shown in Figure 1B. In this case differences in the water structures around the polymeric surface subunits would account for differences in viscosity arising from their shearing against free water. In Figure 1A the water that is structured around the polar groups of the surfactant is qualitatively different from the general term of drag water, to which one attributes the drag viscosity in the canal method and in other methods (Ref. *22*, p. 252; Ref. *24*, p. 104).

Lipid–Lipid and Lipid–Protein Interactions. The DPL–cholesterol and the protein–DPL systems are particularly amenable to interpretation using our membrane model. The high viscosity lattice of DPL can be broken by cholesterol (Figure 9), and the lattice of BSA can be broken by a lipid (*e.g.*, DPL, Figure 10), with a marked loss of surface viscosity. This lattice collapse means formation of independent membrane subunits whose lateral valences are saturated within the subunit, thereby producing a fluid system (Figure 1A); the subunit could be a lipid–lipid system, as with DPL and cholesterol, or a lipid–protein system. The phenomenon of lattice collapse with loss of surface viscosity is impressive in the DPL–albumin system since individually both components have a high surface viscosity.

These observations are consistent with the mosaic model of the membrane that was derived from monolayer studies (*2, 4, 5, 12, 13*). Therein, the structural or bimodal (amphipathic) protein in the membrane (natural or artificial) interacts with the polar peripheries of the polymeric lipid structures alongside the protein. The EPR data of Jost *et al.* (*29*) support this concept, *i.e.*, an appreciable portion of the lipid is in lateral hydrophilic bonding with the protein whereas the other lipid is free, probably within the lipid cluster, and preserves the lipid character.

The surface viscosity data contrasts some EPR and NMR studies of the DPL–cholesterol system. In these studies cholesterol caused fluidity of the alkyl chains of the phospholipid below the transition temperature of the phospholipid; it caused rigidity above the transition temperature (*26, 30, 31, 32*). [Chain rigidity is often interpreted as membrane rigidity (*33*).] However, unless viscosities and microviscosities are distinguished experimentally (they are hypothetical in Figures 1A, B), the data cannot mean that fluidity or rigidity of the alkyl chains is identical with fluidity or rigidity of the membrane. Our surface viscosity experiments were carried out at 25°C, far below the transition temperature of DPL (41°C); no effect from cholesterol was noticed at higher temperatures such as 40°, 41°, 42°, and 45°C. At these temperatures one could expect film rigidity and increased viscosity if the rigidity of the alkyl chains attending the DPL–cholesterol complex were accompanied by rigidity of the membrane. Instead, rigidity of the alkyl chains caused by cholesterol could mean increased rigidity within the DPL–cholesterol bimolecular complex or within the DPL–cholesterol domain; such membrane subunit structures have been recognized on several grounds (*2, 30, 34*). A localized rigidity would not be detected by the surface viscosity probe, which recognizes an extended lattice (Figure 1B); it could, however, be measured as microviscosity within the membrane subunit (circles in Figures 1A, B) by the polarization of fluorescence as it was done with other systems (*14*). Consequently, one can have chain and domain rigidity in a fluid membrane.

Joos' data on distearoyl lecithin (DSL)–cholesterol mixed films (*35*) coincide with data from our DPL–cholesterol system in the sense that cholesterol reduced the viscosity of the lecithin film, and the surface viscosity decreased with increasing cholesterol concentrations. However, the comparison and interpretation of surface viscosity data require caution (*2, 6*). For example, in Joos' experiments the lipid was distearoyl lecithin (DSL), the subphase was distilled water, phospholipid and cholesterol were premixed, and viscosity was measured by the rotational surface Couette method. By the torsion oscillation method, at all film pressures the surface viscosity change of cholesterol was not above the viscosity value of the clean water surface (Figure 9); by the rotational method the surface viscosity of cholesterol increased with the film pressure (*35*). Although surface viscosity should increase with film pressure, two observations refute this assumption. First, despite the large increase in film pressure, the surface viscosity of serum albumin is abolished by DPL additions; DPL itself has a very high viscosity at the pressure where the DPL–alubumin system has no viscosity (Figure 10). Second, viscosity calculated on the basis of inelastic molecular collisions at high film pressures could not account for the high surface viscosities measured in films (*36*). Therefore, a likely molecular model of the surface viscosity is a surface lattice of intermolecular or intersubunit interactions mediated by interfacial water (*2, 6, 13*).

Conclusions

(1) Surface viscosity does not depend generally on the molecular weight of the surfactant. Indeed at all film pressures, the viscosity decreased in the order dihydroceramide (MW 702) > dihydrolactoside (MW 864) >> RNase (MW 14,000) > ganglioside (MW 1,500).

(2) Formation of the lipid–Ca–lipid bidentate complex does not cause an increase in surface viscosity as shown by the low viscosity of the ganglioside–Ca–ganglioside system.

(3) Drag of the subphase water by shearing or lateral diffusion of surfactant molecules may not be the source of surface viscosity for the five highly hydrated sugars of ganglioside should drag more water than ceramide, which has no sugar moiety. Yet, the ceramide film is very viscous; the ganglioside film is not.

(4) "Hydrophobic interactions" (the alleged direct cohesion between the hydrophobic chains of the lipid) may be irrelevant to surface viscosity since cholesterol and dihydrosphingomyelin films are not viscous. Cholesterol, one of the most hydrophobic amphipathic molecules, forms highly incompressible films; this indicates small intermolecular distances. Dihydrosphingomyelin has hydrophobic chains that are as long and saturated as those of DPL and also a smaller cross-section than that of DPL.

(5) Molecular area is important, for proximity favors interaction; however, it cannot be a general criterion. At the same molecular area of 85 A^2, surface viscosity is zero for sphingomyelin and dihydrosphingomyelin and is large for dihydrogalactocerebroside; the molecules of these two lipids are comparable with regard to molecular weight and fatty acid chains. Rather, a combination of small molecular area and specificity of hydration must be needed to produce molecular orientation, intermolecular interaction, and solvent interactions, resulting in surface viscosity.

(6) Surface pressure has an effect insofar as it brings lipid molecules and surface clusters closer. Accordingly, increased film pressures mean greater surface viscosity, but this is not a general conclusion. Rather,

specificity of hydration and clustering is needed to produce viscosity. For example, the highly viscous film of BSA at 21 dynes/cm pressure loses viscosity dramatically but gains pressure as dramatically when DPL is added. Separately, BSA and DPL are highly viscous, but together they lose viscosity.

(7) The major effect therefore must be sought either in (a) the formation of a surface lattice which is generated by hydrophilic lateral interactions and which extends to the entire film area (Figure 1B), or (b) in the specific organization of water around lipid molecules and their clusters such that the resistance to shearing and flow is increased between hydrated surface subunits (Figure 1A) and surface-free water (the solvent). If one breaks the hydrophilic lattice, surface viscosity vanishes —*e.g.*, at 37°C and in the DPL–cholesterol and BSA–DPL systems. Hence the importance of the mosaic structure of the membrane of Figure 1A, B is apparent, and both temperature and lipid–protein association could be nature's devices to make fluid membranes.

Acknowledgments

E. M. Scarpelli is a recipient of a Research Career Development Award from the National Heart and Lung Institute, NIH (5 K03 HL 10613).

Literature Cited

1. Adam, N. K., "The Physics and Chemistry of Surfaces," Oxford University Press, Oxford, 1971.
2. Colacicco, G., in "Biological Horizons in Surface Science," pp. 247–84, Prince, L. M., Sears, D. F., Eds., Academic Press, New York, 1973.
3. Colacicco, G., Hendrickson, R., Joffe, S., *Proc. Nat. Acad. Sci.* (1972) **69,** 1848–50.
4. Colacicco, G., *Ann. N.Y. Acad. Sci.* (1972) **195,** 224–61.
5. Colacicco, G., *J. Colloid Interface Sci.* (1969) **29,** 345–64.
6. Colacicco, G., Scarpelli, E. M., in "Biological Horizons in Surface Science," pp. 368–421, Prince, L. M., Sears, D. F., Eds., Academic Press, New York, 1973.
7. Scarpelli, E. M., in "The Surfactant System of the Lung," Lea and Febiger, Philadelphia, 1968.
8. King, R. J., Klass, D. J., Gikas, E. G., Clements, J. A., *Amer. J. Physiol.* (1973) **224,** 788–95.
9. Reifenrath, R., *Resp. Physiol.* (1973) **19,** 35–46.
10. Kaibara, M., Kikkawa, Y., *Amer. J. Anat.* (1971) **132,** 61–9.
11. Weibel, E. R., Untersee, P., Gil, J., Lulauf, N., *Resp. Physiol.* (1973) **18,** 285–308.
12. Colacicco, G., *Lipids* (1970) **6,** 636–49.
13. Colacicco, G., Buckelew, A. R., Scarpelli, E. M., *J. Colloid Interface Sci.* (1974) **46,** 147–51.
14. Rudy, B., Gitler, C., *Biochim. Biophys. Acta* (1972) **288,** 231–6.
15. Joly, M., in "Progress in Surface Science," Vol. 1, pp. 1–50, Danielli, J. F., Pankhurst, K. G. A., Ridiford, A. C., Eds., Academic Press, New York, 1964.
16. Blank, M., *J. Colloid Interface Sci.* (1969) **29,** 205–9.
17. Shimshick, E. J., McConnell, A. M., *Biochemistry* (1973) **12,** 2351–60.

18. Colacicco, G., Buckelew, A. R., Scarpelli, E. M., *J. Appl. Physiol.* (1973) **34,** 743–99.
19. Condorelli, S., Cosmi, E. V., Scarpelli, E. M., *Amer. J. Obstet. Gynec.* (1974) **118,** 842–8.
20. Colacicco, G., Rapport, M. M., *J. Lipid Res.* (1966) **1,** 258–63.
21. Colacicco, G., *Lipids,* in preparation.
22. Davies, J. T., Rideal, E. K., "Interfacial Phenomena," Academic Press, New York, 1961.
23. Cone, R. A., *Nature (New Biol.)* (1972) **236,** 39–43.
24. Gaines, G. L., "Insoluble Monolayers at Liquid–Gas Interfaces," John Wiley, New York, 1966.
25. Shah, D. O., Schulman, J. H., Advan. Chem. Ser. (1968) **69,** 189–209.
26. Phillips, M. C., "Progress in Surface and Membrane Science," Academic Press, New York, 1972.
27. Deamer, D. W., Cornwell, D. G., *Biochim. Biophys. Acta* (1966) **116,** 555–62.
28. Tien, H. T., Diana, A. L., *Chem. Phys. Lipids* (1968) **2,** 55–101.
29. Jost, P. C., Griffith, O. H., Capaldi, R. A., Vanderkooi, G., *Proc. Nat. Acad. Sci.* (1973) **70,** 480–84.
30. Phillips, M. C., Finer, E. G., *Biochim. Biophys. Acta* (1974) **356,** 199–206.
31. Ladbrooke, B. S., Chapman, D., *Chem. Phys. Lipids* (1969) **3,** 304–67.
32. Trauble, M., Sackman, E., *J. Amer. Chem. Soc.* (1972) **94,** 4499–510.
33. Papahadjopoulos, D., Poste, G., Schaeffer, B. E., *Biochim. Biophys. Acta* (1973) **323,** 23–42.
34. Colacicco, G., Rapport, M. M., Advan. Chem. Ser. (1968) **84,** 157.
35. Joos, P., *Chem. Phys. Lipids* (1970) **4,** 162–8.
36. Blank, M., Britten, J. S., *J. Colloid Interface Sci.* (1965) **20,** 789–800.

Received October 25, 1974. Work supported by the National Heart and Lung Institute, NIH (HL 16137); New York Heart Association; and Health Research Council of the City of New York (U-2091E).

20

Role of the Bovine Serum Albumin Subphase in Relation to Surface Viscosity and Film Transfer

HENRI L. ROSANO, SHU HSIEN CHEN, and JAMES H. WHITTAM

The City College of the City University of New York, New York, N. Y. 10031

The compression or decompression of bovine serum albumin monolayers spread on an aqueous substrate at a pH near the isoelectric point can effect surface tension. The surface pressure changes depend on the distance between the position of the surface pressure measuring device and the compression barrier. This effect is minimal at a pH above or below the isoelectric point and undetected for small molecules (myristic acid and eicosyl sodium sulfate) even when the substrate contains substituted alkyl amines. A theory is proposed which attributes the above observation to surface drag viscosity or the dragging of a substantial amount of substrate with the BSA monolayer. This assertion has been experimentally confirmed by measuring the amount of water dragged per monolayer using the technique of surface distillation.

Many authors (*1, 2, 3*) have compared the surface behavior of macromolecules, especially proteins, with the behavior of low molecular weight monolayers. This paper notes a series of effects that occurred when bovine serum albumin (BSA) was spread on various clean liquid surfaces and was compressed or decompressed. The transfer of the protein monolayer and of some small chain monolayers was also studied using a surface distillation technique.

In general the surface pressures of a monolayer under continuous compression (or decompression) are obtained by measuring the surface tension at one position on the monolayer surface. If the rate of compression (or decompression) is slow enough, it is generally assumed that surface tension is independent of the relative position in the trough. This appears to hold true for most small chain monolayers but not for many protein monolayers. The reasons for this observation are the subject of this paper.

Experimental

The experimental apparatus for measuring surface pressures has been described by Christodoulou and Rosano (*4*). The compression and expansion rates were between 0.009 and 0.03 cm/sec. Isotherms were determined when the system regained equilibrium after short periods of compression.

A second apparatus originally developed by Abribat, Rosano, and Vaillet (*5*) for studying the transfer of film (distillation isotherme superficielle) was used to determine if the subphase was also being dragged with the surface film. A cathetometer was used to measure the change in height of the substrate.

The surface pressures were determined from surface tension measurements which were made by suspending a wettable sand blasted platinum blade from a microforce transducer-amplifier system (model 311 A, The Sanborn Co., Waltham, Mass.). The transducer output was recorded continuously on an x-y recorder (model 370, Keithley Instruments, Inc., Cleveland, Ohio 44139). The surface tensions were reproducible within ±0.1 dyne.

An aqueous protein solution containing 0.5% 1-pentanol was deposited on the aqueous substrate with an Agla micrometer syringe (Burroughs Wellcome Co., Tuckahoe, N. Y.). The substrate and film were retained in an edge-paraffin coated silica trough ($65 \times 14 \times 2$ cm). The surface was cleaned by dusting and aspirating, with a glass tube, calcinated talcum powder from the substrate.

The bovine serum albumin crystals were purchased from the Nutritional Biochemicals Corp., Cleveland, Ohio. A solution was prepared by dissolving about 50 mg protein in a 10-ml solution of 0.5% 1-pentanol (Fisher Scientific Co., Fairlawn, N. J. 07410). The oleic acid and decyl alcohol used in the distillation experiments were purchased from the Hormel Institute, Minnesota. The water for the substrate was distilled from a Stokes still and foamed in a 600-ml medium porosity sintered glass funnel. The foam was removed several times by sweeping the surface to remove surface active impurities.

Results

Compression and Decompression Studies. Bovine serum albumin was spread on an isoelectric substrate (pH 5.3) and initially compressed to a pressure (π) of 5 dynes/cm. The molecular weight of the protein, 66,590 g/mole (pH 5.35 at 26°C and co-surface of 12,000 A^2/molecule) was determined with a surface micromanometer. The moving barrier compressed the monolayer at a constant rate of 1.96×10^{-2} cm/sec for 60 sec and was then stopped while the surface pressure was recorded at various positions in the trough. Figure 1 describes the results of π *vs.* time at various distances from the moving barrier. The surface pressures depend on the distance between the wettable plate and the moving barrier—*i.e.*, the further from the barrier the pressure is recorded, the smaller is the slope ($\partial\pi/\partial T$).

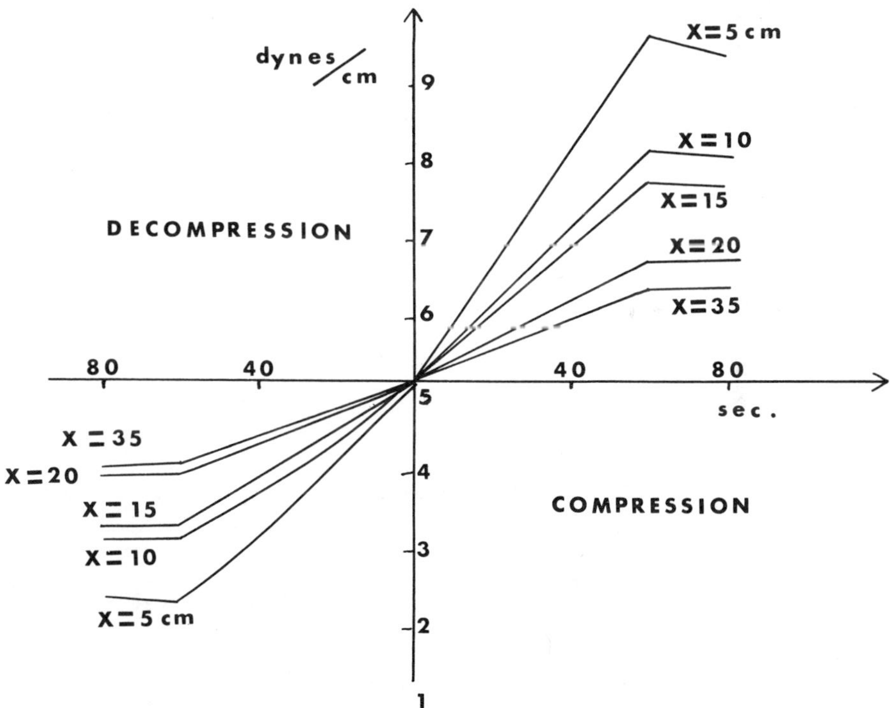

Figure 1. Bovine serum albumin monolayer on water (pH 5.2, 24°C) compressed first at 5.0 dynes/cm. Rate of compression (or compression) 1.5 cm/min.

After the barrier was compressed for 60 sec and then stopped, essentially horizontal lines resulted (Figure 1); thus, the pressure was constant at each position. At small distances from the moving barrier ($x = 5$ and 10 cm) a slight decrease in pressure occurred but shortly leveled out and remained constant during our observations (10–15 min). The system never reached one equilibrium pressure throughout the trough.

A hysteresis phenomenon was observed when the monolayer was expanded after 60 sec of compression. Increasing the rate of compression yields a set of π *vs.* t curves similar to those in Figure 1 but with larger slopes $\partial\pi/\partial T$. The rate of compression had little or no effect on the

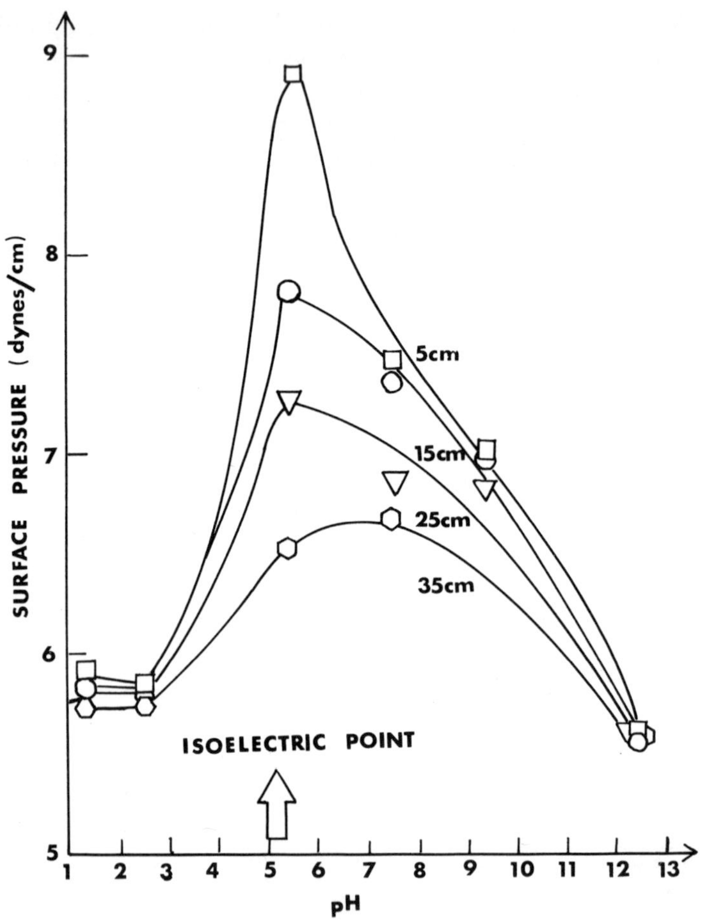

Figure 2. Effect of pH on surface pressure measured at various distances from the compression barrier. Measurements taken after compressing at 1.96 × 10^{-2} cm/sec for 60 sec. Initial pressure 5 dynes/cm.

pressure when measured at a fixed distance providing the monolayer was compressed to the same area.

Figure 2 describes the effect of pH on the surface pressure (π) of bovine serum albumin monolayers at various distances from the moving barrier. Each reading was taken after the barrier was compressed at 1.96×10^{-2} cm/sec for 60 sec. The initial pressure of the system was 5 dynes/cm. The pH was altered with 0.1*N* HCl and 0.1*N* NaOH. Although the pH was varied, the ionic strength was kept constant. When the ionic strength of the aqueous substrate varied from 0.001–0.1*N* (by adding NaCl), the effect was obvious in all cases; the greatest effect appeared at the lowest ionic strength.

When a monolayer of myristic acid spread on 0.01*N* HCl was used over the same pressure range (5 dynes/cm), the change in pressure on compression and expansion was independent of the position of the wettable blade from the moving barrier.

Additional experiments were conducted on eicosyl sodium sulfate monolayers spread over 0.1*N* 2-amino-2-methyl-1-propanol (pH = 9.65, 24.5°C) and 0.1*N* 2-amino-1-butanol (pH 8.5, 24.5°C). In all cases of compression or decompression, the measured surface pressures were independent of the distance between the measuring device and the compression barrier.

Transfer of Monolayer Subphase. The isothermal surface distillation (distillation isotherme superficielle) apparatus is shown in Figure 3. Two troughs of water are connected by a set of glass rods bent at right angles and in parallel contact. Blank and La Mer (*6*) used machined grooved brass rings in their analysis of the transfer process, later Hansen (*7*) described the transfer rate in a rectangular channel. To determine the

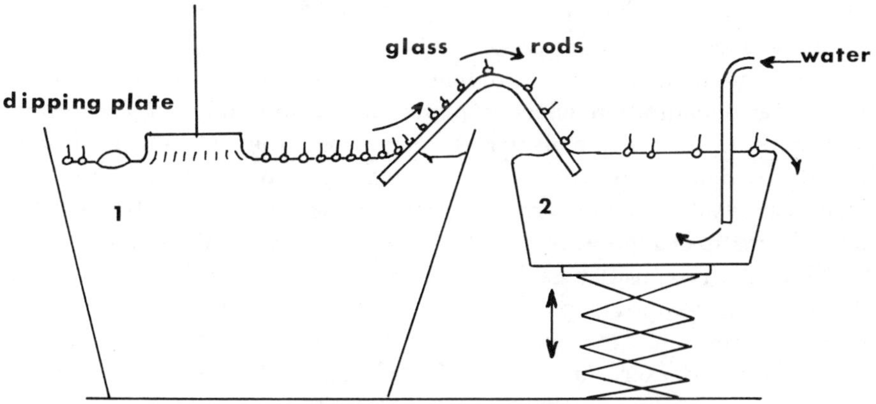

Figure 3. Isothermal surface distillation apparatus

amount of substrate dragged, a monolayer was spread on the substrate in trough 1. For oleic acid, decyl alcohol, and methyl laurate the excess material remained as a lens. For bovine serum albumin, the protein was spread and compressed to approximately 10 dynes/cm. Since trough 2 was continuously flushed with water to clean the surface, a pressure differential existed between the two troughs which caused the monolayer to transfer from the high pressure trough 1 over the glass rods to trough 2 (zero pressure). The surface pressures were determined by the Wilhelmy wettable blade technique previously described. The change in height of the substrate level in trough 1 was measured with a cathetometer which could read to ± 0.02 mm. A correction for the evaporation of substrate was made. Figure 4 describes the changes in surface pressure and substrate level as a function of time for various monolayers. [Four glass rods (od 4 mm) of total length from trough 1 to trough 2 of 6.94 cm were used.]

For oleic acid, decyl alcohol, and methyl laurate the pressure remained constant and equal to the collapse pressure of the monolayer as long as the lens was present. Thereafter the pressure decreased as the monolayer was transferred from trough 1. For the protein monolayer which is somewhat soluble, no plateau region was observed, but the change in surface pressure was gradual and fairly constant.

Interpretation

The surface properties of macromolecules most likely arise from the unfolding of the long chains in the surface region. Flory and Huggins (*8*) developed a theory in 1942 for polymer solution, and Singer (*9*) later applied it to surface films. Since then Frisch and Simha (*10*) and Silberberg (*11*) proposed additional models which account for those portions of the polymer chain that are pushed into the bulk phase, or even lifted off the surface, upon compression.

In proteins this chain unfolding and coiling is known as denaturation. The denaturation effect is most likely responsible for the results in Figure 1. As the monolayer is compressed, the long chains of the protein reorient themselves in the monolayer, thereby effecting the surface viscosity. Since the greatest reorientation is near the barrier, the most dramatic change in the viscosity coefficient is in this region.

For clarity consider each individual protein molecule as an elastic spring and a dashpot in parallel (Kelvin model). As the barrier is compressed a force is developed on the mechanical model. The dashpot compresses and exerts a resisting force—*i.e.*, analogous to the surface viscosity force between the protein and bulk phase. As the piston is compressed the spring will also supply a force—*i.e.*, analogous to the viscosity force developed within the protein molecule as it coils. Finally as the entire "Kelvin solid" is moved, the force will be transferred to the models behind it—*i.e.*, analogous to the viscosity within the monolayer produced by intensity models. With this type of model the pressure will depend on position. This model is also in agreement with the results of Blank (*1*) who observed that BSA monolayers had a plastic component as part of its rheological behavior.

From the above model, it may be assumed that the resistance to the moving barrier is caused by a viscosity effect which is developed in the following ways:

(1) The viscosity within the monolayer itself which can be attributed to the coiling of the molecule and protein–protein interaction.

(2) The viscosity developed between the sheath of water molecules bound to the monolayer and the bulkphase (surface drag viscosity).

Numerous authors studied this phenomena by determining the amount of substrate carried by a moving monolayer—*e.g.,* D. J. Crisp (*12*) and C. Y. Pak and N. L. Gershfeld (*13*). J. H. Schulman and T. Teorell (*14*) estimated that the thickness of the aqueous bulk layer dragged with a moving monolayer of oleic acid is around 0.03 mm. H. L. Rosano (*15*) determined for a 0.07*M* Na_2SO_4/1-butanol/water cell that the diffusion layer at the interface is on the order of 300 microns. With this concept in mind the following mathematical model is proposed to explain if the phenomena is caused by the protein monolayer or the surface drag viscosity effect.

A small and local pressure change is produced in a monolayer by compression at a constant rate. The propagation of the surface pressure is detected at various distances from the point of surface perturbation.

Consider a small element of the monolayer (length M; width ∂x) perpendicular to the ox axis. At the time $t = 0$ the perturbation is produced by the moving barrier. At time t the perturbation has reached the small element of the monolayer $M\partial x$ (where M is also the width of the trough). Let $\partial\pi$ be the difference in surface pressure on each side of the surface element $M\partial x$ and μ the velocity of the perturbation. The forces acting on this small element of surface are: the force caused by the surface pressure,

$$\partial f_1 = \partial\pi M \tag{1}$$

and the force caused by the two-dimensional surface viscosity and the friction between the surface film and the subphase (df_2). These two forces equal the inertial force of the surface element of mass ∂m.

$$\partial f = \partial m \frac{\partial\mu}{\partial t} \tag{2}$$

$t = time.$ Therefore,

$$\partial f_1 - \partial f_2 = \partial f \tag{3}$$

substituting and dividing by $M\partial x$

$$\frac{\partial\pi}{\partial x} - \frac{\partial f_2}{M\partial x} = \frac{\partial m}{M\partial x}\frac{\partial\mu}{\partial t}$$

but $\partial m/M\partial x$ equals the surface density (∂); thus,

$$\frac{\partial \pi}{\partial x} - \frac{\partial f_2}{M\partial x} = \delta \frac{\partial \mu}{\partial t} \tag{4}$$

Let us define $Y = \partial f_2/M\partial x$

$$\frac{\partial \pi}{\partial x} - Y = \delta \frac{\partial \mu}{\partial t} \tag{5}$$

Our first objective is to determine Y. Later we analyze our results to define cases when one of the two viscosity factors is predominant. Equation 4 becomes

$$\frac{\partial \pi}{\partial x} - Y = \delta \frac{\partial \mu}{\partial t} \tag{6}$$

From the continuity equation it can be shown that

$$\frac{\partial \delta}{\partial t} + \frac{\partial(\delta \mu)}{\partial x} = 0$$

$$\frac{\partial \delta}{\partial t} + \frac{\partial \delta}{\partial x}\mu + \delta \frac{\partial \mu}{\partial x} = 0$$

for a small perturbation the surface film density remains practically constant and

$$\frac{\partial \delta}{\partial t} + \delta \frac{\partial \mu}{\partial x} = 0$$

or

$$\frac{\partial \ln \delta}{\partial t} = - \frac{\partial \mu}{\partial x} \tag{7}$$

Gibbs (*16*) noted an elasticity associated with a liquid film if the surface tension varies with the area of the surface; for a thin liquid film of area s, the Gibbs elasticity is given by

$$E = \sigma \frac{\partial \gamma}{\partial \sigma}$$

where E is the film elasticity, σ the molecular area, γ the surface tension, and δ the surface density. Since

$$\sigma = 1/\delta$$

$$\partial\sigma = \frac{1}{\delta^2}\,\partial\delta,$$

$$E = -\frac{1}{\delta}\,\delta^2\,\frac{\partial\gamma}{\partial\delta} = -\,\delta\,\frac{\partial\gamma}{\partial\delta}$$

but

$$\partial\gamma = -\,\partial\pi$$

Therefore

$$E = \partial\pi/\partial\ln\delta$$

By introducing Gibb's elasticity coefficient E, which equals $\partial\pi/\partial\ln\delta$ in Equation 7, one obtains

$$\frac{1}{E}\,\frac{\partial\pi}{\partial t} = -\,\frac{\partial\mu}{\partial x}$$

Therefore

$$\Delta\mu = -\int_{x_0}^{x_1} \frac{1}{E}\,\frac{\partial\pi}{\partial t}\,\partial x$$

Experimental Determination of Y (Table I). The elasticity coefficient (E) and the surface density (δ) are determined directly from the compression isotherm (π *vs.* σ). The monolayer is compressed from a given surface pressure with the position of the wettable blade at distance X from the compression barrier—*i.e.*, compressed from 5.0–5.5 dynes/cm.

Table I. Bovine Serum Albumin on Water (pH 5.2, 24.0°C). Initial Pressure, 9.3 dynes/cm

X (*cm*)	$\frac{\partial\pi}{\partial x}$ $\frac{dyne}{cm^2}$	$\delta\frac{\partial\mu}{\partial t}$ $\frac{dyne}{cm^2}$	Y $\frac{dyne}{cm^2}$	$\eta_{\varepsilon=50A}$ (*cps*)	$\eta_{\varepsilon=5\times10^{-2}\,cm}$ (*cps*)
6.4	-6.16×10^{-2}	10^{-11}	6.16×10^{-2}	5.41×10^{-4}	54.1
10.7	-3.81×10^{-2}	10^{-11}	6.16×10^{-2}	3.44×10^{-4}	34.4
18.5	-7.84×10^{-3}	10^{-11}	7.84×10^{-3}	8.83×10^{-4}	8.83
27.7	-4.45×10^{-3}	10^{-11}	4.45×10^{-3}	7.07×10^{-5}	7.07

Graphs of π *vs.* time are plotted for various positions of the blade (distance X) compressing over the same area. Figure 1 represents a family of curves for bovine serum albumin. Curve fitting of these results leads to

$$\pi(X,t) = (.264) + X^{(-.152)}$$

which can be used to solve for

$$\frac{\partial \pi}{\partial t}, \frac{\partial \pi}{\partial x} \text{ and } \frac{\partial \mu}{\partial t}$$

Solving Equation 6 for the frictional shear gives:

$$Y = -.0401 + X^{(-1.152)} - \frac{\delta(0.264)}{E} \frac{\partial x}{\partial t} X^{-.152}$$

If it is assumed that the local velocity gradient decreases linearly perpendicular to the interface (Newtonian profile at each point X), we can compare the average three-dimensional viscosity for a film thickness of $E = 50$ A and $E = 5 \times 10^6$ A. In the former case we assume that the monolayer and oriented substrate are only 50 A thick; the latter number is based on a much larger oriented subphase (*9, 15*). In the first case the calculated viscosity varies between 7.07×10^{-5} and 5.41×10^{-4} cps; in the second case it varies from 54 to 8 cps (with increasing distance from the barrier). From the calculation of viscosity, the assumption that $E = 5 \times 10^6$ A is more reasonable.

Discussion

In light of the experiment on the compression and expansion of the monolayer and the calculation on the depth of the dragged substrate, we conclude that not only is the coiling of the protein monolayer important in the observed effect but also the structure of the sheath of bound water below the monolayer (subphase) and the extent to which it is bound to the monolayer effect the surface viscosity. Thus we advance the following hypothesis of a non-structured and structured subphase.

For example, in the case of myristic acid and eicosyl sodium sulfate spread on aqueous substrates and substrates containing alkyl amines, the subphase is oriented because of dipole–dipole interactions but is still in the liquid state. This orientation has been accounted for in many monolayer studies concerned with surface potential (*17, 18, 19*). Upon compression, the substrate molecules, although oriented, can slip or be squeezed into the bulk; thus the transmission of surface pressure will be instantaneous and independent of the position of the device for measuring the surface pressure. In the case of the protein monolayers the subphase is bound strongly to the protein (*20*) and produces a "pseudo gel" phase as described by Colacicco *et al.* (*22*). This helps explain our observed phenomenon since on compression the gel will deform and drag along with the monolayer. This idea of a structured subphase explains why protein solutions, although not necessarily good foaming agents, are usually good foam stabilizers. Blank and Lee (*21*) also reported surface tension gradients when studying films of lung extract.

In addition the decrease in the effect with bovine serum albumin above and below the isoelectric point indicates that as the protein becomes ionized, the gel subphase breaks down and becomes similar to that of the myristic acid subphase.

To further substantiate our theory on the role of the subphase structure and surface viscosity, the results of the distillation isotherme superficelle technique must be examined (Figure 3). For oleic acid, decyl alcohol, and methyl laurate more than one monolayer was used (since a lens formed on the surface) to drag the substrate from trough 1 to trough 2. The thickness of the substrate dragged by one monolayer can be determined readily since the total number of molecules placed on the surface and the area per molecule at the collapse pressure are known. Considering this correction, the thickness of substrate-dragged per monolayer is: oleic acid $= 3.28 \times 10^{-3}$ cm, decyl alcohol $= 2.15 \times 10^{-4}$ cm, methyl laurate $= 1.28 \times 10^{-3}$ cm. For the protein the result is

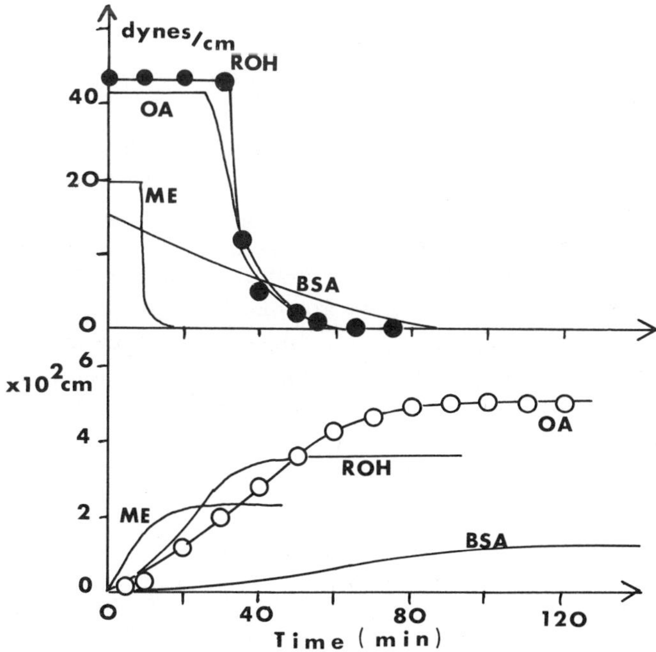

Figure 4. Surface pressure and change in substrate height vs. time using the distillation isotherme superficielle technique. Trough area—201 cm². Bridge—four glass rods, od, = 4 mm, length = 6.94 cm. OA (oleic acid) spread on 0.01N HCl, pH = 1.9, 25°C; COH (decyl alcohol) spread on water, pH 6.45, 22°C; ME (methyl laurate) spread on water, pH 6.45, 22°C; BSA (bovine serum albumin) spread on water, pH 6.45, 22°C.

taken directly from Figure 4: 1.25×10^{-2} cm since only one monolayer was formed and the operating pressure was low enough to prevent desorption. Thus, it is obvious that the protein drags a considerable amount of substrate compared with the smaller molecules studied although each monolayer studied will drag the subphase.

Conclusion

A phenomenon has been observed on only protein monolayers. We hypothesize that not only is the protein monolayer responsible for the effect but the subphase structure also plays an important role.

Film transfer experiments demonstrate that there is always a sheath of subphase associated with the monolayer. However when the monolayer is compressed or decompressed in an enclosed area, there will be a certain degree of slippage at the monolayer/subphase interface. When the subphase water molecules are strongly bound to the monolayer, as for a protein monolayer spread on an aqueous substrate at its isoelectric point, little slippage exists, and the phenomenon in Figure 1 is observed. For monolayer systems where there is substantial slippage at the monolayer/subphase interface the effect will not be observed during compression or decompression experiments. This effect helps explain why proteins are used with other surfactants to stabilize emulsion and foam formulations (*1, 2*).

Nevertheless, no matter how one interprets the theory, the implications of this experiment are far more important and should be understood and used for future work on macromolecular monolayer systems. We urge the use of an additional variable—the distance that the surface pressure measuring device is from the point of perturbation; this must be included in future work to ensure accuracy and reproducibility of monolayer studies.

Literature Cited

1. Blank, M., *J. Colloid Interface Sci.* (1964) 29-32.
2. Joly, M., *Recent Progr. Surface Sci.* (1963) **18,** 555.
3. Tachibana, T., Inokuchi, K., *J. Colloid Interface Sci.* (1953) **8,** 341.
4. Christodoulou, A. P., Rosano, H. L., ADVAN. CHEM. SER. (1968) **84.**
5. Abribat, M., Rosano, H. L., Vaillet, G., *C. R. Acad. Sci., Paris* (1954) **238,** 1219.
6. La Mer, V. K., Blank, M., *J. Colloid Sci.* (1956) **11,** 608.
7. Hansen, R. S., *J. Phys. Chem.* (1959) **63,** 637.
8. Flory, P. J., "Principles of Polymer Chemistry," Cornell University Press, 1953.
9. Singer, S. J., *J. Chem. Phys.* (1948) **16,** 872.
10. Frisch, H. L., Simha, R., *J. Chem. Phys.* (1957) **27,** 702.
11. Silberberg, A., *J. Phys. Chem.* (1962) **66,** 1872.
12. Crisp, D. J., *Trans. Faraday Soc.* (1946) **42,** 619.

13. Pak, C. Y., Gershfeld, N. L., *Nature* (1967) **214,** 888.
14. Schulman, J. H., Teorell, T., *Trans. Faraday Soc.* (1938) **34,** 1337.
15. Rosano, H. L., *J. Colloid Interface Sci.* (1967) **23,** 73.
16. Gibbs, J. W., "Collected Works," p. 301.
17. Gaines, G. L., "Interscience Monographs on Physical Chemistry," Chap. 4, Interscience, 1966.
18. Davies, J. T., Rideal, E. K., "Interfacial Phenomena," Chap. 5, Academic, 1963.
19. Adamson, A. W., "Physical Chemistry of Surfaces," 2nd ed., pp. 123-136, Interscience, 1967.
20. Karel, M., *J. Food Sci.* (1973) **38,** 756.
21. Blank, M., Lee, B., *J. Colloid Interface Sci.* (1971) **36,** 1.
22. Colacicco, G., Buckelew, Jr., A. R., Scarpelli, E. M., *J. Colloid Interface Sci.* (1974) **46,** (1).
23. Adam, N. K., "The Physics and Chemistry of Interfaces," p. 148, Oxford University Press, London, 1944.
24. Becker, P., *Amer. Perfumer* (1962) 7721.

RECEIVED October 17, 1974.

21

Dynamic Surface Measurements as a Tool to Obtain Equation-of-State Data for Soluble Monolayers

E. H. LUCASSEN-REYNDERS

6 Kingsway, Heswall, Merseyside, England

J. LUCASSEN, P. R. GARRETT, D. GILES, and F. HOLLWAY

Unilever Research Port Sunlight Laboratory, Wirral, Merseyside, England

A dynamic technique is described for obtaining surface elasticity (ϵ_0) vs. *surface pressure (π) curves which can be transformed into accurate π–A curves for soluble monolayers. Small amplitude periodic area variations are used with a sufficiently high frequency to make monolayers effectively insoluble in the time of the experiment even though they behave as soluble in equilibrium measurements. ϵ_0–π plots are given for some nonionic surfactants. Straight line portions in these plots illustrate that surface interactions are too complex to be described by a Frumkin isotherm. In the limit of very low surface pressures there is no trace of an "ideal gaseous" region. Some examples show the implications of particular ϵ_0–π curves for equilibrium and dynamic surface behavior.*

Interactions at surfaces have long been at the center of interest in the study of surfactant monolayers and have been thought to influence both static and dynamic surface properties considerably (*1, 2*). Although the theoretical interpretation and even the definition of surface interactions may be controversial, the experimental method has not been in doubt. Invariably, the equilibrium surface pressure *vs.* molar area relationship has been used as a criterion for assessing interactions in monolayers since interactions, no matter what their precise definition, must appear in the measurable quantity of surface tension (γ) or surface pressure ($\pi = \gamma^\circ - \gamma$) at a given surface concentration (Γ) or molar area ($\bar{A} = 1/\Gamma$). This classical way to assess surface interactions is very suitable for insoluble monolayers where the molar area is a directly measurable variable. For soluble monolayers, however, where the adsorption can only be arrived at indirectly through Gibbs' adsorption theorem, the static method cannot be applied with much accuracy.

In this paper we propose a dynamic method for tackling the problem of interactions in soluble monolayers. The surface dilational modulus, ϵ, is a dynamic surface quantity very sensitive to interactions. It is defined as the decrease in local surface pressure per unit relative increase in surface area A in an oscillatory experiment:

$$\varepsilon = \frac{-d\pi}{d \ln A} \tag{1}$$

For the simple case of insoluble monolayer behavior the total number of surfactant molecules at the surface is constant, and the total surface area A is directly proportional to the molar area $\bar{A}$; in this case the dilational modulus, ϵ, is simply related to the slope of the π–$\bar{A}$ curve. The dilational modulus, therefore, can be expected to be even more sensitive to anything that affects the surface pressure at given $\bar{A}$ than the surface pressure itself is. In fact the greater sensitivity of ϵ to surface interactions has been demonstrated by a number of numerical examples for a given simple concept of surface non-ideality (*3*). A more important advantage, however, of the dilational modulus as an interaction criterion is that it can be measured accurately for many soluble monolayers. By adjusting the frequency of the oscillatory experiment, many soluble monolayers can be made to behave effectively as though they were insoluble since for each bulk surfactant concentration there exists a frequency above which surface–bulk diffusional interchange is negligible (*4*). At some higher frequency range one must expect a region corresponding to the time scale of molecular reorientation processes within the monolayer. This frequency range is usually well separated from that of surface–bulk interchange (*5*)—a conclusion which can be checked for any specific system by experimentally determining the frequency dependence of the modulus. In the intermediate range of high (but not too high) frequencies the modulus is frequency independent, and the surface pressure is in equilibrium with the local value of the adsorption. At such frequencies the product of Γ and A is constant during the experiment, and the modulus, ϵ, approaches a limiting value, ϵ_0, given by:

$$\varepsilon_0 = + \frac{d\pi}{d \ln \Gamma} \tag{2}$$

Measurement of the high frequency modulus, ϵ_0, as a function of the equilibrium surface pressure, π, should provide a sensitive criterion for interaction for monolayers that are quite soluble by normal standards, which involve much longer time spans than the inverse frequency of the compression/expansion experiment. A numerical example of the greater sensitivity of an ϵ_0 *vs.* π plot, compared with that of the π *vs.* log c relationship is shown in Figure 1 for a hypothetical case. The specific definition of surface interactions used here to arrive at numerical values includes all mechanisms that produce deviations from Szyszkowski-Langmuir adsorption behavior. Ideal behavior, with zero surface interactions, then is represented by zero values of $\ln f_1{}^s$ in the equation of state:

$$\frac{\pi}{RT\Gamma^\infty} = -\ln\left(1 - \frac{\Gamma}{\Gamma^\infty}\right) - \ln f_1{}^s \tag{3}$$

where Γ^∞ is the saturation value of the surfactant adsorption, and $f_1{}^s$ is the solvent's interfacial activity coefficient. Thermodynamically, this activity coefficient is defined (*3*) from the equations for matter in a field of force rather than from equations implying that the monolayer is an autonomous region. Ideal surface behavior ($f_1{}^s = 1$ for all Γ values) according to this definition results in ϵ_0 *vs.* π and π *vs.* log c relationships represented by Curves A in Figure 1. Curves B and C show the effect of increasing deviations from ideality as expressed in the simple interaction model that leads to Frumkin adsorption:

$$\ln f_1{}^s = H^s \left(\frac{\Gamma}{\Gamma^\infty}\right)^2 \tag{4}$$

where H^s is the partial molar free energy of interaction at infinite dilution. Both sets of curves were computed by obtaining ϵ_0 and π at given values of Γ/Γ^∞ from Equations 2–4, while the log c values corresponding to these Γ/Γ^∞ values followed from the Frumkin isotherm equation:

$$\ln c/a = \ln \frac{\Gamma/\Gamma^\infty}{1 - \Gamma/\Gamma^\infty} - \frac{2H^s}{RT} \cdot \frac{\Gamma}{\Gamma^\infty} \tag{5}$$

The numerical value of the bulk–surface distribution coefficient a in this expression does not influence the shape of the π *vs.* log c curve. It is obvious that the shape of the ϵ_0 *vs.* π curve is affected by interactions to a far greater extent than is the shape of the π *vs.* log c curve. Also, the position of the ϵ_0 *vs.* π curve is fixed, but that of the π *vs.* log c curve floats, depending on the bulk–surface distribution coefficient, which has no direct relationship with surface interactions. In fact, the three π–log c curves in Figure 1 have been anchored to the same concentration in the high surface pressure region to illustrate the almost identical shape in this region, where measurements usually are the most accurate. Thus, qualitative assessment of interactions in soluble monolayers requires accurate determination of the entire π–log c curve, where in principle a single point on the ϵ_0 *vs.* π curve suffices. The argument which has been presented here for interactions in single surfactant monolayers can be extended to cover mixed monolayers (*3*) since interactions between surfactants are also reflected in ln f_1^s in Equation 3.

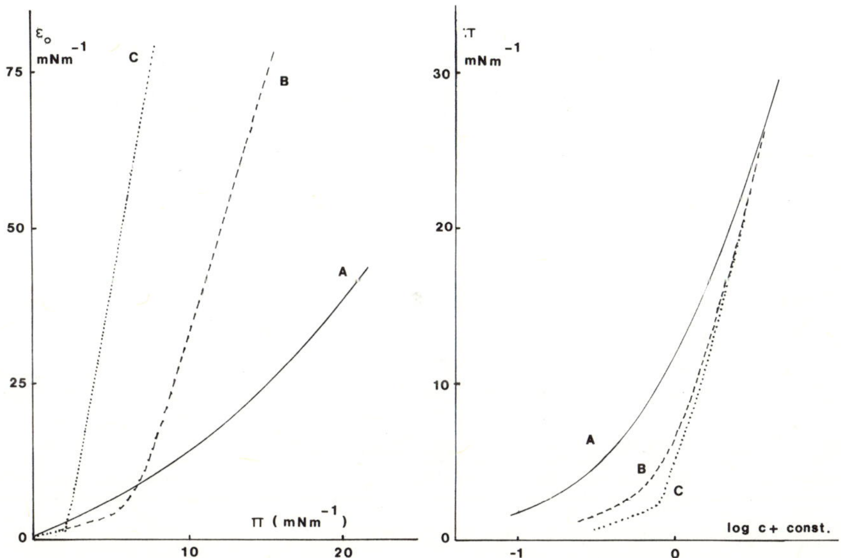

Figure 1. Effect of interactions on ε_0 vs. π *and* π vs. *log* c *curves. Quadratic non-ideality term: A:* $H^s = 0$; *B:* $H^s = 1.55$ RT; *C:* $H^s = 2.175$ RT; $RT\Gamma^\infty = 17.15$.

Measured ϵ_0 *vs.* π curves are given below for some single surfactant systems, and the results are interpreted tentatively in terms of surface interactions. Some implications of specific ϵ_0 *vs.* π curves for interpreting equilibrium and non-equilibrium surface behavior are also discussed.

Experimental

The techniques used to measure dilational surface properties have been described in detail (*4, 6*). Surfaces of surfactant solutions are subjected to small amplitude, sinusoidal area variations, and the dilational modulus is given by the ratio between surface tension change measured and fractional area change applied. The modulus generally depends on the frequency at which the experiment is carried out, and this frequency dependence appears generally to be determined by diffusional exchange between bulk and surface (*4, 6*). To obtain the required value of ϵ_0—the limiting value of the modulus at high frequency—it is necessary either to extrapolate to infinite frequency or to carry out the experiment at such high frequency that diffusion transport of surfactant between surface and bulk is negligible.

For the extrapolation technique, the dilational modulus must be measured over a wide enough range in the frequency spectrum. Figure 2 shows examples of such experiments. It can be shown (*4*) that if ϵ is plotted against frequency on a bilogarithmical basis, all solutions containing one single surfactant should give curves of the same shape if the only relaxation mechanism is diffusion. A simple matching procedure then provides one extrapolated ϵ_0 value from each series of ϵ *vs.* frequency experiments. In its present stage, this extrapolation technique, which

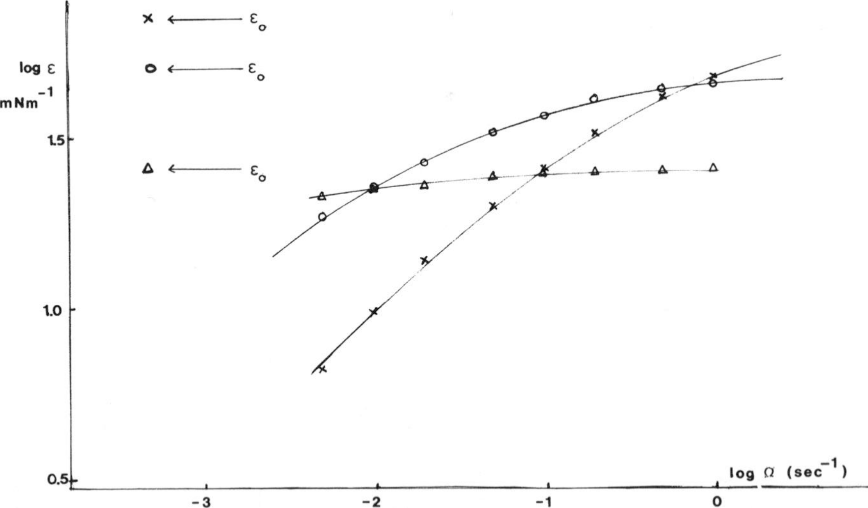

Figure 2. Example of surface dilational modulus vs. *frequency curves. Dodecyl triethylene glycol ($C_{12}E_3$) solutions. △ = 10^{-9} mole/cm³; ○ = 5 × 10^{-9} mole/cm³; X = 2 × 10^{-8} mole/cm³. Abscissa: log ω (ω is angular frequency in Hertz). Ordinate: $\log_{10}\epsilon$ (mNm⁻¹). Drawn lines: theory below CMC.*

provides a single point in an ϵ_0 *vs.* π plot from a series of modulus measurements, cannot be applied to surfactant mixtures. In that case (*7*) the frequency dependence of the dilational modulus is so complex that no simple recipe for extrapolation can be given. Soluble surfactant mixtures thus can only be tackled by measuring within the high frequency range where both surfactants behave as though they were insoluble.

In a second technique, a freshly swept surface of a dilute surfactant solution is subjected to periodic compression–expansion cycles. With the right choice of surfactant and concentration, a rapid periodic surface tension change will be superimposed on a slow transient surface tension change owing to adsorption onto the swept surface. Figure 3 gives an example of such a "tensio-elastogram." Experiments of this kind offer a change of surface pressure with time as well as a change of dilational modulus with time. From these data ϵ *vs.* π plots can be constructed. Moreover, if care is taken to ensure that the inverse frequency of the oscillatory disturbance is small compared with the characteristic time of the adsorption process, ϵ can be equated to ϵ_0.

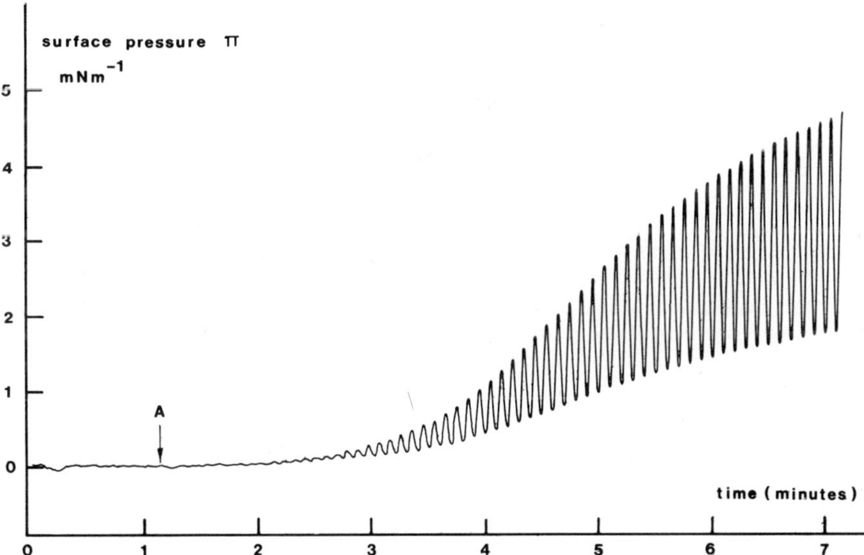

Figure 3. Example of surface tensio-elastogram, giving change of surface tension and dilational modulus with time (schematic). A: start of periodic surface deformation. Abscissa: time after sweeping of surface (minutes). Ordinate: surface pressure (mNm^{-1}).

Interpretation of the ϵ_0 *vs.* π curves obtained from tensio-elastograms is, at present, only feasible for systems containing a single surfactant. For multicomponent systems, it will be difficult to obtain the composition and the composition changes of the surface during the adsorption process.

The ϵ_0 *vs.* π curves can be converted into π *vs.* $\bar{A}$ or π *vs.* Γ curves by integration starting at high adsorption:

$$\ln \Gamma_2/\Gamma_1 = \int_{\pi_1}^{\pi_2} \frac{1}{\varepsilon_0} \, d\pi \tag{6}$$

and finding one absolute Γ value from the π–log c curve since ϵ_0 only gives information on relative adsorption changes.

Materials

Samples of the nonionic and zwitterionic surfactants used were prepared in this laboratory and had a purity better than 99.5%. Sodium dodecyl sulfate was a sample prepared and purified in this laboratory. No minimum in the surface tension was found near the CMC. The dodecyl trimethylammonium bromide was a sample from Palmer Research Co. Ltd., showing a minimum of about 1 mNm^{-1} at the CMC. Water distilled from alkaline permanganate was used to prepare all solutions. All experiments were done at 25°C where water had a surface tension of 72.25 mNm^{-1}.

Results and Discussion

Figures 4 to 6 give experimental ϵ_0 *vs.* π curves for different single surfactant systems. The 1:1 cationic–anionic mixture effectively behaves as a single surfactant, DTA–DS (*8*). The π vs. $\bar{A}$ curve obtained by

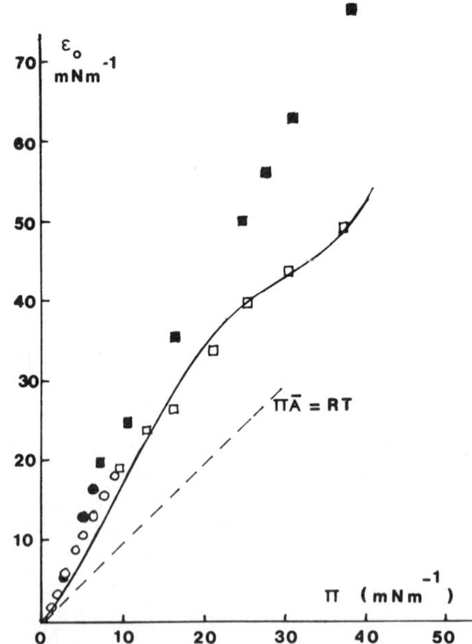

Figure 4. Dependence of extrapolated dilational modulus, ε_0, on surface pressure π for dodecyl triethylene glycol (■ and ●) for dodecyl hexaethylene glycol (□ and ○). Dashed line: $\pi\bar{A}$ = RT. Drawn line: predicted curve for RT ln $f_1^s = 1.77\ \theta^2 - 8.96\ \theta^3 + 9.48\ \theta^4$. $\theta = \Gamma/\Gamma^\infty$; $RT\Gamma^\infty = 10.5\ mNm^{-1}$. Abscissa: surface pressure, mNm^{-1}. Ordinate: ε_0, mNm^{-1}.

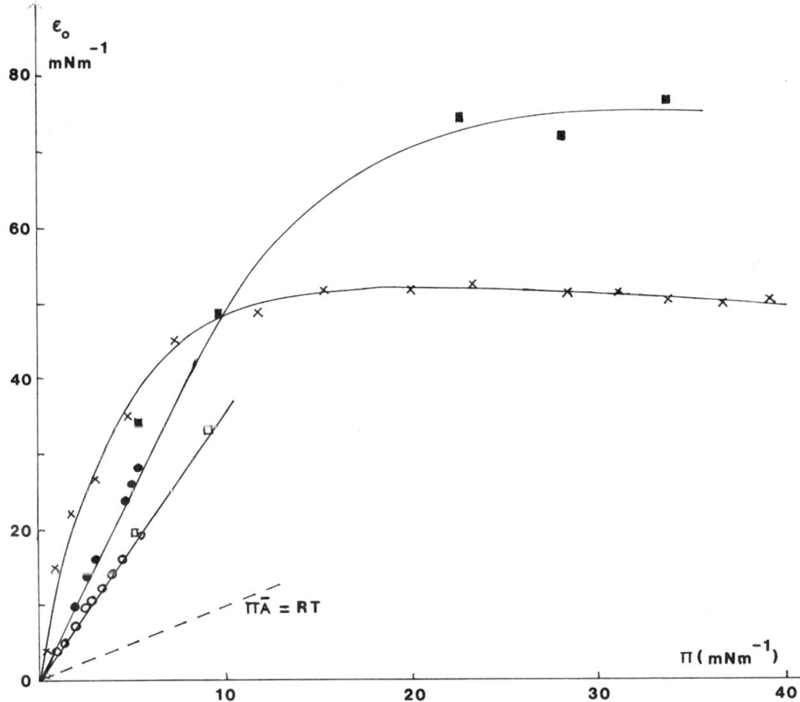

Figure 5. Dependence of extrapolated dilational modulus, ε_0, on surface pressure π for tetradecyl dimethylammoniopropane sulfonate (TDPS) (□ and ○) and hexadecyl dimethylammoniopropane sulfonate (HDPS) (■ and ●). X: C_{20} analog [from data of Goddard and Kung (9)].

integrating the ϵ_0 *vs.* π curve with Equation 6 is given for this compound in Figure 7. The squares always refer to ϵ_0 values obtained by extrapolation, and the circles represent tensio-elastogram data. In Figure 5, ϵ_0–π data for the C_{20} sulfobetaine, obtained from the π–A curve published by Goddard and Kung (9), are given as well. A regular sequence is apparent only in the low pressure region of the C_{14}–C_{16} and C_{20} data.

A striking aspect of the ϵ_0–π curves shown is the total apparent absence of a low pressure region with unit slope in the limit of zero surface pressure where Equation 3, like all other surface equation of state formulations, reduces to $\pi\bar{A} = RT$. Qualitative comparison of the experimental ϵ_0–π curves in Figures 4–6 with the theoretical ones in Figure 1, reveals that a single quadratic non-ideality term in Equation 3 is unable to obscure the $\pi\bar{A} = RT$ region to the extent shown by the experimental results. Introduction of higher order terms into Equation 3 appears to be a means to produce ϵ_0 *vs.* π curves in which the limiting $\epsilon_0 = \pi$ region is obscured by higher order terms even at very low surface pressures. Figure

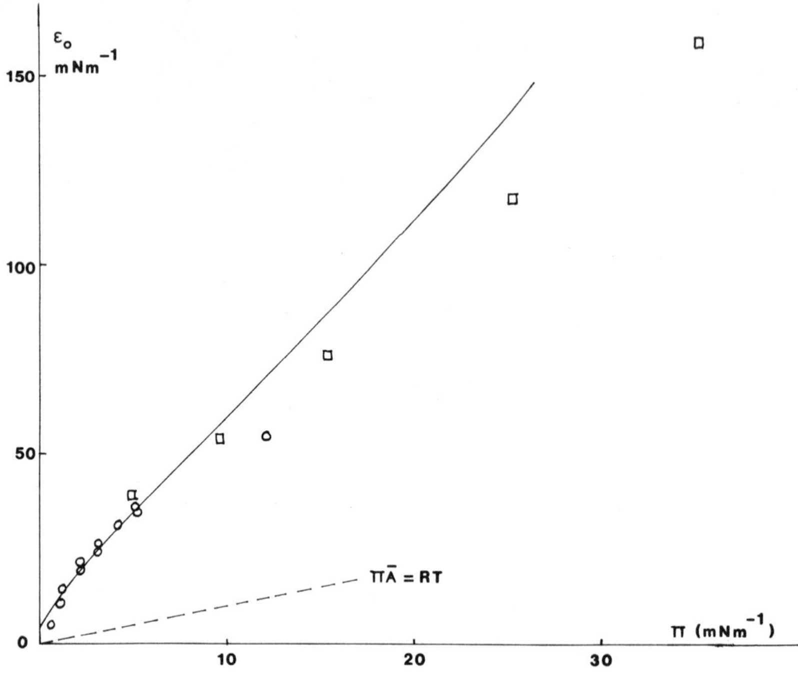

Figure 6. Dependence of extrapolated dilational modulus, ε_0, on surface pressure π for equimolar mixture of sodium dodecyl sulfate and dodecyl trimethylammonium bromide. Drawn line: predicted curve for RT *ln* $f_1^s = 5.80\ \theta^2 - 6.17\ \theta^3$. ($\theta = \Gamma/\Gamma^\infty$) $RT\Gamma^\infty = 16.9$ *(per chain).*

6 shows an example of the effect of a cubic term in Equation 3, which gives fair agreement for surface pressures up to 6 mNm^{-1}, for the cationic–anionic mixture, but it gives too high ε_0 values at higher surface pressures, where apparently terms with still higher order become important.

By using a series expansion for the $\ln f_1^s$ term in Equation 3, it is always possible, in principle, to curve-fit any experimental data in the same way that non-ideality in bulk mixtures can be expressed in a series expansion. For surface mixtures, however, the problem is aggravated by the inevitable arbitrariness in defining the contents of a surface phase. This problem is not restricted to any particular model used to describe surface behavior. In the gas model approach the numerical value of the limiting area A_0 becomes a critical factor in the low area region while in the surface solution model used here it is not obvious which value should be used for Γ^∞ (the limiting surfactant adsorption) since the extrapolation beyond the measured Γ range depends on the analytical form of the equation of state. Inclusion of Γ^∞ as an adjustable parameter, together with the coefficients in the series development for $\ln f_1^s$, would seem to provide a solution, but convergence in the minimization procedure then often turns out to be poor.

An example of a curve-fitted equation of state is shown in Figure 4 for the nonionic $C_{12}E_6$ while Figure 8 shows the fit for the surface pressure–log concentration curve for the same compound. A second- and third-order negative ("repulsion") term and a fourth-order positive ("attraction") term in the series expansion for $\ln f_1^s$ provide a good description of the experiments.

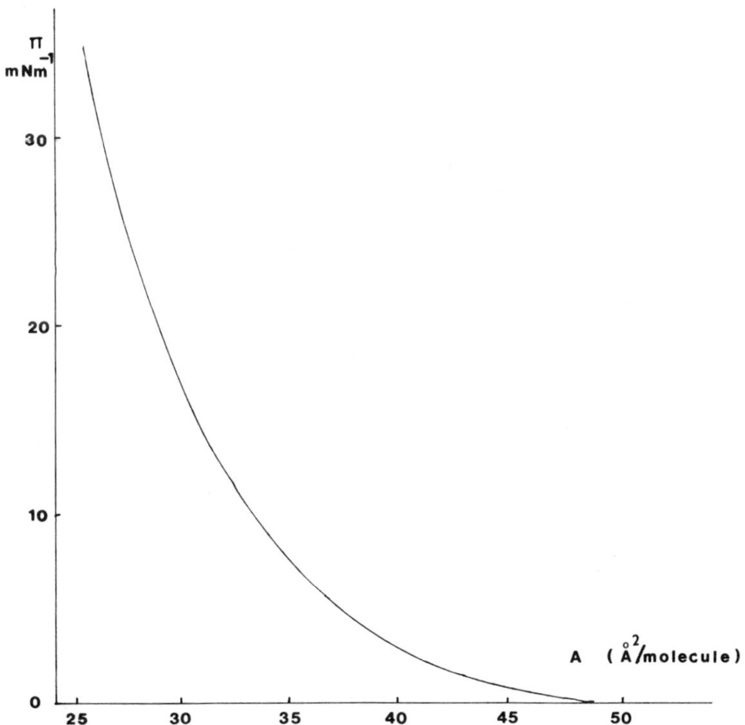

Figure 7. π vs. $\bar{A}$ curve, obtained by integrating ε_0 vs. π curve in Figure 6, anchored to surface coverage obtained from π vs. log c in saturation region. Abscissa: area per aliphatic chain (A^2/molecule). Ordinate: surface pressure in mNm^{-1}.

Not too much physical significance should be attached to the numerical values of the parameters used to describe $\ln f_1^s$ in view of the above mentioned arbitrariness. What can be concluded with certainty, however, is that the description of surface non-ideality in all systems investigated requires at least two or three terms in a series expansion.

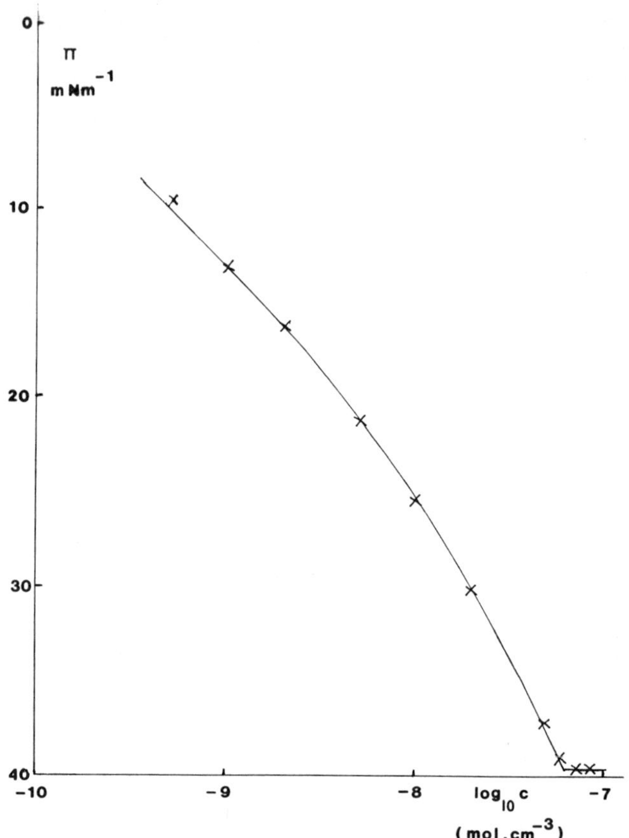

Figure 8. Surface pressure vs. *log concentration curve for dodecyl hexaethylene glycol. Drawn line: predicted curve with parameters given for Figure 4 and with* a = *7.5* × *10*$^{-11}$ *mole/cm*^{3} *and* RTΓ^∞ = *10.5 mNm*$^{-1}$.

The information on surface behavior obtained from this dynamic method could not be obtained from the equilibrium π–log c curve, where the effect of most interaction mechanisms tends to be swamped in the very steep, nearly linear ascent of π at near saturation.

This aspect can be illustrated analytically if, as for sulfobetaines, non-ionics [and also for certain protein solutions (*10*)], the initial part of the ϵ_0 *vs.* π curve can be described, albeit roughly, by:

$$\varepsilon_0 \left(= \frac{d\pi}{d \ln \Gamma}\right) = n\,\pi \tag{7}$$

where n is larger than 1. Combination of this expression with Gibbs' adsorption law:

$$d\pi = RT\ \Gamma\ d\ \ln\ c \tag{8}$$

and integration yields:

$$RT\ \ln\ c = \frac{n}{n\text{-}1}\ k^{\ (1/n)}\ \pi^{\ n\text{-}1/n} + k_2 \tag{9}$$

When n is much larger than unity, as for hexadecylsulfobetaine where it is about 5, this equation predicts a practically linear relationship between π and ln c.

The accurate information about the surface tension–surface coverage relationship which for soluble monolayers is contained in the ϵ_0–π curves, can be used, for example, to interpret rate of adsorption measurements. For fluid–fluid interfaces, adsorption onto an initially "clean" surface can be assessed only indirectly by measuring the changing interfacial tension. An example is shown in Figure 9, giving the change in surface pressure

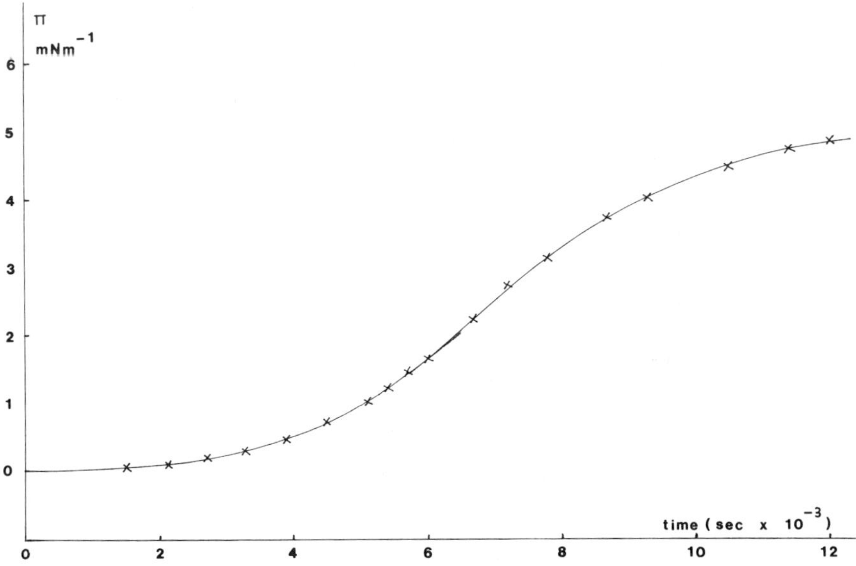

Figure 9. Surface pressure vs. *time for a 5 × 10^{-10} mole/cm^3 solution of HDPS after sweeping of surface. Abscissa: time in seconds. Ordinate: surface pressure in mNm^{-1}.*

with time for a dilute solution of the hexadecylsulfobetaine following sweeping of the surface. These π *vs.* t data show an induction period, during which the surface pressure stays close to zero. Induction periods of this type which have been observed for various surface-active substances, might at first suggest the presence of an adsorption barrier (*11*) or a slow rearrangement in the surface.

The ϵ_0–π curve for the same compound enables us to check the validity of such interpretations by transforming the π *vs.* t curve into an adsorption *vs.* time relationship. For the case under consideration, the Γ *vs.* $\sqrt{t}$ curve is given in Figure 10. Here the induction period has completely disappeared, and the linearity of the initial part of the Γ vs. $\sqrt{t}$ plot is consistent with a diffusion-controlled mechanism, rather than with the presence of an adsorption barrier. From the Ward-Tordai (*12*) equation for the initial part of the adsorption:

$$\Gamma \approx 2c \sqrt{\frac{Dt}{\pi}} \tag{10}$$

onto a clean surface, a diffusion coefficient $D = 1.2 \times 10^{-5}$ cm^2/sec is found, which is of the right order of magnitude.

This example emphasizes the danger of using the ideal Langmuir-Szyszkowski equation of state in converting surface tension–time data into adsorption–time data even for very dilute monolayers. Also it clearly shows that any conclusion about the existence or non-existence of an

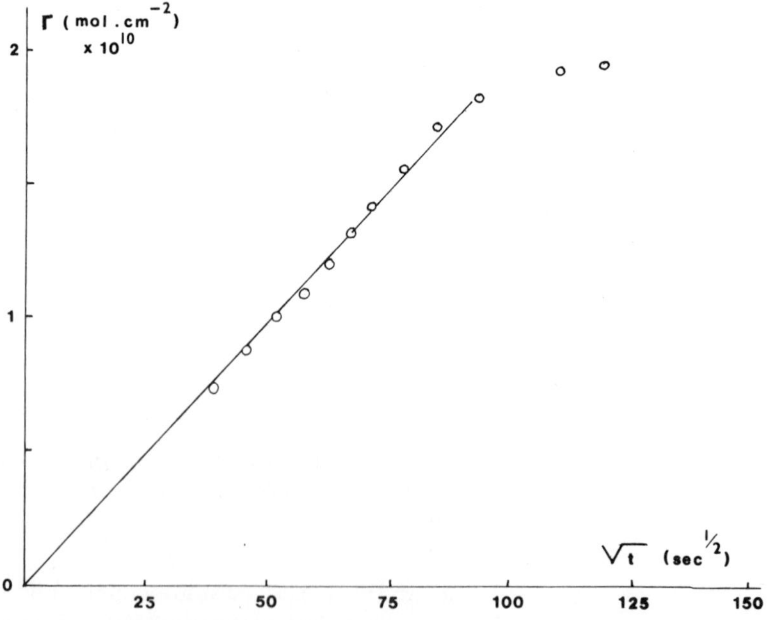

Figure 10. Data of Figure 9 converted into a Γ vs. $\sqrt{t}$ *plot. Ordinate: surfactant adsorption in mole/cm²*.

adsorption barrier should be based on an accurate knowledge of the surface equation of state.

Although no direct relationship between apparent induction periods and adsorption barriers or rearrangement processes is shown to exist, the phenomenon is related to interactions in the surface in an indirect way since the shape of the ϵ_0 *vs.* π curves depends on such interactions. It is also clear that induction periods can be of great practical significance for systems in which surface or interfacial tensions move towards equilibrium after an initial rapid extension as occurs during foaming or emulsification. Once again this emphasizes the need for a better knowledge of the factors determining equilibrium surface behavior as expressed in a surface equation of state for various surface-active materials.

Acknowledgment

Thanks are due to E. D. Goddard for providing us with tabulated data for the π–A curve of the C_{20} sulfobetaine.

Literature Cited

1. Davies, J. T., Rideal, E. K., "Interfacial Phenomena," 2nd ed., Academic Press, New York, London, 1963.
2. Goodrich, F. C., *Proc. Intern. Congr. Surface Activity, 2nd, London* (1957) **I,** 85.
3. Lucassen-Reynders, E. H., *J. Colloid Interface Sci.* (1973) **42,** 563, 573.
4. Lucassen, J., van den Tempel, M., *Chem. Eng. Sci.* (1972) **27,** 1283.
5. Lucassen, J., Hansen, R. S., *J. Colloid Interface Sci.* (1967) **23,** 319.
6. Lucassen, J., Giles, D., *J. Chem. Soc., Faraday Trans. I,* (1975) **71,** 217.
7. Garrett, P. R. *et al.,* unpublished data.
8. Lucassen-Reynders, E. H., *Kolloid-Z., Z. Polymere* (1972) **250,** 356.
9. Goddard, E. D., Kung, H. C., *J. Colloid Interface Sci.* (1971) **37,** 585.
10. Blank, M., Lucassen, J., van den Tempel, M., *J. Colloid Interface Sci.* (1970) **33,** 94.
11. Tsonopoulos, C., Newman, J., Prausnitz, M., *Chem. Eng .Sci.* (1971) **26,** 817.
12. Ward, A. F. H., Tordai, L., *J. Chem. Phys.* (1946) **14,** 453.

RECEIVED September 24, 1974.

22

Electron Microscope Studies of Monolayers of Lecithin

H. E. RIES, JR.,[1] M. MATSUMOTO, N. UYEDA, and E. SUITO

Institute for Chemical Research, Kyoto University, Kyoto, Japan

Electron micrographs of the monolayers of a synthetic lecithin, β,γ-dipalmitoyl-DL-α-glycerylphosphorylcholine, are markedly different from those of fatty acids and cholesterol. However, the overall thin film properties of these materials, as indicated by pressure–area isotherms, are quite similar. At low surface pressures, the island structures of lecithin films are far less regular in contour than those of fatty acid or cholesterol films. The lecithin films also have an unusual microporosity. This perforated structure persists at intermediate and high pressures. After monolayer collapse, the long, flat, ribbonlike structures formed by lecithin are less regular than the well defined collapse structures of fatty acids and cholesterol. Such similarities and differences are related to molecular geometry and the location and strength of polar groups.

Lecithin, one of the principal lipids in cell membranes, controls many important biological processes. Nevertheless, little is known about the structure of its films (*1, 2, 3, 4*). Electron micrographs now show remarkable properties for the thin film or monolayer of dipalmitoyl lecithin transferred quantitatively from a water surface. In many respects the water corresponds to the aqueous phases that bound cell membranes.

Experimental

Pressure–area isotherms were determined for synthetic lecithin and compared with those for related compounds. Electron micrographs were obtained for samples transferred from representative films before and after collapse. Basic apparatus and techniques have been described (*1, 5, 6, 7*).

[1] Present address: Whitman Laboratory, Department of Biology, University of Chicago, Chicago, Ill. 60637.

Materials. All materials had the highest purity available. The synthetic lecithin, β,γ-dipalmitoyl-DL-α-glycerylphosphorylcholine, was obtained from the Sigma Chemical Co. Chloroform, the principal volatile solvent used for spreading, was obtained from the Bojin Pharmaceutical Chemical Laboratories. Twice-distilled benzene was used for some of the experiments with related materials. Film-balance measurements with the solvents alone showed that remaining impurities were negligible.

The water used as a substrate was twice distilled and had a pH slightly below 7.0 and a specific conductivity of about 0.5 μmho. Because the pH was always below 7.0, its variation had a negligible effect on the lecithin isotherm. Thorough testing of the water surface on the film balance ensured freedom from significant capillary-active contamination.

Surface-Pressure Measurements. Both the Wilhelmy and the Langmuir-Adam-Harkins techniques have been used for pressure–area measurements (*1*). For most of the experiments a long, shallow trough coated with Teflon contained the twice-distilled water on which the monolayer was spread (*6*). The apparatus was housed in a glass-walled cabinet that could be kept closed during pressure and area adjustments and during the transfer of samples to electron microscope screens. The entire apparatus was mounted on a concrete base that was essentially free of vibration as indicated by a long optical lever reflected from the water surface. Glass weighing pipets were used to spread a few drops of the dilute solutions of film-forming materials in volatile solvents.

To minimize contamination, all parts of the system were thoroughly cleaned before each experiment; the water surface was swept many times with small Teflon coated barriers before the film was spread. During pressure–area experiments, the film area was reduced in small decrements and pressures were measured at 2-min intervals. The compression continued until collapse; this was indicated by a constant or falling pressure. Temperatures rarely varied more than 0.1°C during individual experiments.

Electron Microscopy. A modified Langmuir-Blodgett method was used to transfer monolayer samples to electron microscope screens that were sandwiched between Formvar and a glass plate (*5, 6*). A motor drive raised the plate slowly through the water–air interface, and a variable-speed motor drive moved the compressing barrier at a rate that maintained constant surface pressure during the transfer. Many samples were transferred at low surface pressures because the lipids in membranes are undoubtedly subjected to relatively small "horizontal" or surface pressures.

Following transfer, film samples were placed in a high vacuum and were shadowcast with germanium at an angle of approximately 10°. The samples were then examined at a direct magnification of at least 5000 times in a modified JEM-Type 7 electron microscope.

There are many difficulties in transferring monolayer samples for electron microscope studies—the evaporation or removal of the interposed water film between the monolayer and the Formvar, vibration, mechanical problems, and various other strains. Therefore, we do not claim a one-to-one correspondence between the state of a film on the water surface and the structures observed in the micrographs. Nevertheless, it is of interest to compare the *sequence* of changes that occur during compression on the water surface with the *sequence* of changes observed in the microscope.

Results and Discussion

Pressure–Area Isotherms. In spite of the apparently bulky structure of the lecithin molecule (Figure 1), the isotherms presented in Figure 2 indicate strong and well-behaved monolayers. The isotherm at 20.0°C has an initial pressure rise near 70 A^2/molecule, a pronounced inflection at 50 A^2 and 4 dynes/cm, and a steep high-pressure portion that extrapo-

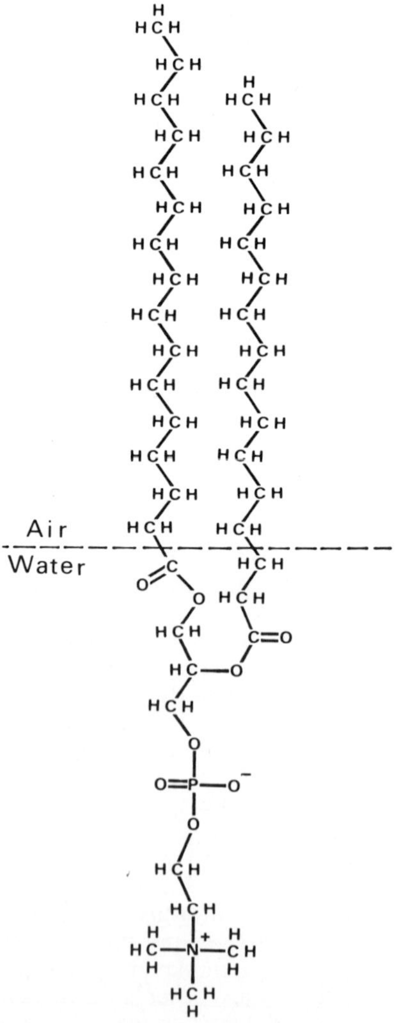

Figure 1. Schematic of a synthetic lecithin molecule

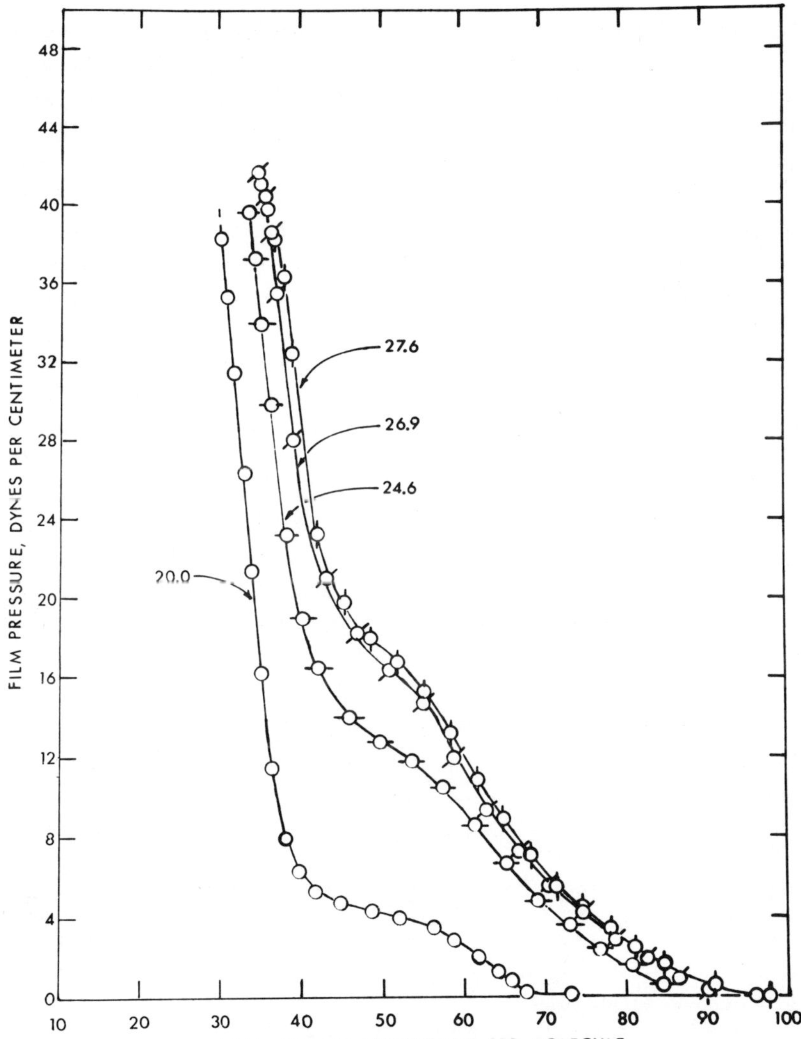

Figure 2. Pressure–area isotherms for a synthetic lecithin at several temperatures (°C)

lates to 39.5 A^2/molecule. The collapse pressure is relatively high, close to 40 dynes/cm, and the compressibility, 0.0060, is relatively low for such a complex structure. These data support the approximate orientation shown in Figure 1 although details are not yet established (*2, 3, 4*).

The effects of increasing temperature are demonstrated in Figure 2 and Table I. The extrapolated area increases linearly with the temperature. Moreover, the inflection-point pressure at 50 A^2 also increases linearly with increasing temperature. A marked reorganization or repacking of the molecules, intra- and/or intermolecular, evidently takes place at 50 A^2. Staggered packing that involves a vertical shift of the lecithin molecules might account for such a phase change. This vertical shift could greatly reduce the effect of repulsive forces between the large polar groups. As expected, the phase change or inflection-point plateau is less pronounced at higher temperatures. Also, as might be anticipated if collapse is a mechanical phenomenon, the collapse pressures are essentially the same at the four temperatures studied. Compressibility values for the films in the upper pressure region are also remarkably similar.

The isotherm for cholesterol (*1*) is surprisingly similar to that for lecithin (Table II); both are important components of many cell membranes. The extrapolated area of 39.0 A^2/molecule approximates that of a double-chain system, and the cohesive and adhesive forces give a collapse pressure of 43 dynes/cm, close to the 40 dynes/cm for lecithin. In Table II cholesterol is also compared with stearic acid, the classic compound in monolayer research. Collapse pressures and compressibilities are similar. The rigid packing of the complex cholesterol molecules is clearly demonstrated by its monolayer compressibility—0.0012, the smallest value in these studies.

Table I. Monolayer Properties of a Synthetic Lecithin

Temperature, °C	*Area, A^2/molecule*[a]	*Inflection Point, dynes/cm*[b]	*Collapse Pressure, dynes/cm*	*Compressibility, cm/dyne*[c]
20.0	39.5	4.0	(39)	0.0060
24.6	45.0	12.5	(40)	0.0064
26.9	47.5	16.5	(42)	0.0061
27.6	48.5	17.5	(41)	0.0058

[a] Extrapolated area at zero pressure.
[b] Effectively at 50 A^2/molecule.
[c] Compressibility is $(a_o\text{-}a_1)/a_of_1$, where a_o is the extrapolated area at zero pressure and a_1 is a smaller area at pressure f_1.

Table II. Comparison of Monolayer Properties for Stearic Acid, Cholesterol and Lecithin at 25°C

	Area, A^2/molecule[a]	*Collapse Pressure, dynes/cm*	*Compressibility, cm/dyne*[b]
Stearic acid	20.6	(41)	0.0015
Cholesterol	39.0	(43)	0.0012
Lecithin	45.0	(40)	0.0064

[a] Extrapolated area at zero pressure.
[b] Compressibility is $(a_o\text{-}a_1)/a_of_1$, where a_o is the extrapolated area at zero pressure and a_1 is a smaller area at pressure f_1.

Electron Micrographs. Representative electron micrographs are shown in Figure 3. Arrows indicate the direction of shadowcasting, and the magnification is shown by the 1μ scale.

The blank sample of Figure 3A was obtained by raising the Formvar-covered screen through a clean water–air interface before the film was spread. Such micrographs establish both the flat smooth surface of the Formvar as well as the fine texture of the vapor-deposited germanium.

A typical island or cluster structure for the lecithin film transferred

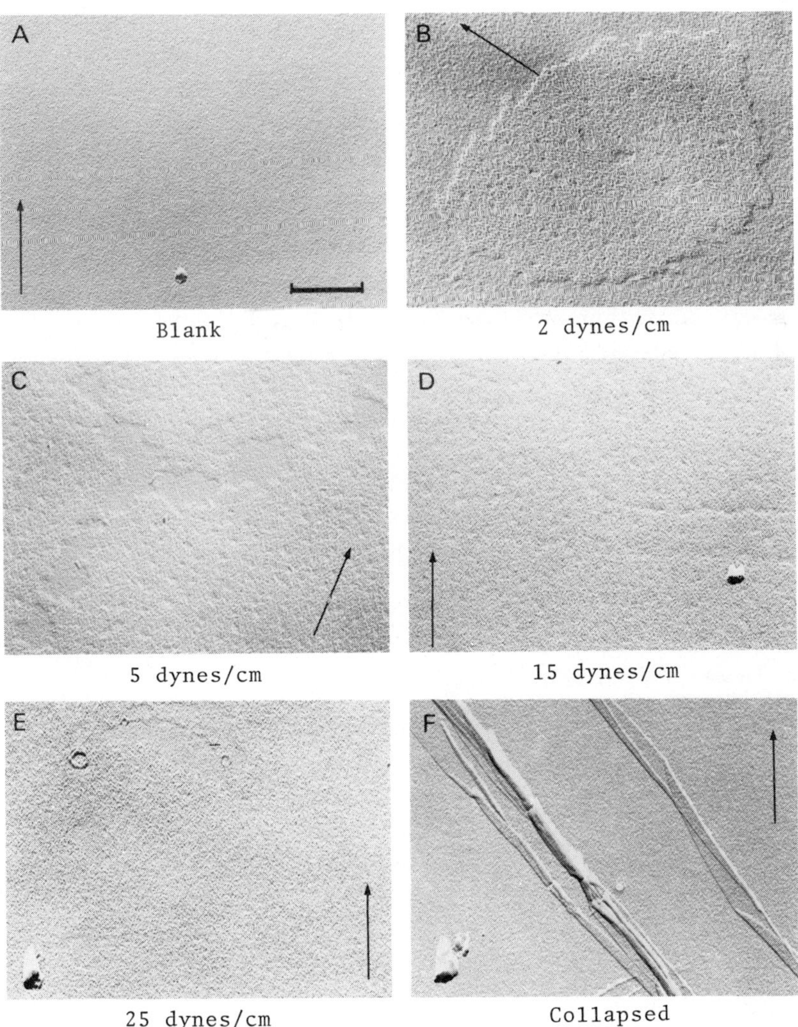

Figure 3. Electron micrographs of lecithin films transferred at various surface pressures. The scale shows 1μ, and the arrows indicate the direction of shadowcasting.

at a low surface pressure (2 dynes/cm) is shown in Figure 3B. Monolayer thickness (*ca.* 40 A) as observed in the electron microscope suggests an elongated molecular structure with extended polar groups (Figure 1). Lecithin islands contain more small holes and have edge contours that are considerably more irregular than those for *n*-hexatriacontanoic acid (C-36) (*5, 6*) and cholesterol.

At 5 dynes/cm (Figure 3C), the micrographs show large areas of a two-dimensional continuous phase with discontinuous so-called uncovered areas. Micro-islands and individual molecules beyond the limit of resolution of the microscope may be present in the so-called uncovered area. At 15 dynes/cm (3D), further compression increases the ratio of the covered to the uncovered area, but the small holes persist.

At 25 dynes/cm (3E), much of the so-called uncovered area evidently disappears, and large areas are covered homogeneously. Following film collapse (3F), long, flat, ribbonlike structures, apparently two molecules thick, appear. This collapsed material is less regular in structure than that formed in the collapse of C-36 acid films.

There are three principal differences between the film structures observed for lecithin and those for the C-36 acid (as well as for cholesterol): (a) lecithin islands have many more holes and less regular edges; (b) the small holes in lecithin films persist at much higher surface pressures; (c) the collapse fragments formed by lecithin are less structured. Such differences are, of course, related to molecular geometry and polarity. The hydrocarbon chains of lecithin are much shorter than those of the C-36 acid; the forces of cohesion are thus smaller. The polar extremity of lecithin is larger and stronger (high dipole moment) than the carboxy group of the acid (or the hydroxy group of cholesterol), and may, therefore, interfere with close packing.

Conclusion

Combined film balance and electron microscope studies reveal some remarkable properties of lecithin monolayers. A microporosity of the film, which is observed at low surface pressures and persists into the intermediate and high pressure regions, might, if present in membrane structures, be related to the penetration of proteins and other materials into cell membranes. Films of related materials and their mixtures warrant detailed study.

Acknowledgment

We thank Tohru Takenaka for helpful suggestions and the use of some of his film balance apparatus and Yoshio Saito and Tsunoru Yoshida for valuable assistance in the experimental work.

Literature Cited

1. Gaines, G. L., Jr., "Insoluble Monolayers at Liquid–Gas Interfaces," Interscience Publishers, New York, 1966.
2. Cadenhead, D. A., "Recent Progress in Surface Science," J. F. Danielli, A. C. Riddiford, M. D. Rosenberg, Eds., Vol. 3, p. 169, Academic Press, New York, 1970.
3. Shah, D. O., Schulman, J. H., *J. Lipid Res.* (1967) **8,** 227.
4. Standish, M. M., Pethica, B. A., *Trans. Faraday Soc.* (1968) **64,** 1113.
5. Ries, H. E., Jr., Kimball, W. A., "Proceedings of the Second International Congress on Surface Activity," Vol. 1, p. 75, Butterworths Scientific Publications, London, 1957.
6. Ries, H. E., Jr., Walker, D. C., *J. Colloid Sci.* (1961) **16,** 361.
7. Adamson, A. W., "Physical Chemistry of Surfaces," 2nd ed., Interscience, New York, 1967.

RECEIVED September 23, 1974.

23

Monolayer Studies of Pure Nitroxide Fatty Acid Spin–Label Probes

D. A. CADENHEAD and F. MÜLLER-LANDAU

Departments of Chemistry and Biochemistry, State University of New York, Buffalo, N.Y. 14214

Isotherms for monomolecular films of a series of nitroxide stearic acid and nitroxide palmitate and stearate (methyl ester) films are reported for the temperature range 9°–50°C. The isotherms depend to a significant degree on the location of the nitroxide (oxazolidine) ring. For films of 8-nitroxide methyl palmitate and 12- or 16-nitroxide stearic acids or esters the bipolar nature of these molecules predominates, and gaseous expanded isotherms are obtained. For films of 5-nitroxide stearic acid or ester the molecule behaves more like a monopolar amphiphatic molecule. The isotherm shapes and their negative temperature expansion dependence are explained in terms of the ability of these molecules to adopt two molecular conformations at the air–water interface: erect, with only the primary carboxyl or methyl ester group in the substrate, and bent, with both the primary and the oxazolidine ring group in the substrate. Implications for spin–label studies of lipid bilayers and cell membranes are pointed out.

The use of spin–label probes to investigate cell membrane structure and function clearly demonstrates the fluidity of membrane lipid structures (*1, 2, 3, 4*); however, a spin–label probe sees only its immediate environment. Predictions (*5, 6, 7, 8*) and data (*9, 10, 11, 12*) show that the introduction, for example, of a substituted oxazolidine ring as part of a typical amphiphatic lipid molecule can also significantly perturb a "normal" lipid environment. Consequently, some quantitative observations that used spin–label techniques need revision while others may be reduced to the level of qualitative predictions.

To evaluate possible perturbation or impurity membrane effects and to understand the behavior of spin-labeled lipids in a membrane-like environment, we used monomolecular films as model membrane systems. Of course we realize that the energetics of a monolayer and a membrane are not precisely identical (*13*); however, a monolayer closely approximates half a bilayer, and a bilayer appears to be a general, if not universal, major component of membrane structure (*14, 15*). One further problem with such an approach is that under typical spin–labeling conditions, molar ratios of $\geq$100:1 (host lipid:lipid probe) are usually used. At those concentrations the macroscopic properties (surface pressure, π; surface potential, ΔV; and surface viscosity, η) cannot be accurately distinguished from those of the pure host lipid. Concentrations of $\leq$50:1 molar ratio (host:probe) are required for a reasonable resolution. However, since the probe only sees its immediate molecular neighbors, the *effective* concentration is closer to a 6:1 (host:probe).

To understand the behavior of the mixed host–probe films, it is essential to understand the behavior of the pure films of both components. We previously investigated in detail the behavior of the pure films of 12-nitroxide stearate [2-(10-carboxydecyl)-2-hexyl-4,4′-dimethyl-3-oxazolidinyloxyl] including its extraordinary temperature dependence (*7, 8, 16*). In this paper we extend our investigations to the pure films of other nitroxide stearic acid (and methyl ester) probes where the oxazolidine ring is attached to various carbon atoms of the stearic acid (or ester) hydrocarbon chain. This series of spin–label probes is one of those most extensively used to study cell membrane structure. It was used to define, among other things, an order parameter establishing the fluidity (*17*) and polarity profiles (*18*) of lipid bilayers.

This presentation, based solely on pure film studies, shows that the validity of fluidity and polarity profiles as determined from spin–label probes, is questionable. More important, however, is the fascinating insight into the unique surface chemistry of these compounds.

Experimental and Results

The 12-nitroxide stearic acid and the 8-nitroxide methyl palmitate were donated by J. D. Morrisett (John Baylor College of Medicine, Houston, Tex.). All other nitroxide stearic acid and stearate probes were purchased from Syva (California). All probe molecules were subsequently purified by preparative thin layer chromatography (TLC) until a single spot was obtained under all development conditions. TLC was carried out in a dry nitrogen atmosphere to avoid possible oxidation or hydrolysis during purification. Ultrapure silica gel, free from any plasticizers, was used to avoid possible further contamination.

All films were studied in an automated Wilhelmy film balance. The balance and film-handling techniques are described in a previous publication (*19*). The films were spread on a distilled water substrate (pH 6).

The water was distilled twice from glass (initially from alkaline permanganate). It was subsequently twice distilled from quartz. Surface areas at most pressures were reproducible within ±0.5 A²/molecule; however, at low pressures, for gaseous isotherms, the error was somewhat greater. Surface potentials were also obtained but are not reported here. Surface potential values, however, confirm observations made with surface pressure measurements (*8*).

The results are shown in Figures 1 through 7. Figures 1 through 6 show isotherms over a range of temperatures for the 5-, 12-, and 16-nitroxide stearic acids, for the 5- and 16-nitroxide methyl stearates, and for the 8-nitroxide methyl palmitate over the range 9°–50°C. [We use 5-NS, 5-NS(Me), and 8-NP(Me) for the trivial names 5-nitroxide stearic acid, 5-nitroxide methyl stearate, and 8-nitroxide methyl palmitate, with

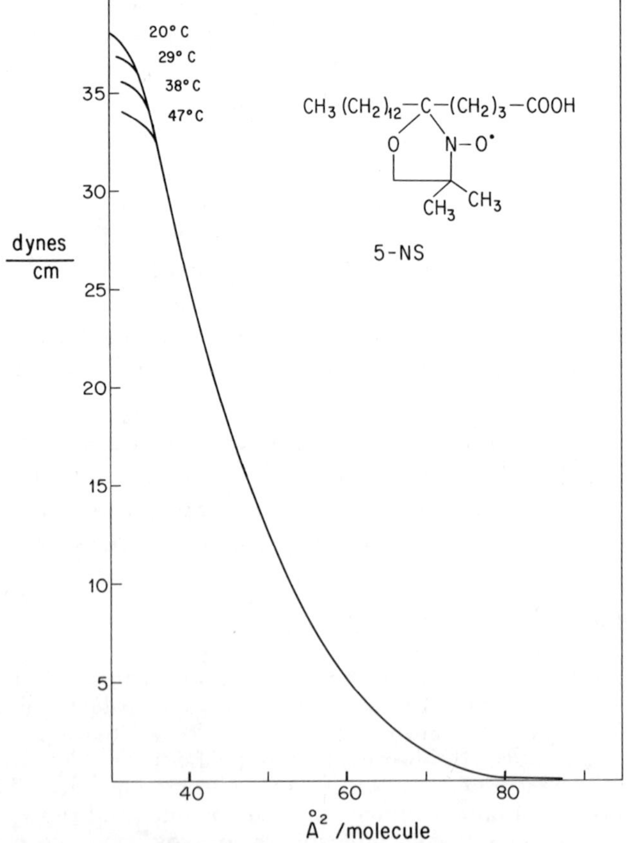

Figure 1. Surface pressure (dynes/cm) vs. area/molecule (A²) for 5-nitroxide stearic acid [5-NS] at 20°, 29°, 38°, and 47°C. The isotherms at these temperatures are identical except for the small region near collapse where the separate isotherms are labelled.

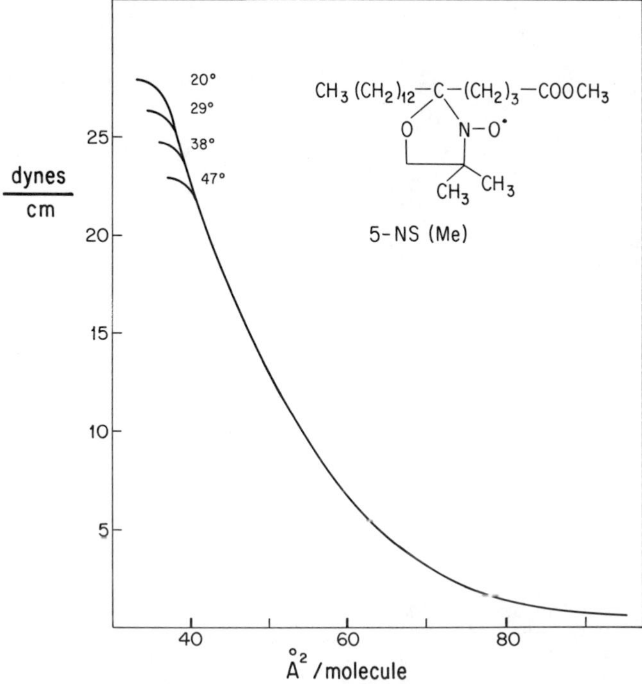

Figure 2. Surface pressure (dynes/cm) vs. area/molecule (A^2) for 5-nitroxide methyl stearate [5-NS(Me)] at 20°, 29°, 38°, and 47°C. The isotherms at these temperatures are identical except for the small region near collapse where the separate isotherms are labelled.

similar abbreviations for the other compounds.] Figure 7 compares the 20°C isotherms for the complete series of spin–label probes.

Discussion

Isotherm Shape and Interpretation. The substitution of an oxazolidine ring for two hydrogen atoms on the stearic acid (or methyl ester chain) results in a significant expansion when the isotherm is compared with the parent stearic acid isotherm (Figure 7). This is most evident for compounds where the oxazolidine ring is attached at the 12-position on the carbon chain or beyond. A highly expanded form is also evident for 8-NP(Me) *vs.* either methyl stearate or palmitate. This suggests strongly that the corresponding 8-NS(Me) would also exhibit a highly expanded form since the expansive effect of the oxazolidine ring is significantly greater than the condensation induced by two additional carbon atoms. When this is the case, the isotherm is in a gaseous expanded form. For 5-NS and 5-NS(Me), however, the isotherm shape approaches that of a liquid expanded form with the ester, as expected, being slightly more expanded. For 5-NS there appears to be a lift–off, *i.e.*, a fairly sharp rise

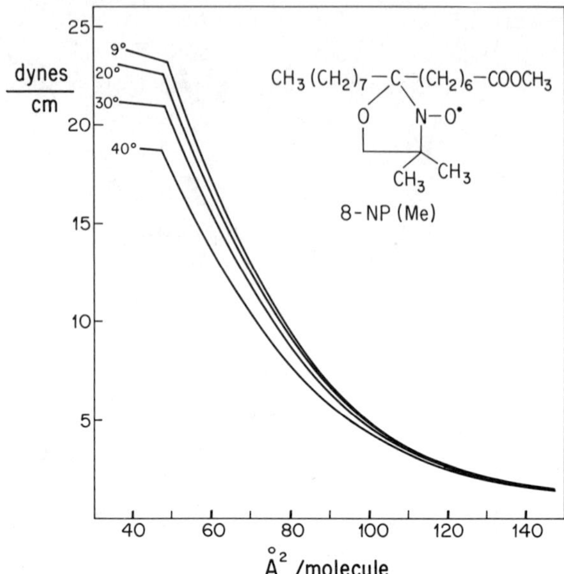

Figure 3. Surface pressure (dynes/cm) vs. *area/molecule (A^2) for 8-nitroxide methyl palmitate [8-NP(Me)] at 9°, 20°, 30°, and 40°C. Broken line indicates a near liquid-like film collapse.*

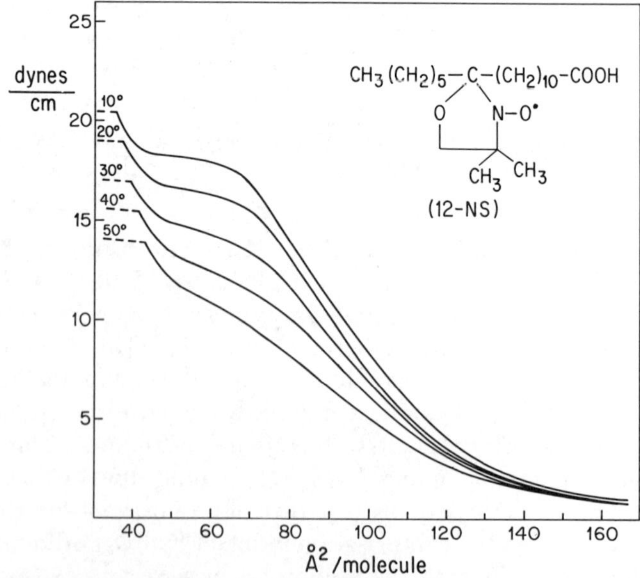

Figure 4. Surface pressure (dynes/cm) vs. *area/molecule (A^2) for 12-nitroxide stearic acid [12-NS] at 10°, 20°, 38°, 40°, and 50°C. Broken line indicates a near liquid-like film collapse.*

in surface pressure at low surface pressures. In addition, the surface potential seems to vary at areas/molecule higher than lift–off in a somewhat irregular fashion, indicating a two-phase system. The data, however, are not clear enough to identify the physical state definitely.

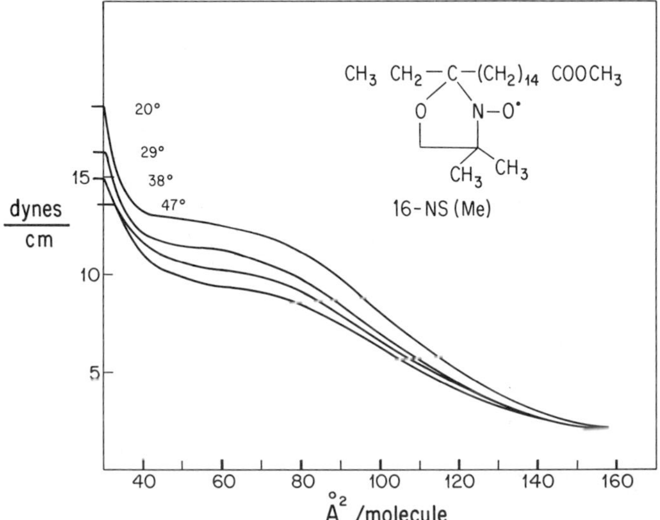

Figure 5. Surface pressure (dynes/cm) vs. area/molecule (A²) for 16-nitroxide methyl stearate [16-NS(Me)] at 20°, 29°, 38°, and 47°C. The isotherms have the same order (top to bottom) as the indicated temperatures.

From N. K. Adam's observation that many bipolar amphiphatic molecules form gaseous expanded films, it is evident that these nitroxide probe films, with the possible exception of 5-NS and 5-NS(Me), behave as typical bipolar amphiphatic films. The shift to lower areas/molecule for 5-NS and 5-NS(Me) shows that, as the oxazolidine ring moves closer to the primary polar group, the bipolar nature of the molecule diminishes. Thus the carboxyl group and oxazolidine ring for these compounds effectively constitute one polar group.

In previous publications, we discussed the isotherm shape for 12-NS (*7, 8, 16*). We postulated (*see* Figure 8) that in the gaseous state both polar groups are immersed in the aqueous substrate. With compression, a close–packed, bent conformation is attained; any further compression results in the weaker polar group (oxazolidine) being forced out of the interface and an erection of the probe molecule. These molecular conformational changes are reflected in the isotherm by the inflection at approximately 70 A²/molecule, followed by a region of high compressibility and a short condensed region (Figure 4). The α,ω dibasic alcohols

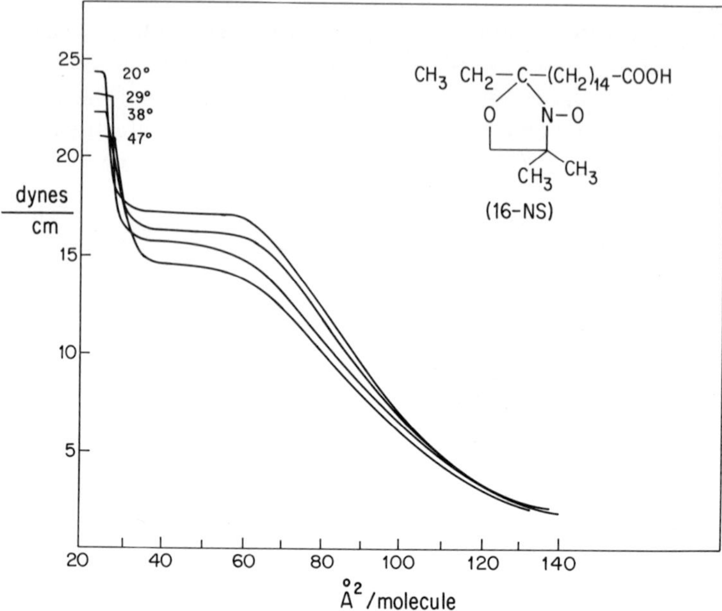

Figure 6. Surface pressure (dynes/cm) vs. *area/molecule (A^2) for 16-nitroxide stearic acid [16-NS] at 20°, 29°, 38°, and 47°C. The order of temperatures shown (top to bottom) is for the isotherms over all but the close-packed region (<40 A^2/molecule).*

and aldoximes examined by Adam *et al.* (*20*) have similar gaseous expanded isotherms but exhibit sharp breaks or phase changes rather than inflections as do the corresponding acids (*21*) even though the two polar groups are identical and the isotherm breaks do not always correspond to twice the close–packed area. The isotherm for a rigid, bipolar amphiphatic molecule, β-estradiol diacetate (*22*) has an even sharper transition, suggesting that the more gradual inflection for 12-NS results partly from the flexible hydrogen chain's taking up part of the compressional strain and partly from the difference in polarity between the two polar groups. In addition, the gradual inflection, rather than a sharp phase change, indicates that unlike other cases (*20, 21, 22*), the transition may not be from one pure conformation to another, but rather from a situation where one conformation predominates to another situation where the reverse is true. A similar comparative examination of surface potentials (not shown) also indicates a gradual rather than an abrupt transition (*8*).

The relatively limited condensed region of the 12-NS isotherm indicates perturbation of alkane chain packing when the oxazolidine ring is forced into the hydrophobic region. The stability of this region increases as the oxazolidine group shifts towards the terminal methyl group. This indicates that the oxazolidine ring induces less of a perturbation in this position. In contrast, it appears that a close–packed bent conformation, as indicated by the absence of an inflection in the isotherm, is destabilized as the oxazolidine ring moves from the 12- to 8-position, but it becomes increasingly stable at the 5-position. Apparently as the oxazolidine and carboxylic groups approach one another, stearic hindrance inhibits a close–packed bent conformation. The higher collapse pressure of 8-NP over 12-NS suggests that a bent conformation (but not a close-packed bent conformation) is more stable for the former molecule. However, by the time the oxazolidine ring attains the 5-position, a considerably enhanced stability is achieved, presumably because in the bent conformation the two polar groups can now fully immerse the intervening three methylene groups. In other words, we expect for 5-NS that the oxazolidine ring spends more time in the aqueous substrate rather than that water is then capable of greater penetration of the bilayer. Note again that as

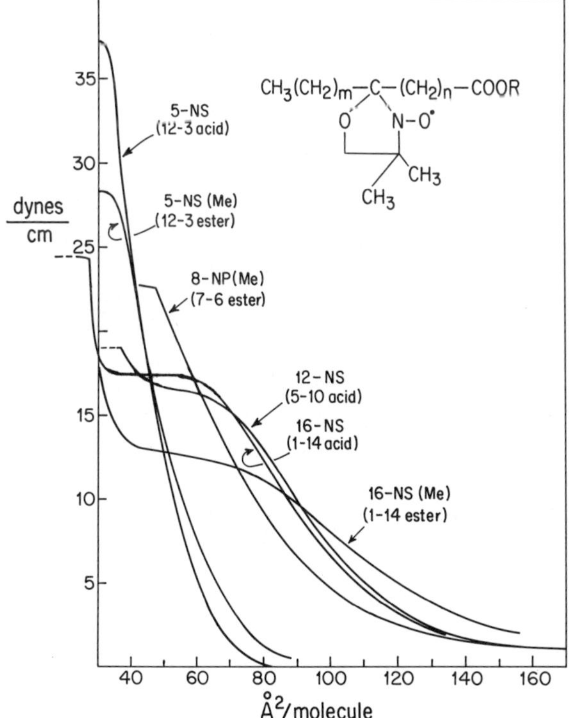

Figure 7. Surface pressure (dynes/cm) vs. area/molecule (A²) for 5-NS, 5-NS(Me), 8-NP(Me), 12-NS, 16-NS(Me), 16-NS and stearic acid at 20°C. Terminology in parenthesis refers to the number of methylene groups between the terminal methyl group and the oxazolidine ring (m) and between the oxazolidine ring and the primary polar group (n).

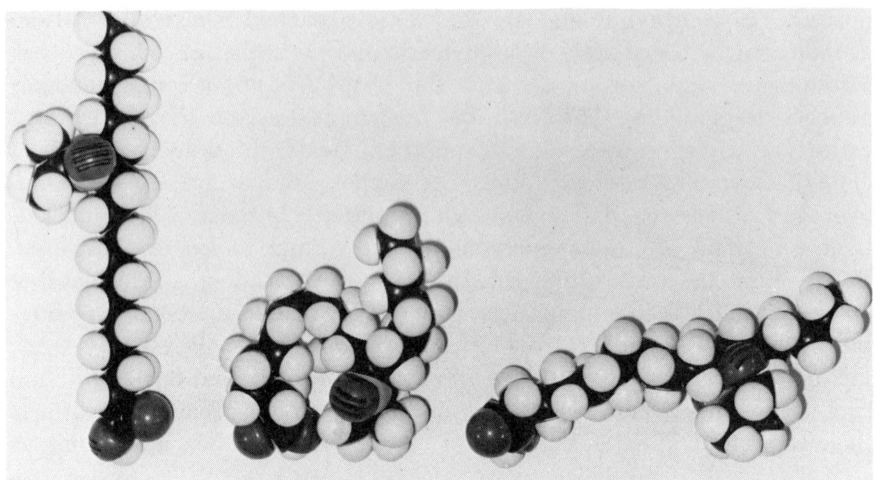

Figure 8. Conformational changes for 12-NS using C.P.K. molecular models.

(a) Erect conformation: only the carboxyl group is in the aqueous substrate (left).
(b) Close-packed bent conformation: both the carboxyl and the oxazolidine ring (nitroxide group) are depicted in the aqueous substate. The arrangement shows a minimum area/molecule (~70 A²/molecule) in this conformation (center).
(c) Gaseous bent conformation: both polar groups are in the aqueous substrate. A continuous change between this arrangement and that depicted in (b) is possible (right).

the two polar groups approach one another, the pre-inflection portion of the isotherm changes from a highly expanded to an expanded film, reflecting the shorter distance between the two polar groups.

Temperature Dependence. The extraordinary film temperature dependence exhibited by most members of this series was discussed previously for 12-NS (*16*). Enhanced film loss does not account for a film condensation with increasing temperature, presumably through increasing solubility, with increasing temperature. There are two reasons why this is clearly not the case. First, such a film loss would simply shift the entire isotherm to lower areas/molecule without any alteration in shape. Here, however, they clearly do change (*see* Figure 4). Second, even though the film was unstable at low pressures, an instability presumably arising from substrate attack on the oxazolidine ring, this instability disappeared at higher pressures where the molecule was in an erect conformation (*7*). This stability pattern is the reverse of what we would expect if solubility were the problem. Indeed, when 12-NS and myristic acid were compared at 15 dynes/cm, the latter appeared slightly more soluble.

The only other examples of a film which condenses with increasing temperature are those of hexadecyl and of octadecyl urea (*23*). In these cases, the explanation suggested was that intermolecular hydrogen bonding between the polar head groups preserves an expanded but rigid film structure. With increasing temperature this structure breaks up, producing a condensed but more fluid film. This explanation does not apply to films of 12-NS for several reasons. First, it is difficult to envisage a suitable polar head conformation (using CPK models) which would facilitate strong multiple hydrogen-bond formation. Second, the pre-inflection region gives no indication of the rigidity or viscosity expected for a hydrogen-bonded network of polar head groups. Third, although both films condense with increasing temperature, the isotherm shapes and their variations with temperature are quite different. Thus, the octadecyl urea films show a similar sharp phase change at all temperatures with only the surface pressure altered. In contrast, the inflection for 12-NS clearly diminishes with increasing temperature; at 50°C it is barely prceeptible.

We explain this strange temperature dependence in the following way (*16*): as the temperature increases, the oxazolidine ring–substrate interactions are overcome, thus destabilizing the close–packed bent conformation. We envisage a dynamic equilibrium between two conformations: an erect and a bent 12-NS, the latter having both polar groups in the substrate. With increasing temperature the stability of the close-packed bent conformation diminishes, and the isotherm inflection disappears, indicating a continuous transition from a near horizontal to near vertical orientation. When both polar groups are identical, as for the α,ω compounds of Adams *et al.* (*20*), Jeffers and Dean (*21*), and our β-estradiol diacetate (*22*), the break in the isotherm is sharper. This indicates a well defined phase change when the film is only in a bent conformation (*i.e.*, both polar groups in the substrate) at pressures below the phase transition. Since only the spin–label probes of all these compounds (*21, 22*) show a condensing film with increasing temperature, an equilibrium must exist between two conformations, with the one of lower area being more stable at higher pressures and temperatures.

On shifting the oxazolidine ring from the 12 to the 16 position, we find further supporting evidence for this explanation (Figures 5 and 6). In Figure 6 the expansion of 16-NS shows a large negative temperature dependence at areas/molecule greater than the isotherm inflection and a small positive temperature dependence in the condensed region where only the erect conformation is possible. The corresponding isotherms for 16-NS(Me) have a similar negative temperature dependence at large areas/molecule and indicate an isotherm crossover for only the 38° and

47°C isotherms. The diminished stability of the condensed region coincides with the use of the less polar ester group as the terminal (or primary) polar group.

The only other region where expansion might occur with increasing temperature is at the very high areas/molecule, which correspond to a gaseous film. Here the molecule must have both polar groups in the substrate, and thus only one basic molecular conformation appears possible. In this region, however, the isotherms virtually coincide, and precision is low; there is little hope of demonstrating a positive expansion temperature dependence here.

Films of 8-NP(Me) still show a negative expansion temperature dependence; the magnitude of the effect, however, is reduced presumably because of the decreased stability of the erect and close-packed bent conformations. [The sharp break at higher pressures indicates film collapse before the inflection, and a completely erect film is not attainable.] CPK molecular models clearly indicate considerable stearic hindrance in attaining a close-packed bent conformation when only six methylene groups intervene between the primary polar group and the oxazolidine ring. A similar argument pertains to 8-NS(Me).

Films of 5-NS and 5-NS(Me) have a remarkable coincidence of isotherms over a 30°C range; a negative expansion temperature dependence only appears in the inflection as the films approach collapse. The inflection seems to be real and can be differentiated clearly from the collapse behavior for 8-NP(Me). Thus the plots of surface pressure *vs.* area/molecule and surface potential *vs.* area/molecule (not shown) for 5-NS, 5-NS(Me), and 2-NP(Me) differ from each other. Moreover, the inflections for 5-NS and 5-NS(Me) remain invariant with a varying rate of compression whereas the collapse pressure of 8-NP(Me) is slightly pressure sensitive. It appears, therefore, that films of 5-NS and 5-NS(Me) exhibit a zero temperature dependence over most of the pressure range but a negative temperature dependence as they approach collapse.

Even if 5-NS and 5-NS(Me) did not have a small pressure region where a negative temperature expansion dependence was observed, the zero temperature dependence over most of the isotherm is still remarkable and calls for an explanation. For this region, a decrease in area/molecule with temperature caused by a shift in the equilibrium between the two primary conformations (erect and bent) apparently balances the normal expansion of the films with increasing temperature. Moreover, since 5-NS and 5-NS(Me) can apparently attain close-packed conformations (as indicated by the inflection), a close-packed bent conformation is more stable then with 8-NS(Me). Nevertheless, 8-NP(Me) reveals a more extensive negative area–temperature dependence, presumably because the decrease in area/molecule is greater for 8-NP(Me) when the molecule is erected. For 5-NS and 5-NS(Me) a bent conformation (both polar groups in the interface) can occupy only a slightly greater area/molecule than the erect conformation and must involve the immersion of the first three methylene groups. Thus, while our interpretation of isotherm shape for 5-NS and 5-NS(Me) indicates an effective single polar group, this could be achieved either on the basis of water penetration to the fifth carbon atom on the chain or with the oxazolidine ring being pulled down into the interface. The temperature dependence clearly supports the latter explanation.

Biological Significance. The evidence here shows that as the oxa-oxazolidine ring moves along the alkane chain from the carboxyl to the terminal methyl group, the film behavior changes drastically, especially when the ring is near to the carboxyl group. In studies with mixed films (*24*) this behavior is somewhat modified but by no means eliminated. One of the assumptions made in ESR studies where the position of the oxazolidine ring is varied, is that the behavior of the molecule is essentially unchanged. Clearly, this is not the case. This means that the order parameter and fluidity gradient calculated by McConnell (*17*) are probably wrong, particularly the values for carbon atoms near the primary polar group. Recent studies by Seelig *et al.* (*9, 10, 11, 12*) confirm this by comparing spin-label and deuterium resonance spectra. Also the polarity profile deduced by Griffith *et al* (*18*) should predict too high a polarity for the first few carbon atoms on either side of the lipid bilayer. When allowance is made for this, it may be unnecessary to invoke unlikely and extensive water penetration. It should be noted that McConnell (*17*) and Griffith (*18*) used not only fatty acid spin labels but also phospholipid spin labels. We have also carried out preliminary studies using these latter compounds, however results indicate similar though smaller effects.

The picture is fairly clear for the sharply altered behavior of 5-NS, 5-NS(Me), and the rest of the series in view of the ready stabilzation of the bent conformation, it is less clear for the differences between other positions of the oxazolidine ring on the chain. In particular, a detailed evaluation of changes in the role that a bent conformation might play for 12-NS and 16-NS requires further study of mixed lipid films. While it seems likely that under typical membrane conditions a bent conformation will play only a minor role, present indications are that it is not negligible and will vary with the shift from the 12 to the 16 position (*24*).

At least one report based on observed ESR spectra indicates that a bent conformation for 12-NS(Me) is likely in a micellar environment (*25*). The work concerned 12-NS(Me) in aqueous suspensions of egg lecithin, lysolecithin, and sodium dodecyl sulfate (SDS). Since the coupling constant for the probe depends on the polarity of its environment, this was evaluated for a wide range of solvents, and the calibration was then used to establish the polarity of the oxazolidine ring environment in the different suspensions. The Z values (*26*) for these suspensions were 65.0, 74.0, and 94.1. Consequent proposals were: an erect conformation for the probe in egg lecithin, an erect conformation plus water penetration for the probe in lysolecithin, and a bent conformation for the probe in SDS.

We prefer an explanation based on a dynamic equilibrium between the erect and bent conformations. If a nonpolar environment has a Z of 60 [isooctane (*25*)] and a polar environment has one of 94.3 [water (*26*)], we obtain the conformational distributions shown in Table I.

Table I. Conformations of 12-NS(Me) in a Host Lipid

Host Lipid	*Z Value*	*% Erect*	*% Bent*
Egg lecithin	65.0	85.4	14.6
Lysolecithin	74.0	59.2	40.8
SDS	94.1	0.6	99.4

This simplified treatment clearly depends on the choice of Z for a nonpolar environment. We also assume that 12-NS(Me) can only exist in one or the other conformation and that water penetration may be neglected. If our explanation is correct, ESR studies cannot distinguish the two conformations and report only an average; hence, the time for such a conformation change cannot greatly exceed 10^{-7} sec. Alternatively, the contribution of a bent conformation is detectable, but interpretation of the spectra have so far ignored this possibility. Also, in neglecting all water penetration at the fourth or higher carbon atoms on the alkane chain, we clearly overstate the situation. Water will permeate a monolayer (particularly when expanded) or a bilayer. Nevertheless, we believe that existing polarity or water-penetration profiles overstate the case for water penetration, *i.e.*, they assume that the probe always maintains an erect conformation, and they neglect the possibility that the oxazolidine ring spends substantial time in the interface. The precise time spent is, of course, a function of ring position on the probe molecule chain.

Acknowledgments

We wish to express our appreciation to M. D. Barratt for bringing Ref. 25 to our attention.

Literature Cited

1. Griffith, O. H., Waggoner, A. S., *Accounts Chem. Res.* (1969) **2**, 17.
2. McConnell, H. M., McFarland, B. G., *Quart. Rev. Biophys.* (1970) **3**, 91.

3. Jost, P., Waggoner, A. S., Griffith, O. H., *in* "The Structure and Function of Biological Membranes," L. Rothfield, Ed., Chap. 3, Academic Press, New York, 1971.
4. Smith, I. C. P., *in* "Biological Applications of Electron Spin Resonance," J. D. Bolton, D. Bory, H. Swartz, Eds., Wiley-Interscience, New York, 1972.
5. Cadenhead, D. A., Katti, S. S., *Biochim. Biophys. Acta* (1972) **307,** 279.
6. Cadenhead, D. A., Demchak, R. J., Müller-Landau, F., *Ann. N.Y. Acad. Sci.* (1972) **195,** 218.
7. Cadenhead, D. A., Müller-Landau, F., *Biochim. Biophys. Acta* (1973) **307,** 279.
8. Cadenhead, D. A., Müller-Landau, F., *in* "Proceedings of the XXIst Colloquim: Protides of the Biological Fluids," H. Peeters, Ed., pp. 175-182, Pergamon Press, London, 1973.
9. Seelig, J., Seelig, A., *Biochim. Biophys. Res. Comm.* (1974) **57,** 406.
10. Seelig, J., Niederberger, W., *J. Amer. Chem. Soc.* (1974) **96,** 2069.
11. Seelig, J., Niederberger, W., *Biochemistry* (1974) **13,** 1585.
12. Seelig, A., Seelig, J., *Biochemistry* (1974) **13,** 4839.
13. Phillips, M. C., Chapman, D., *Biochim. Biophys. Acta* (1968) **163,** 301.
14. Singer, S. J., Nicolson, G. L., *Science* (1972) **175,** 720.
15. Bretcher, M., *Science* (1973) **181,** 622.
16. Cadenhead, D. A., Müller-Landau, F., *J. Coll. Interface Sci.* (1974) **49,** 131.
17. McConnell, H. M., McFarland, B. G., *Ann. N.Y. Acad. Sci.* (1972) **195,** 207.
18. Griffith, O. H., Dehlinger, P. J., Van, S. P., *J. Membrane Biol.* (1974) **15,** 159.
19. Cadenhead, D. A., *Ind. Eng. Chem.* (1969) **61,** 22.
20. Adam, N. K., Danielli, J. F., Harding, J. B., *Proc. Roy. Soc. (London) Ser. A* (1934) **147,** 491.
21. Jeffers, P. M., Dean, J., *J. Phys. Chem.* (1965) **69,** 2368.
22. Cadenhead, D. A., Phillips, M. C., *J. Coll. Interface Sci.* (1967) **24,** 491.
23. Glazer, J., Alexander, A. E., *Trans. Faraday Soc.* (1951) **47,** 401.
24. Cadenhead, D. A., Müller-Landau, F., to be published.
25. Dodd, G. H., Barratt, M. D., Rayner, L., *F.E.B.S. Letters* (1970) **8,** 286.
26. Kosower, E. M., *J. Amer. Chem. Soc.* (1958) **80,** 3253.

RECEIVED September 17, 1974. Work supported in part by grant HL-12760-06 from the U.S. Heart and Lung Institute.

24

Molecular Motion of Surfactant Molecules at the Air–Water Interface

ESR Spin Label Results

THOMAS R. McGREGOR, WYONA CRUZ, CANDACE I. FENANDER, and J. ADIN MANN, JR.[1]

Biophysics Department, University of Hawaii, School of Medicine, 1997 East-West Rd., Honolulu, Hawaii 96822

The complementary application of film balance and electron spin resonance techniques has given information regarding molecular motion in monolayers. The molecules used, 3-nitroxide androstan 17β-ol and 3-nitroxide cholestane, have the molecular geometry required for a test of the Cooper-Mann theory of surface viscosity. These ellipsoidal molecules show distinctly different surface behavior as shown by the complete collapse of the hyperfine lines in the case of the cholestane derivative and the apparent lack of exchange broadening for the androstane derivative. These results are modeled in terms of transitional surface regions.

Over the last few years spin labeled compounds have been used as structure probes of biological membranes and model systems. The nitroxide labeled molecules have included the phospholipids (*1, 2*), the fatty acids (*3*), and the molecules pertinent to this work, the steroids (*4, 5*). The use of such probes has provided valuable information concerning the structure and dynamics of the biomembrane as well as raising controversy over the perturbing effects of the label (*6*).

In general the labeled compound is an analog of a compound found naturally in the membrane system, and thus it is assumed that the motional behavior of the labeled species is very similar to that of its analog. It is surprising, therefore, that the surface properties of the spin labeled compounds have not been studied as extensively as those of the unlabeled

[1] Present address: Department of Chemical Engineering, Case Western Reserve University, Cleveland, Ohio 44106.

species. [Cadenhead and Muller-Landau give data pertinent to this point (*7*). They also arrive at some very interesting conclusions pertaining to the perturbation effect of the oxizolidine ring on the surface properties of the fatty acids.] The assumption of motional similarity is a very serious one and one that has been discussed in the context of the membrane problem by other authors (*6*). We believe that our technique and results are pertinent to an examination of the motional similarity problem.

Further, Mann and McGregor (*12*) suggested that the study of the ESR spectral characteristics of labeled molecules in monolayers could lead to an experimental determination of surface viscosity numbers for expanded monolayers. While we do not report surface viscosity numbers here, we do report the spectra obtained using two ellipsoidal molecules in monomolecular films spread on water. We also report the low surface pressure isotherms for these systems and comment on the extent to which isotherms in general reflect molecular motion in monolayers.

The molecules used in this study were 4′,4′-dimethylspiro[5-androstan-1β-ol-3-2′-(1′,3′-oxazolidine-3-oxyl)] and 4′,4′-dimethylspiro[cholestane-3-2′-(1′,3′-oxazolidine-3′-oxyl)]. To simplify identification we refer to them as 612 and 611 respectively, their source (*8*) catalog numbers. Figure 1 shows the structure of these compounds along with their identifying numbers. Note that the major parts of the two molecules—namely the steroid skeleton and nitroxide group—are the same. The difference in these molecules occurs at the 17 position at one end of the "ellipsoid."

Theory

The theory of electron spin resonance (ESR) is much the same as proton magnetic resonance (PMR) and consequently can be couched in similar terms. The difference in the Hamiltonians for the two cases lies mainly in certain constants such as the Bohr magnetron as well as in the use of electron spin operators in the first case for spin-spin interactions.

The Hamiltonian for the ESR case may be written with the important contributions as

$$\hat{\mathcal{H}} = \beta H^{t}\cdot\mathbf{g}\cdot\hat{S} + \hat{S}^{t}\cdot\mathbf{A}\cdot\hat{I} - \sum_{i\neq j} J_{ij}\hat{S}_i{}^{t}\cdot\hat{S}_j + g^2\beta^2 \sum_{i\neq j} \frac{(3\cos^2\theta_{ij}-1)}{r^3{}_{ij}} (\hat{S}_i{}^{t}\cdot\hat{S}_j - 3\hat{S}_{z,i}\hat{S}_{z,j})$$

where H = external magnetic field, $\hat{S},\hat{I}$ = spin operators for electron and nucleus respectively. The first term in this expression is the Zeeman term and is responsible for most of the energy of the spin system. The second term represents the hyperfine interaction between the nuclear and elec-

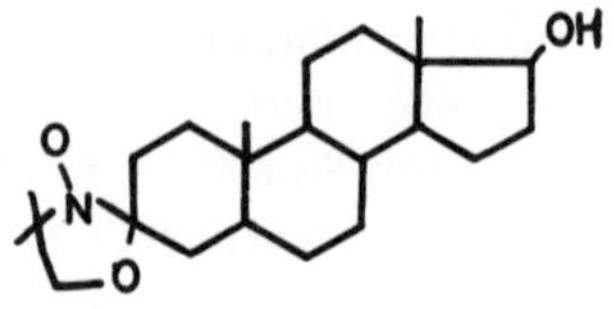

Figure 1. Structures of 4′,4′-dimethylspiro[5-androstan-17β-ol-3-2′-(1′,3′-oxazolidine-3-oxyl)] and 4′,4′-dimethylspiro[cholestane-3-2′-(1′,3′-oxazolidine-3-oxyl)], i.e., *612 and 611, respectively*

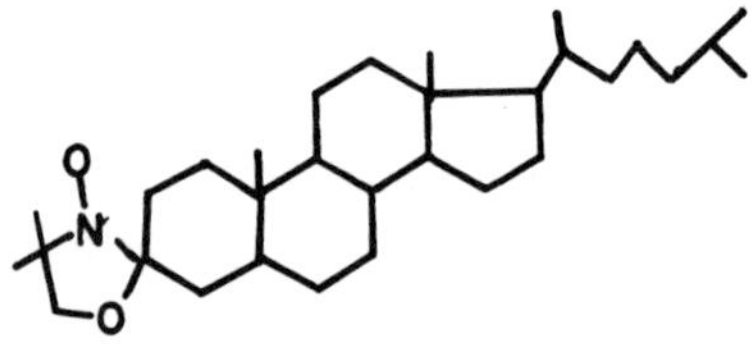

tron spins. The quantities **g** and **A** are rank two tensors, the matrix representation of both being diagonal to reasonable accuracy (*9*) in the molecular frame of reference. Since the magnetic field defines the coordinate system for the experiment—*i.e.*, to a first approximation, electron spin is quantized along the external field—the elements of the **g** and **A** tensor will be orientation dependent. Use of this fact has indeed been made in determining the values of the diagonal elements in the molecular frame (*9*). In general, the two terms in the Hamiltonian discussed so far are important in any ESR experiment in which information concerning orientation to external magnetic fields is important. We expand this point later in reference to the nitroxide spin labeled compounds.

The third term in the Hamiltonian is the exchange term and represents the quantum mechanical exchange interaction. The quantity J_{ij} is, in fact, the exchange integral. Finally, the last term is a dipole interaction term and represents the magnetic dipole-dipole interaction of two spins in close proximity. These last two terms, in contrast to the first two, do not depend upon orientation relative to an external coordinate system but rather depend upon the relative orientation of more than one spin quantity. These interactions furthermore are concentration dependent, a fact which is important for our experiments.

The ESR spectra of nitroxide radicals in solution depends upon the rate of tumbling and upon the concentration of the radical. If the nitroxide moiety is undergoing a tumbling which is rapid on the ESR time scale ($\sim 10^{10}$ sec^{-1}), the spectrometer senses only averaged motion. In this case, the first two terms in the Hamiltonian may be written as

$$g\,\beta\,\hat{H}\cdot\hat{S} \text{ and } a_{H}\hat{S}\cdot\hat{I}$$

where $g = 1/3\ Tr\ \mathbf{g}$ and a_H = hyperfine splitting constant $= 1/3\ Tr\ \mathbf{A}$. Parenthetically, the traces of the interaction tensors **g** and **A** are tensor invarients and consequently are independent of the state of molecular motion. For nitroxides, rapid tumbling results in a three-line spectrum with a separation between lines of about 14 gauss (a_H) and centered at a field corresponding to about $g = 2.006$. These numbers vary somewhat from system to system. This variation is mainly the result of solvent polarities or slightly different bonding situations and is usually only a few percent at most. The effect of increasing radical concentration on the three-line spectrum is an equal broadening of each of the three lines until the lines collapse to a single broad line which then narrows upon further increase in concentration.

Another instance of molecular tumbling is of interest when the possibility of ordered systems exists. Should the ordered system be aligned with the external field such that the field is along the molecular y axis (for example), the rapid tumbling about the y axis averages out the other two principal tensor elements. Then the spectrum obtained depends upon the orientation of the sample relative to the magnetic field.

At the other extreme of molecular motion is the situation where the radical is tumbling very slowly on the ESR time scale. This sort of behavior is obtained by using low temperatures and solvents of high viscosity. Here there is no averaging of tensorial elements. The resulting spectrum is a sum of spectra of radicals in every possible orientation. As temperature is decreased and/or viscosity is increased, the lines of the nitroxide spectrum are not all broadened equally. Because of the larger value of A_{zz} relative to the other elements of the diagonalized **A** tensor, the high field line on the spectrum is broadened more rapidly than are the others. Although this situation is not germane to the present work, we include it for completeness.

In the application of ESR spectroscopy to the study of monolayers, one is particularly interested in comparing the resulting spectra with the known state of the monolayer. To this end, it is necessary to consider the surface pressure–area curve of the system. The expected spectra should depend upon whether the monolayer being examined spectroscopically was in a gaseous, liquid expanded, liquid, or any of the states insoluble monolayers have traditionally shown. Since we are interested in gaseous systems and since one expects the dilute gaseous film to exhibit well known patterns, the question arises as to what the apparent state function which fits the experimental isotherm means on a dynamic-molecular level. Two-dimensional gas laws constructed from kinetic arguments which ignore substrate interaction with polar head groups are obviously suspect (*10*), yet, the expectation that monolayers in the "gaseous" state will obey a gas equation seems to persist. We suggest that there is no *a priori* reason to expect a 2-D gas equation to be followed at ordinary surface pressures and areas. Even if the expanded monolayer does follow a 2-D isotherm, the interpretation of that result should involve possible delocalization.

Experimental

All ESR spectra were taken on the Varian E-4 spectrometer using the E-231 resonant cavity. The cavity was used with a waveguide modified such that the cavity stacks were aligned horizontally. This horizontal arrangement permitted the entrance and the proper orientation of the miniature trough on which the monolayers were spread. The miniature troughs were made to order by the James F. Scanlon Co., 2428 Baseline Ave., Salvang, Calif. 93463. These sample cells were of quartz, and during a run they were covered by a quartz cover slip to ensure a closed system. The overall length of the cell was 16 cm. Of this, 6 cm was a flat portion which contained the trough. This portion was 9 mm wide and 3 mm thick. The trough itself was in the center of the 6-cm portion. Centered on the 6-cm portion was a rectangular region 8 mm wide, 3 cm long, and 1 mm deep. Centered within this region was the originally designated trough. This trough was 7 mm wide, 2 cm long, and 0.5 mm deep. Early experiments were done using this section with paraffined sides; however, the spreading solvent tended to dissolve and spread the paraffin. Consquently, the entire depressed region without paraffin was used as a trough. The remaining 10 cm of the cell consisted of a quartz rod 5 mm in diameter which fitted into a cell holder. The cell holder was mounted on a Wetzlar micromanipulator in such a way that there could be pivotal motion of the cell about a fulcrum on the holder. This, together with the degrees of freedom of the micromanipulator, allowed good alignment of the cell with the nodal E plane of the cavity.

The cell, when situated in the cavity, had its flat surfaces parallel to the external magnetic field. Since the magnets on the E-4 were not movable, it was not possible to obtain spectra with the field perpendicular to the surfaces of the cell and hence the surface of the water.

The monolayers in the cell were deposited on doubly distilled water. The method of deposition consisted of running out a small drop of spreading solution and allowing it to touch the surface. The volume of the drop was 1 μl and was delivered to the surface with an accuracy of $\pm 1\%$. The spreading solvent was cyclohexane which was chosen for its solvent properties and its boiling point. Other solvents were tried. These included petroleum ether, chloroform, and cyclopentane. Cyclopentane and petroleum ether evaporated very rapidly from the water surface in the cell and thus gave rise to the possibility of precipitation of the spin labeled compound. While these solvents were appropriate for a conventional trough, the same solvent was desired for the monolayer deposition on both the conventional trough and the miniature trough. Chloroform had the desired solvent properties and boiling point but owing to its density was difficult to use. Cyclohexane, then, was the spreading solvent. Although spectral grade cyclohexane was used, it was routinely run through a silica gel column before use. The monolayers of both 611 and 612 were deposited in the same way with the same solvent. When monolayers with decreased areas per molecule were desired, more spreading solution was added to the previous system. Experiments were also done using only a single deposition to achieve a desired area per molecule, and no detectable difference in results could be noted.

For 612 monolayers, it was necessary to show that the surfactant molecules were actually confined to a surface region. The simplest way to do this was to use standard monolayer removal techniques or modifications thereof and monitor the signal of the residual monolayer. The method adopted as being most suitable for the experimental arrangement was a "touching" technique (*10*). The monolayer was spread on the miniature water surface with the proper calculated area per molecule, and the spectrum was taken. The cell was then removed from the cavity, and a wet quartz slide was brought into contact with the surface for 3–5 sec. The cell was then replaced into the cavity, and a spectrum was taken under the same instrument settings as the previous one. A reduction in intensity of signal indicated a smaller number of spin systems in the active region of the cavity.

The small size and the particular design of the sample cells rendered the determination of pressure area characteristics in them impractical. Consequently, these characteristics were determined with a conventional trough (*11*). The solvent for the spin label solution was cyclohexane, and the volumes of spreading solutions used were 0.2–0.3 ml. These volumes were applied with a pipet by placing drops at various locations on the water surface. Techniques were checked by determining well known pressure–area curves such as steric acid.

At pressures above about 2 dynes/cm for 612 and about 3 dynes/cm for 611, the original records show some time dependence. There may be several reasons for this time dependence, one of which could be leakage around the barriers. To check this possibility, talcum powder was sprinkled on the film side of the barriers, and the presence of talcum on the clean surface side of the barriers was determined. Since no talcum flow was noted, the possibility of leakage was rejected.

All substrates had a pH of about 6.5. All experiments were done at room temperature, 23° ± 1°C.

Results

The results of the film balance measurements at low pressures are shown in Figures 2 and 3 in the form of $\Pi\Sigma$ *vs.* Π plots. Here, Π is the

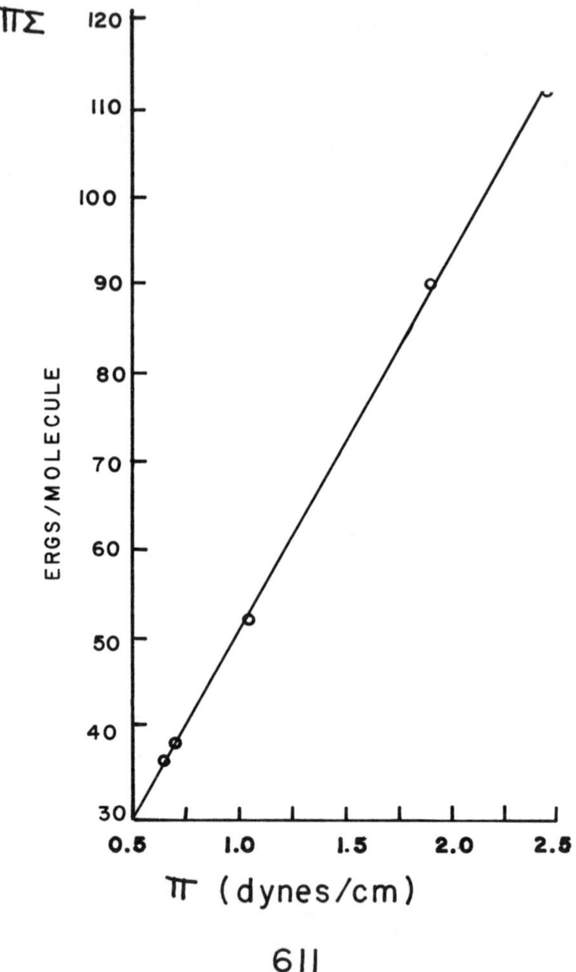

Figure 2. Surface behavior of 611 monolayer at low surface pressures. In this pressure regime, the film follows the equation $\Pi\Sigma = C + B\Pi$. B, *the slope of the line, can be interpreted as the effective co-area of the molecule.*

surface pressure, and Σ is the area per molecule. Plots of this type are useful in determining how far a monolayer system is deviating from a slightly nonideal type of state equation. If straight lines are obtained, as in these cases, the slope of those lines should correspond to excluded areas per molecule or effective co-areas. It should be emphasized once again that these are low pressure plots. At much higher pressures, negative deviation becomes pronounced. This is especially true for the 612 monolayers which did not support a surface pressure of more than a few dynes/cm before collapse. Table I summarizes the film balance data for both monolayers. One immediately notices the low values of the intercepts, indicating that these terms are not kT types. In addition, the coarea of 612 seems abnormally low for a steroid while that of 611 seems to be about right. It is possible that the hydrocarbon on 611 chains contributes the excess area over that of 612.

Figures 4 and 5 give the typical ESR spectra of the two monolayers. The ESR spectra of the 612 monolayer in Figure 5 were obtained by successive application of the touching technique. In both figures, the problem of signal–noise ratio is evident. The figures are the results of single scans, but we have since used a time averaging computer to enhance the signal–noise ratio. That data will be published in a subsequent paper. The signal–noise ratio in Figure 5 is somewhat better than it should be since some overmodulation (by 2.5 gauss) was done to enhance the signal.

Discussion

The most interesting feature of the pressure–area data in light of ESR results is that there is nothing particularly striking about them. The

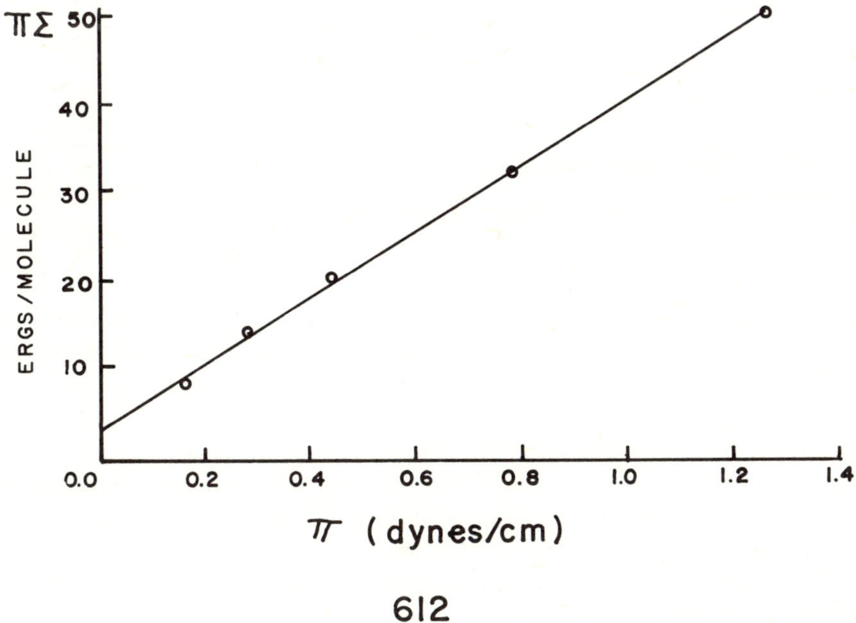

Figure 3. Surface behavior of 612 monolayer at low pressures. An equation of the same form as that describing the 611 monolayer fits the 612 data. Note the low limit of pressure.

Table I. Summary of Film Balance Data[a]

	Parameters	
Monolayer	$\Sigma_o (A^2/m)$	C *(ergs/m)*
612	35.7	4.5
611	41.2	10.0

[a] Equation of state: $\Pi(\Sigma - \Sigma_o) = C$.

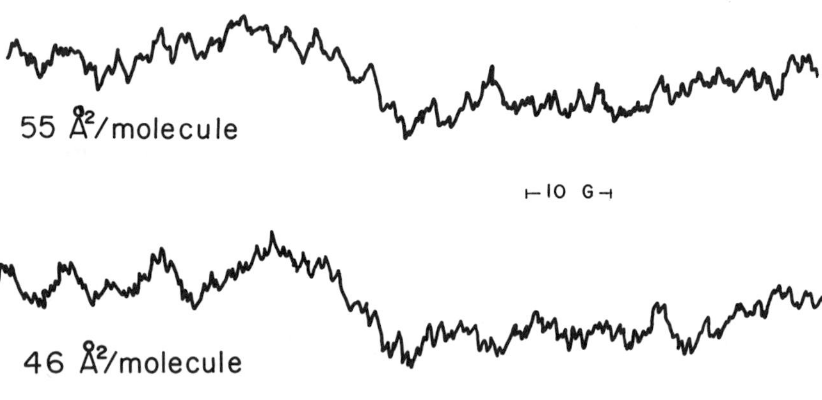

611

Figure 4. Typical spectra of 611 monolayer. Two spectra taken with a single scan at different surface densities. Complete collapse of the hyperfine is evident although low signal–noise ratio renders accurate linewidth measurements impossible. Scale is the same in Figure 5.

two systems formed by the 611 and the 612 molecules at the air–water interface are behaving classically at low pressures. One may comment at this point on the equations which fit the data. The form $\Pi\Sigma = \Sigma_0\Pi + C$ or $\Pi(\Sigma - \Sigma_0) = C$ is that of a slightly imperfect gas provided C is a kT term and provided Σ_0 is the co-area of the molecule. As stated previously, the constant C in both cases is much too low to be kT. Consequently, the equation ought not to be construed as a gas equation. One should be aware that the form of the equation which fits the data does not necessarily imply the physical state of the system. In any case, a temperature study of these isotherms would be interesting but was not attempted. It may be possible by this means to determine how much of the term C is caused by kT and what the form of the monolayer substrate interaction may be.

Although the film balance measurements at low pressure indicate that the systems in question are behaving similarly, the ESR results show that the molecular motion in these systems is very different. The spectra for the 611 system shows a total collapse of the hyperfine splitting and indicates an almost crystalline spectrum. We discount the possibiity of extensive patch or island formation at the pressures where ESR measurements were made. First, the film balance data indicate a more expanded film than expected if patch formation were important. Secondly, if patches were present, they would be in equilibrium with freely moving molecules, and these would give rise to a triplet superimposed on the single line. This was not observed. We conclude that the ESR spectra of the 611 monolayer are those of a system in which the molecules are undergoing rapid motion laterally and thus are in frequent collision. The frequent collisions are accompanied by the exchange process which is responsible for the complete collapse of the hyperfine structure.

The spectral characteristics of the 612 monolayer indicate that the molecules are colliding much less frequently, and the apparent lack of hyperfine collapse indicates that the time between effective collisions is long on the ESR time scale. The problem to be considered, then, is how two systems can show such similar classical surface behavior and such dissimilar motional behavior.

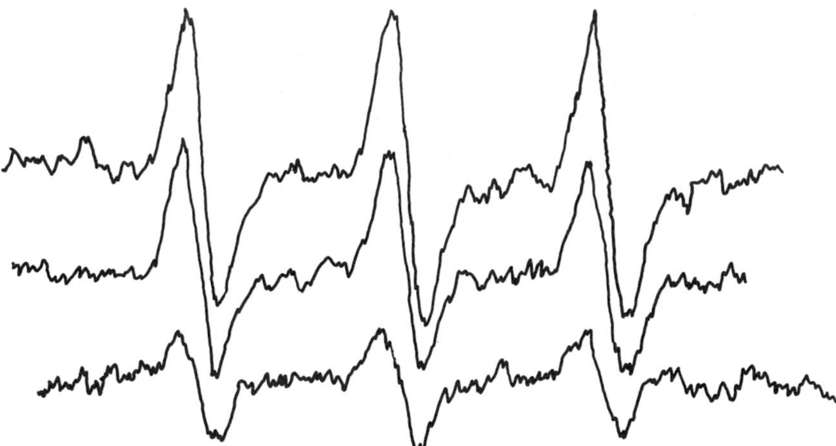

Removal of 612 monolayer at

constant total area

Figure 5. Typical spectra of 612 monolayer. Again the signal noise problem is apparent. In addition, some distortion of spectra was introduced by overmodulation required to enhance signal. The decrease in signal amplitude upon successive application of the "touching" technique suggests localization of the 612 molecules to the surface region.

Since the salient features of the ESR spectrum depend upon the nitroxide–nitroxide interaction, one must examine questions of molecular orientation at the interface. The 611 molecule has only one polar group —*viz.*, the oxazolidine system. Assuming the existence of a substrate whose only function is to receive the polar group of the surfactant molecule, one would expect the 611 molecule to be oriented toward the vertical with the nitroxide moiety in the substrate. The 612 molecule, however, has two polar groups—the oxazolidine ring and the hydroxyl group. The hydroxyl group being more polar is likely to be the one in the substrate. This molecule will also tend to be oriented vertically. The experimental co-areas confirm vertical orientation for both molecules. Further, at very low pressures the parts of the molecules not in the substrate will execute motion such that each would have a cone of excluded volume associated with it. One can reasonably assume that the trajectory about the axis of the cone occurs on a time scale of the order of magnitude of the rotational correlation time since these molecules are rigid by virtue of the steroid skeleton. So the end of the molecule not in the substrate must be rotating with a frequency of about 10^{10} sec^{-1} and encountering other molecules executing similar motion. The encounter frequency must be considerably less. The maximum that this encounter frequency would be is about 10^9 sec^{-1}, corresponding to the number of collisions a fairly small molecule would suffer in the gas phase. Thus, for every encounter of the groups in the substrate there would be a minimum of 10 encounters for the tails. On this basis the nitroxide moieties of the 612 should be in collision 10 times as often as the corresponding moiety of the 611. In such a situation, exchange broadening would be expected from the 612 system rather than from the 611 system. This is opposite to what is found experimentally.

We believe that the answer to this apparent anomaly may be found in the concept of three-dimensional interfaces. This concept is associated with the existence of a transition region between two liquid phases such that physical properties vary continuously through the interface from one bulk phase to another. This implies the existence of at least two environments in which a molecule may find itself. In the particular case of the air–water interface, the two extreme environments would correspond to a gas and a liquid. The 612 molecule, having two polar groups, would tend to be found in the more liquid regions of the interface because of the tendency for solvation of the polar group. The 611 molecule, on the other hand, has a hydrocarbon tail and the less polar group. This molecule would be found in the energetically favorable gaseous regions rather than the deeper regions favored by 612. These different regions will certainly offer different local viscosities to the attendant molecules, and thus these molecules will exhibit the motional behavior appropriate to those viscosities.

One of the more puzzling aspects of the work just completed has been the notable absence of a concentration dependence upon the spectral characteristics of the monolayer systems (*5*). It is suggested that the three-dimensional interface concept also handles the apparently anomolous behavior nicely. One notes that there can be a meaningful quantity known as concentration only if the volume of the system is well defined. Since the surface concentration depends only on number of molecules per unit area, one would expect that if the surfactant molecules were confined to a surface, there would have been a surface concentration dependence. However, if one is dealing with a three-dimensional interface, one must consider a "surface" concentration function which depends upon thickness as well as upon area. If this thickness depends upon the number of molecules present in some unknown manner, then one has not defined concentration in any meaningful way. Thus if no surface concentration dependence of the spectral characteristics of these systems is noted, one must admit the possibility of delocalization. Specifically, at the surface pressures indicated, it is suggested that the air–water interface is a capacitive one for the molecules which were studied.

The three-dimensional interface can rationalize the difference in the ESR spectra of the two systems. If one accepts the argument, one is led to some disturbing conclusions concerning the film balance data in general. Here, these data indicate that we have been dealing with very similar systems. The obvious explanation for the lack of resolution in film balance data is that the film balance measurement is a macroscopic measurement and is thus much too gross to differentiate between thick and thin interfaces. Yet, spreading pressure–area curves are often interpreted in terms of molecular configurations strongly localized to monolayers. Assumptions concerning molecular motion are also made when the data are fit to gas law equations. It is not clear to us that the range of validity of these assumptions concerning orientational and motional behavior of surfactant molecules has been thoroughly explored. The ESR technique, however, is intimately concerned with the effects just mentioned and may very well be the tool required for the exploration. At this point, we must regard the interpretation of film balance work, only in the terms of a highly localized monolayer, with skepticism.

Acknowledgment

We are indebted to Laurence Piette and the personnel of the Biophysics Laboratory for helpful suggestions and fruitful discussions during this work.

Literature Cited

1. Devaux, P., Scandella, C. J., McConnell, H. M., *J. Mag. Res.* (1973) **9**, 474-485.
2. Gaffney, B. J., McConnell, H. M., *Chem. Phys. Letters* (1974) **24**, 310.
3. Jost, P., Libertini, L. J., Hebert, V. C., Griffith, O. H., *J. Mol. Biol.* (1971) **59**, 77.
4. Hubbell, W. I., McConnell, H. M., *Proc. Nat. Acad. Sci.* (1969) **63**, 16.
5. Sackmann, E., Träuble, H., *J. Amer. Chem. Soc.* (1972) **94**, 4482.
6. Seelig, J., Niederberger, W., *Biochem.* (1974) **13**, 1585.
7. Cadenhead, D. A., Muller-Landau, F., ADVAN. CHEM. SER. (1975) **144**, 294.
8. Syva Corp., 3181 Porter Dr., Palo Alto, Calif. 94304.
9. Libertini, L. J., Griffith, O. H., *J. Chem. Phys.* (1970) **53**, 1359.
10. Gaines, Jr., G. L., "Insoluble Monolayers at Liquid-Gas Interfaces," p. 159, Interscience, New York, 1966.
11. Mann, Jr., J. A., Hansen, R. S., *Rev. Sci. Instr.* (1963) **34**, 702.
12. Mann, Jr., J. A., McGregor, T. R., ADVAN. CHEM. SER. (1975) **144**, 321.

RECEIVED October 22, 1974. Work supported by Petroleum Research Fund Grant No. 6689-AC5, 6.

25

Molecular Motion of Surfactant Molecules at the Air–Water Interface

ESR Spin Exchange Relaxation as a Means of Measuring Surface Viscosity

J. ADIN MANN, JR.[1] and THOMAS R. McGREGOR

Chemistry and Biochemistry and Biophysics Departments, University of Hawaii, School of Medicine, 1997 East-West Rd., Honolulu, Hawaii 96822

Statistical theories of surface viscosity predict values of 10^{-10} g/sec or smaller. ESR spin exchange effects provide a method that may be unique for measuring such small values. Block equations can be constructed that describe the macroscopic magnetization of the spin system including the relaxation mechanism involving spin exchange caused by molecular collisions. These equations were solved as a function of the exchange frequency, ω_{ex}, for the nitroxide free radical case. Spectra were computed by assuming the degrees of motional freedom observed with the labeled androstane or cholestane molecules localized at the air–water interface. Crude estimates for computing collision frequencies and surface viscosities were constructed to outline the concepts involved. Even though line shape calculations show that surface viscosities $< 10^{-14}$ g/sec can be measured, the signal-to-noise ratio becomes troublesome as the surface density approaches 10^{13} spins/cm^2. The effects of substrate viscosity on surface viscosity and the exchange frequency are discussed.

It has been difficult to analyze accurately the fluid mechanics of the various classical devices used to measure surface viscosity (*1, 2*) until recently (*3, 4, 5*). Unfortunately, the most sensitive device (*4, 5*) could measure to only about 10^{-5} g/sec. This sensitivity is sufficient for measur-

[1] Present address: Department of Chemical Engineering, Case Western Reserve University, Cleveland, Ohio 44106.

ing viscosities of dense monolayers but cannot detect surface viscosity of dilute gas monolayers.

Theoretical predictions of surface viscosity have been made by Blank and Britten (*6*) who constructed a formalism based on fluctuations. Cooper and Mann (*7*) posed a dilute gas kinetic theory of monolayer transport that involved a modified Boltzmann equation which could be solved in detail. The substrate surface coupling could be easily accounted for quantitatively. The results of these calculations were similar in the sense that surface viscosities of the order of 10^{-10} g/sec were computed for typical monolayer systems. Even when the extensions involving the dense gas region were formulated and computed, the expected surface viscosity numbers were around 10^{-8} g/sec.

The Cooper–Mann theory of monolayer transport was based on the model of a sharply localized interfacial region in which ellipsoidal molecules were constrained to move. The surfactant molecules were assumed to be massive compared with the solvent molecules that made up the substrate and a proportionate part of the interfacial region. It was assumed that the surfactant molecules had many collisions with solvent molecules for each collision between surfactant molecules. A Boltzmann equation for the singlet distribution function of the surfactant molecules was proposed in which the interactions between the massive surfactant molecules and the substrate molecules were included in a Fokker–Planck term that involved a friction coefficient. This two-dimensional Boltzmann equation was solved using the documented techniques of kinetic theory. Surface viscosities were then calculated as a function of the relevant molecular parameters of the surfactant and the friction coefficient. Clearly the formalism considers the effect of collisions on the momentum transport of the surfactant molecules.

For dilute monolayers, the intermolecular potential between two surfactant molecules was not needed to construct the kinetic formulas. Obviously there are solvent–surfactant interactions since the entire system is a liquid. However, it was possible to incorporate such interactions into a friction coefficient that can be computed in principle or left as a parameter; details are given by Cooper and Mann (*7*). The extension to dense gas monolayers requires a potential function and a radial distribution function for the surfactant molecules. The formulation based on the Rice–Allnatt approach was developed by Cooper and Mann (*7*).

We have been searching for experimental methods that can measure surface viscosities as low as 10^{-10} g/sec or measure the collisional dynamics that should correspond to the Mann–Cooper model. To qualify, the experimental method must respond to dilute monolayers having densities less than 10^{14} molecules/cm^2. From our experience with the ESR spin label technique for measuring bulk viscosity effects in ultrathin films (*8*), we developed analogous methods for monomolecular films deposited on aqueous substrates.

The ESR spin label technique is a well established way to study molecular motion in anisotropic media such as biological membranes and their models. A spin label is a molecule that contains a stabilized free radical moiety whose ESR spectrum is sensitive to orientation of label molecules and their concentration. Extensive literature covers the rotational motion of molecular probes using such techniques. Three useful references are the review by McConnell and McFarland (*9*), the book edited by Muus and Atkins (*10*), and the recent review article by Atkins (*11*) on relaxation processes and line widths. Even though certain technical problems are unsolved, the line shape analysis for the rotating nitroxide probe is relatively well understood. In the sense that the rotating probe is inhibited by interactions with its neighbors, including solvent molecules, one can construct a method for measuring local viscosity effects by observing the ESR line shape distortion of hindered probe molecules. For example, the rotational correlation time of a probe molecule is often quoted to be related to the viscosity of the medium through

$$\tau_c = \frac{4\pi r^3}{3\,kT}\,\eta^* \tag{1}$$

where r is the effective radius of the spin label, T is the temperature and η^* is the effective viscosity. Experiments (*12*) with gels and certain solids show that η^* need not be the macroscopic viscosity of the fluid. Also, the appropriate probe molecules for testing the Cooper–Mann theory are ellipsoidal; thus more than one rotational correlation time is required in principle. We wish to emphasize that it is just these properties that make spin labeling an important tool in studying biological membrane structure and dynamics. Such structures are complicated mixtures of lipid and protein molecules that interact strongly. Even though probe molecules may disrupt the structure somewhat, their ESR spectra can give strong indications of local ordering and fluidity in membrane systems. Similar studies can be done with mixed monolayer systems in which spin label molecules are incorporated as an additional component. Preliminary experiments along these lines are reported in a second paper of this series (*13*) (Chapter 24). However, the nature of surface momentum transport for a multicomponent system is not well understood. Since we preferred to work with a single surfactant component, a second method was investigated and is described in the remainder of the paper.

The character of spin label ESR spectra changes dramatically as a function of concentration. Even though accurate modeling of the concentration effect depends upon sorting out a number of possible mechanisms, collisions are strongly involved in the change in spectral line shapes at higher spin label concentrations. If the spin label molecules can be constrained to an interfacial region, it is possible to monitor the collision dynamics in a situation closely resembling the model for the kinetic theory calculation of Cooper and Mann (*7*).

This paper predicts spectra for a dilute gas of spherical particles, each containing a spin label moiety. The formalism developed by Cooper and Mann is not used although the generalization is certainly possible. Rather, we used the simplest model that retains the collisional character of the dilute gas kinetic theory.

Theory

The formalism necessary to calculate ESR spectra that arise from collisional exchange interaction is outlined briefly, and detailed equations used in our calculation are given. A simple but naive model for the collision frequency and its connection to the spin label, exchange collision frequency is developed. The model is instructive and indicates the limit of sensitivity of the spin label method in measuring surface viscosities.

The Macroscopic Theory of Exchange Contributions to the ESR Line Shape. Following Sackman and Träuble (*14, 15, 16*) and Devaux, Scandella, and McConnell (*17*), we adopted the Block equation formalism for spin exchange effects in computing ESR spectra. (Consult preceding references for the validity problem in using the macroscopic equations). Because reasonable diffusion coefficients for lateral spin label transport in biological membrane structures can be computed using the Block equation formalism, our choice is supported on a pragmatic level. The Block equations are satisfactory for our purposes. Nevertheless, a microscopic theory of ESR line shapes is possible and might be useful in connection with the surface transport problem.

Recall that the spin Hamiltonian for the system is composed of the following terms:

$$\hat{\mathcal{H}} = \hat{\mathcal{H}}_{\mathfrak{x}} + \hat{\mathcal{H}}_{\mathrm{ex}} + \hat{\mathcal{H}}_{\mathrm{dipole}} \tag{2}$$

$$\mathcal{H}_{\mathfrak{x}} = \beta\, \hat{H}^{t}\cdot\mathbf{g}\cdot\hat{S} + \hat{S}^{t}\cdot\mathbf{A}\cdot\hat{I}$$

We assume that the dipole–dipole interaction term can be ignored for this calculation. However Sackman and Traüble (*14, 15, 16*) and Devaux and McConnell (*17*) believe there is a contribution from this term in real spectra. The problem appears tractable if measurements are made at temperatures high enough to average out the dipole–dipole terms.

The first term of the spin Hamiltonian predominates strongly in the low spin label concentration regime. For nitroxide spin labels, the tensor, g, involved in the Zeeman term that couples the magnetic field to the electron spin operator is almost of cylinder symmetry and not isotropic. Similarly, the tensor, **A**, that couples the electron spin operator to the nuclear spin operator of the nitrogen atom is almost of cylinder symmetry and, again, not isotropic. From experimental results, these two tensors are diagonal in almost the same molecular coordinate system; the Zeeman and hyperfine splitting depend on the orientation of the molecular coordinate system to the magnetic field direction. The dependence on orientation is easily measured when nitroxide molecules are incorporated in the appropriate host lattice. The ESR spectrum of these molecules is a sharp triplet with splittings that depend on orientation with respect to the magnetic field. When the spin label molecules are free to tumble rapidly, compared with the ESR time scale, sharp triplets are again observed at a splitting intermediate between the maximum and minimum splittings seen in crystal orientation studies. There are stochastic models that can produce these spectra with good precision (*9, 10, 11*). When the spin label molecules are locked into a structure, but with random orientations, a greatly distorted triplet spectrum is observed which can be rationalized on the basis of a stochastic model. Computation of spectra for correlation times between the free tumbling and the frozen situations is much harder to accomplish and involves disputes as to the validities of the models invoked. Nevertheless, spectra for these situations can be simulated relatively well and are improving rapidly. Experimental data reported in a subsequent paper suggest that spin label molecules that were used to form ellipsoid like gaseous monolayers tumbled quite freely. The appropriate line widths for the hyperfine lines are available in the literature and were used in our calculations.

The second term in the spin Hamiltonian relates to collision dynamics. Specifically, $\hat{\mathcal{H}}_{ex}$ is expressed as

$$\hat{\mathcal{H}}_{ex} = - \sum_{i \neq j} J_{ij} \, \hat{S}_i \cdot \hat{S}_j \qquad (3)$$

where J_{ij} is the exchange integral for two adjacent spins. Since the exchange integral depends strongly on the separation of the two spins, the exchange term becomes important only when two molecules approach each other close enough for physical collision. However, not every collision involves a spin exchange. For example, if the colliding spin label molecules have parallel spins, spin exchange will not be seen. Also, collision configurations can occur with spin label moieties at opposite ends of each molecule, *i.e.*, sufficiently separated so that J_{ij} remains small. These properties are built into the model by the relation,

$$\omega_{ex} = f\ \omega_{coll} \tag{4}$$

where f is the probability that a collision will be productive.

We first show how to calculate the spectra as a function of the exchange frequency. Without spin exchange, the Block equation reads:

$$\frac{d}{dt} G_j + \left[\frac{2\pi}{T_{2j}} + i\ (\omega_j - \omega)\right] G_j = i\ Mo_j\ \gamma\ H_1 \tag{5}$$

where $\{G_j\}$ are the complex magnetizations; $\{Mo_j\}$ are the static magnetizations; γ is the gyromagnetic ratio, H_1 is the microwave magnetic field taken to be a constant in the rotating frame assumed for this problem, T_{2j} are the relaxation times associated with the widths of the hyperfine lines, and ω_j are the Larmor frequencies of the hyperfine lines. The index j runs over the three nuclear spin numbers -1, 0, 1. Equation 5 is modified to include exchange by assuming that the interaction should follow a first-order kinetic scheme:

$$\frac{d}{dt} G_A = -P_{AB}G_A + P_{BA}G_B \tag{6}$$

where $P_{AB}\ \Delta t$ is the probability that an exchange will occur between spins A and B in the period Δt. $G_A P_{AB}\ \Delta t$ spins will go from state A to state B while $G_B P_{BA}\ \Delta t$ spins will go from state B to state A.

Adding the appropriate terms of Equation 6 to Equation 5 gives the starting equation for the line shape analysis:

$$\frac{d}{dt} G_j + \left[\frac{2\pi}{T_{2j}} + i\ (\omega_j - \omega)\right] G_j + 2\pi \sum_k (G_j P_{jk} - G_k P_{kj}) = i\ Mo_j\ \gamma\ H_1 \tag{7}$$

Write G as the sum of real and imaginary functions, $G_j = u_j + iv_j$ and then separate Equation 7. Since most experiments are done under "slow passage conditions," set $(d/dt)G_j = +0$. Finally, scale u_j and v_j by the factor $Mo_j\gamma H_1$ to arrive at the working equations:

$$\frac{2\pi}{T_{2j}} u_j - (\omega_j - \omega)\ v_j + 2\pi \sum_k (u_j P_{jk} - u_k P_{kj}) = 0$$

$$\frac{2\pi}{T_{2j}} v_j + (\omega_j - \omega)\ u_j + 2\pi \sum_k (v_j P_{jk} - v_k P_{kj}) = 1 \tag{8}$$

The set of coefficients $(2\pi P_{jk})$ are the exchange frequencies $\{\omega_{jk}\}$. Further assume that each exchange is equally probable so that $\omega_{jk} = \omega_{ex}$ for all jk.

Equation 8 can be put into the form of a matrix equation for numerical work:

$$\mathbf{M}\, \boldsymbol{u} = \boldsymbol{c} \tag{9}$$

where **M** is obvious after specifying $\boldsymbol{u}^t = (u_{-1},u_0,u_1,v_{-1},v_0,v_1)$ and $\boldsymbol{c}^t = (0,0,0,1,1,1)$. Taking $\omega = \gamma H_o$ where $\gamma = g\beta/\hbar$ then

$$\{\omega_j\} = (\omega_o - \left(\frac{g\beta}{\hbar}\right) a_H),\ \omega_o,\ \omega_o + \left(\frac{g\beta}{\hbar}\right) a_H$$

where a_H is the coupling coefficient for the hyperfine lines. When the spin label is tumbling rapidly compared with the ESR "time scale," $a_H = 1/3$ trace $\mathbf{A} = 14$ gauss, and this is the case of immediate interest (*13*). Solution of the matrix Equation 9 for $\boldsymbol{u}$ as a function of ω proceeds easily by numerical methods; the observed spectrum is computed from the imaginary components as:

$$S\ (\omega) - v_{-1} + v_o \mid v_1 \tag{10}$$

The conventional methods for detecting ESR signals actually produce the derivative $dS/d\omega$. Differentiating Equation 8 after ω and reorganization gives

$$\mathbf{M}\, \frac{d\boldsymbol{u}}{d\omega} = \boldsymbol{D} \tag{11}$$

where **M** is unchanged from Equation 9 and

$$\boldsymbol{D}^t = (-v_{-1}\ -v_o\ v_1\ u_{-1}\ u_o\ u_1).$$

The numerical method involved first solving Equation 9 for $\boldsymbol{u}$ for given values of ω_{ex} and ω from which $\boldsymbol{D}$ was defined and then solving Equation 11 for $d\boldsymbol{u}/d\omega$. The observed derivative spectrum is given by:

$$\frac{dS}{d\omega} = \frac{dv_{-1}}{d\omega} + \frac{dv_o}{d\omega} + \frac{dv_1}{d\omega} \tag{12}$$

The computation of the $2\pi/T_{2j}$ terms in **M** was done by the relation:

$$\frac{2\pi}{T_{2j}} = \frac{\sqrt{3}}{2}\,\gamma\ \Delta H_j \tag{13}$$

where ΔH_j is the appropriate line width of the hyperfine lines that obtain in the absence of exchange.

The computations were done very simply in time sharing mode using the IBM APL/360 language. APL is a relatively new time sharing language in the sense that it has been commercially available for about five years although Iverson (*18*) originated the scheme some 15 years ago. The language has a distinct advantage in conciseness and in the power of the primitive functions for manipulating arrays. The APL functions for our computations are reproduced in the appendix (Figure A1). Only seven lines of coding were used to define the functions that computed both u and $du/d\omega$ as a function of the array of given frequencies, ω. Execution was rapid; only several minutes of observer time was required to compute a 50-point spectrum. Actual CPU time was not measured accurately but seemed to be less than 1 sec for each spectrum.

An Ultra-Simplified Model for the Collision Frequency Estimate. A naive but instructive model for the collision frequency, ω_{coll}, can be constructed following an argument by Hirschfelder, Curtis, and Bird (HCB) (*20*) for the three-dimensional situation. Even though a considerably deeper theory for the collision frequency can be constructed along the lines of the Cooper–Mann (*7*) kinetic theory of surface transport, all that is needed now is a crude model to provide information on the potential usefulness of spin label methodology for measuring the very small surface viscosities predicted by the kinetic theory of localized monolayers.

Imagine a sharply localized set of spherical molecules of diameter d_s moving with velocities $\pm\Omega\hat{e}_x$ and $\pm\Omega\hat{e}_y$ on the (x, y) surface of a substrate fluid composed of small molecules. The substrate molecules interact with the larger surfactant molecules through frequent collisions which, in the Cooper–Mann model, are accounted for by a term involving a friction coefficient added to the two-dimensional Boltzmann equation. This complication is finessed by asserting that substrate coupling modifies the molecular speed, Ω, which is assumed constant. The relative velocity of particles moving head on is 2Ω, and $\sqrt{2}\Omega$ is for those particles moving at right angles.

Collisions in a time interval, Δt, for particles moving in the $\pm x$ directions relative to a particle moving in the $+x$ direction will occur when the particle centers are inside the rectangle $2d_s$ by $2\Omega\Delta t$. There are n molecules per unit area; one-fourth go in the $-x$ direction, so that $nd_s\Omega$ collisions per unit time will occur.

The number of collisions in a time interval, Δt, for particles moving in the $\pm y$ direction relative to a particle moving in the $+x$ direction is $(2\sqrt{2})/2\ nd_s\Omega$. An estimate of the total number of collision in Δt is an estimate of the collision frequency:

$$\nu_c = (1 + \sqrt{2})\ n\ d_s\Omega = \xi\ n\ d_s\Omega \tag{14}$$

The mean free path is approximately

$$l = \frac{\Omega}{\nu_c} \tag{15}$$

Following HCB, the net flux in the y direction of the x component of the momentum (nmv_x)is the $\Pi_{xy} = \Pi_{yx}$ component of the surface pressure tensor, Π. Suppose that a velocity field is produced such that on the line $y = -l$ molecules are moving in the ($-x$) direction, on $y = 0$ their velocity is zero, and they are moving in the ($+x$) direction on $y = l$. Thus a gradient is imagined in the x component of the momentum, $p_x = nmv_x$. Note that the molecules have their last collisions at $y = \pm l$. The net flux of momentum, Π_{xy}, across the line $y = 0$ is taken to be $1/4\ \Omega[p_x^{(-)} - p_x^{(+)}]$ where $p_x^{(\pm)} = \pm l(dp_x/dy)$; all are parallel with HCB. Therefore,

$$\Pi_{xy} = -\ \tfrac{1}{2}\ \Omega\ lnm\ \frac{dv_x}{dy} \tag{16}$$

so that

$$\bar{\eta} = \frac{1}{2}\ \Omega\ lnm \tag{17}$$

Note that

$$n\ l = \frac{1}{\xi\ d_s} \tag{18}$$

depends only on the molecular dimensions; thus $\bar{\eta}$ is independent of n as expected in the dilute gas regime.

The final result of this crude model is

$$\bar{\eta} = \frac{m}{2\ \xi^2\ d^2_s}\ \frac{\nu_c}{n} \tag{19}$$

where the ratio ν_c/n depends only on the product $\xi d_s \Omega$ and is independent of n. Notice that as the substrate viscosity increases, the friction coefficient that couples the surfactant molecules to the substrate increases, restricting the velocity distribution so that in the limit $\Omega \rightarrow 0$. In that case ν_c and $\bar{\eta}$ both go to zero. While perhaps against one's three-dimensional intuition the result is nevertheless true and is obtained in our much more detailed theory of surface transport (7).

Discussion

The distortion of the initial triplet spectrum from that for $\omega_{ex} = 0$ proceeds in two ranges (Figure 1). First, the triplets broaden and overlap;

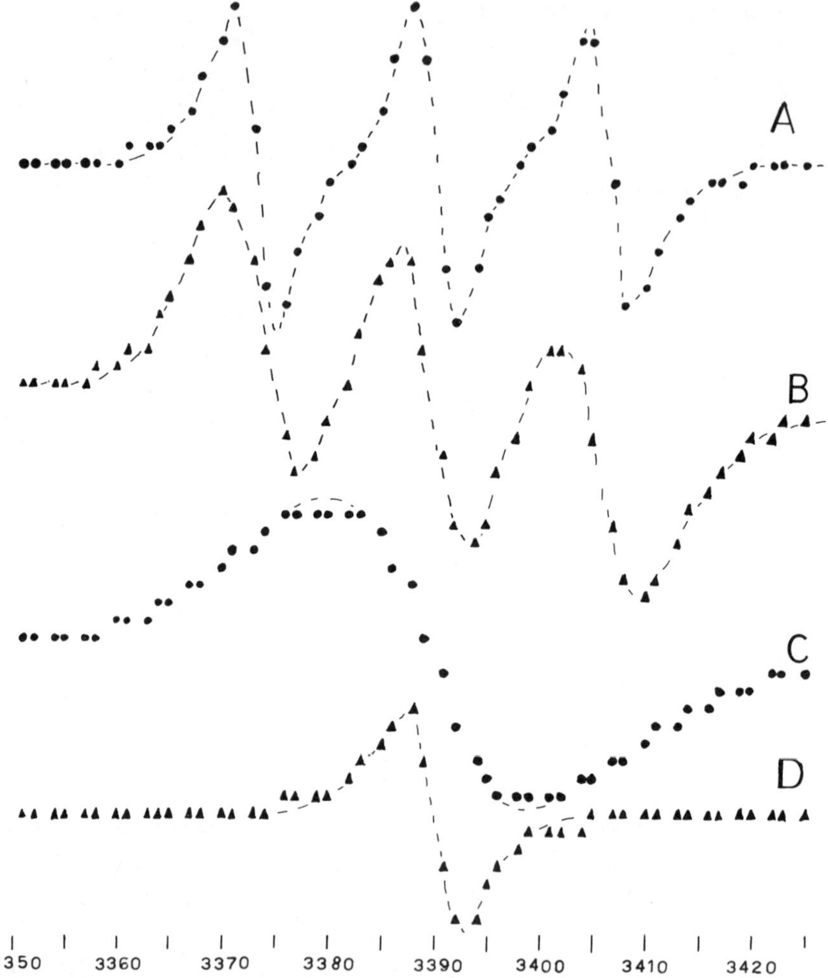

Figure 1. Computed derivative spectra for different exchange frequencies. For the spectrum labeled A, $W_{ex} = 0$; B, $W_{ex} = 3 \times 10^7 Hz$; C, $W_{ex} = 10^8 Hz$; D, $W_{ex} = 10^9 Hz$. The magnetic field, in gauss, is plotted, rather than W. *The microwave frequency is assumed as 9.5 GHz for this conversion.*

second, a single broadened band narrows. Figures 1 and 3 show that the spectra change immeasurably for ω_{ex} less than 10^5 Hz and that the triplet structure of the spectrum is lost above about 10^7 Hz. The spectrum narrows measurably until ω_{ex} is about 10^{11} Hz; beyond this, further narrowing is not observed. An exchange frequency range $10^5 \leq \omega_{ex} \leq 10^{11}$ Hz is available, in principle, for studying two-dimensional collision effects.

Since the viscosity coefficients are more fundamental, we computed these material coefficients from the collision frequencies. The computations (Figures 1 and 2) were based on the published (*14, 15, 16*) $\omega_{ex} = 0$ line shape and frequency coefficients for the androstane molecule. This molecule was also used in a separate paper in this series (*13*). Unfortunately, a shape factor could not be built into the naive model; thus only the molecular weight of androstane is relevant, which means that $m \cong 6.3 \times 10^{-22}$ g. With d_{eff} as the effective diameter of the androstane molecule in A units and Σ in units of A^2/molecule, the collision frequency is estimated by

$$\omega_{coll} = 9.1 \times 10^{22} \times \bar{\eta} \times \frac{d^2_{eff}}{\Sigma} \tag{20}$$

The effective diameter of a rapidly rotating ellipsoidal molecule can be taken as the length of the major axis, about 15 A for the androstane molecule. The surface density, $1/\Sigma$, corresponding to $d^2_{eff}/\Sigma = 1$, is 4×10^{13} molecules/cm^2.

The exchange frequency is computed from f in Equation 3. First, we assume that the spin exchange collisions are "strong," *i.e.*, $[J\tau_{coll}] >> 1$ where J is the exchange integral expressed in appropriate units, and τ_{coll} is the characteristic lifetime of the collision complex. Second, Devaux *et al.* (*17*) showed that only one-third of these collisions could be productive since the colliding nitroxide moieties have different nitrogen nuclear and electron spins. Therefore $f \leq 1/3$, and for maximum sensitivity in the detection of surface viscosity, $f = 1/3$. For this approximation

$$\omega_{ex} = 3 \times 10^{22} \times \bar{\eta} \tag{21}$$

from which it follows the range of surface viscosities observable in principle is

$$10^{-17} \leq \bar{\eta} \leq 10^{-11} \text{ g/sec} \tag{22}$$

Probably f is less than 1/3 since the molecular geometry prevents certain collisions from being productive. For example, since the nitroxide moiety is on one end of the molecular ellipsoid of androstane, such molecules can collide with the NO groups separated by nearly 30 A; this would lead to a nonproductive collision if τ_{coll} is short enough. The work of Sackmann and Tröuble (*14, 15, 16*) with androstane suggests that f is close to its maximum for this case. In addition, solvent properties can result in small values of f. According to Eastman *et al.* (*22*), the variation of the pH or ionic strength of the solution can lessen the effects from the strong collisions.

More generally, Eastman *et al.* (*21, 22*) give the exchange frequency as

$$\omega_{ex} = \frac{1}{\tau_2}\left[1 + \frac{1}{(2J\ \tau_{coll})^2}\right]^{-1} \qquad (23)$$

where J and τ_{coll} have been defined, and τ_2 is the time between collisions. Obviously for strong exchange, $\omega_{ex} = 1/\tau_2$, but a correction function should be inserted to complete the correspondence with Equation 4. Several possibilities are suggested by Equation 23 for modifying ω_{ex} when τ_2 is constant: reduce τ_{coll} by using a low viscosity medium, enclose the

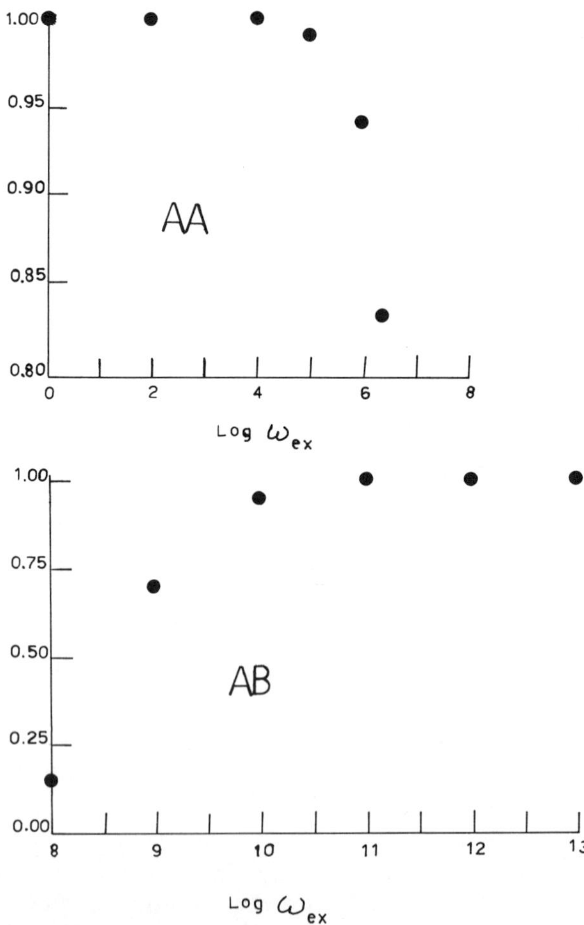

Figure 2. Measure of the change in peak heights as a function of exchange frequency and the corresponding but crude approximate values of the surface viscosity: AA. The triplet is resolved. The peak-to-peak height of the center peak as a function of W_{ex} *is compared with* $W_{ex} = 0$ *case. AB. Only a single*

NO moiety by bulky side groups to reduce J, or include the NO moiety as part of a charged species to increase the distance of closest approach. Unfortunately, τ_{coll} is greatly decreased by other solvents since water has a small viscosity. Also it appears that charge effects tend to be modest (*22*) and will not allow the orders of magnitude diminution of ω_{ex} required to handle dilute monolayers that have $\bar{\eta} > 10^{-11}$ g/sec. Steric protection of the NO group appears to be a better way of adjusting ω_{ex} for a given $\bar{\eta}$ range.

Even if the various ways of adjusting the $J\tau_{coll}$ factor are exploited, the range of $\bar{\eta}$ values may not extend into the range of values that would

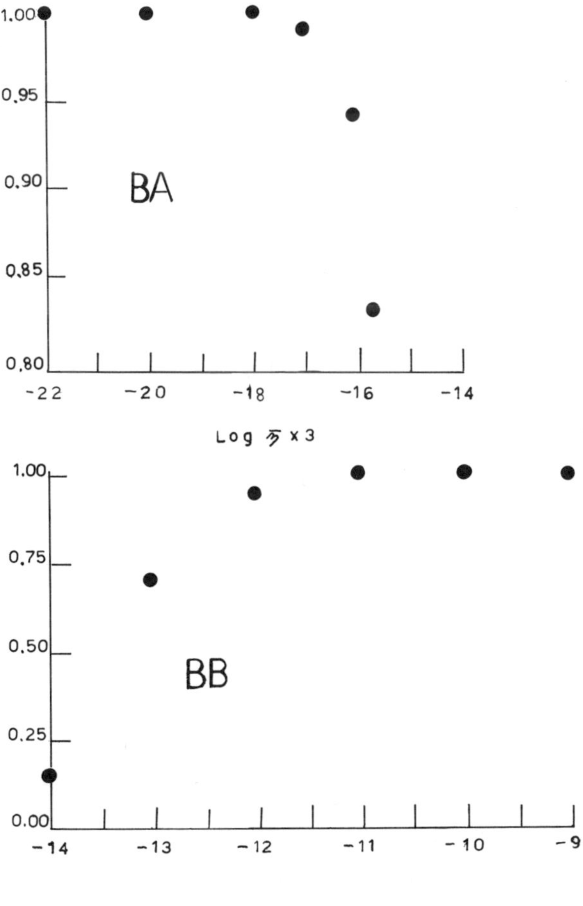

band is observed. The peak-to-peak height is compared with the peak-to-peak height of the fully narrowed band that obtains for $W_{ex} = 10^{13}Hz$. *BA and BB duplicate AA and AB except that the surface viscosity corresponding to* W_{ex} *is plotted.*

match the measurements made by mechanical devices. It may be necessary to work with mixed films involving reduced nitroxides mixed with the active form. The value of f could be carefully controlled, and the two species of surface molecules would be nearly identical with respect to the collision dynamics. Even so, it is unlikely that f could be diminished by more than a factor of 1000 by dilution before analytical uncertainties become intolerable.

Even though τ_{coll} would increase, one could explore the effect of increased substrate viscosity on τ_2 since for dilute gas monolayers $\bar{\eta}$ decreases as the friction coefficient coupling the substrate and monolayer increases. However, the substrate macroscopic viscosity should match the microscopic viscosity; gels would be difficult to use since the two viscosities are not matched (*12*).

There is a striking contrast between our surface collision model in Ref. *7* and the formulas (*e.g.* Ref. *22*) used to relate τ_2 to the three-dimensional transport of spin label molecules. In the three-dimensional model, the spin labels are diffused by Brownian motion, in which case (*18*)

$$\frac{1}{\tau_2} = f^* \times 4\pi \; nd \; D \tag{24}$$

where d is the "interaction distance," n is the density of radicals, f^* is a correction factor defined to take into account spin label–spin label interactions, and D is the diffusion coefficient. The Stokes–Einstein relation is

$$D = \frac{kT}{6\pi \; a \; \eta} \tag{25}$$

where a is the radius of the diffusing particle and η is the viscosity of the medium in which the particles are diffusing. In this case,

$$\frac{1}{\tau_2} = f^* \times \frac{2}{3} \, mk \times \frac{d}{a} \times \frac{T}{\eta} \tag{26}$$

On the other hand, Equation 19 has the collision frequency, ν_c, identified here as $1/\tau_2$, proportional to the surface visocity, but the substrate viscosity is not displayed directly. Certainly Equation 19 represents the two-dimensional collision frequency expected from a gaseous monolayer, but the substrate coupling is not explicit.

It is certainly possible to retrace the derivation of Equations 24 and 25 in a context of a set of Brownian diffusion particles moving on a fluid substrate. The result apparently is that the collision frequency for a low density of surfactant molecules will be proportional to T/η; η will not be involved at all. The substrate effect dominates the surface transport in such a model. In a sense the Copper–Mann theory is a generalization of this model that takes into account the lateral interactions of the Brownian particles.

The construction of Cooper and Mann (*7*) for the surface viscosity includes the substrate effect by a model that represents the result of very frequent molecular collisions between the small substrate molecules and the larger molecules of the monolayer. This was done by adding a term to the Boltzmann equation for the 2D singlet distribution function that is equivalent to the friction coefficient term of the Fokker-Planck equation from which Equations 24 and 25 can be constructed. Thus a Brownian motion aspect was introduced into the kinetic theory of surface viscosity. It would be interesting to derive the collision frequency of Equation 19 using the better model (*7*) and observe how the T/η variable of Equation 26 emerges.

Chapter 24 discusses the experimental details for measuring collision distorted spectra by the spin label method. A satisfactory signal-to-noise ratio when the number of spins is in the range of 10^{13} radicals is difficult to attain. Time averaging techniques can effectively improve the signal-to-noise ratio, but we still have not been able to work with surface densities less than about 2×10^{13} molecules/cm^2 of spin label surfactant when the signal is broadened. Because the Varian E4 system that we used for these studies had a heavy work load, it was impossible to optimize the intrinsic stability of the system for our particular application. However, on the basis of our experience (*13*), it is possible to work with densities to perhaps 0.5×10^{13} molecules/cm^2 when time averaging is coupled with a precise cell design.

We assume that conditions can be controlled to minimize additional relaxation effects such as magnetic dipole–dipole interactions. As the number of relaxation mechanisms decreases, the information necessary for a line shape analysis of the spectra also decreases. Thus, a poorer signal-to-noise ratio can be tolerated, and the signal can be smoothed by curve fitting techniques. Since little is known at the molecular level about two-dimensional transport coefficients, such as the surface viscosity, large uncertainties can be tolerated. In this sense, we believe that much can be learned from monolayer experiments using spin label surfactants.

It is less certain that the combination of instrumental noise at such low concentrations and the existence of multiple relaxation mechanisms will allow the determination of collision frequencies accurate enough to allow a critical test of the Cooper–Mann model and its generalization to delocalized monolayers. We are exploring this possibility.

Appendix

APL functions for computing the arrays $\boldsymbol{u}$ and $d\boldsymbol{u}/d\omega$ are shown in Figure A1 as they were typed by an IBM 2741 terminal. A function was used to do the conversion of units and the construction of the $\boldsymbol{P}$ vector composed of the elements

$$P \leftarrow W_{\mathrm{ex}},\ \{T_j\},\ \{W_j\}.$$

This vector was the left domain of the spectrum function G_{ex}. The right domain of G_{ex} was an array of frequencies, W. W was usually composed of 50 numbers to obtain a reasonable representation of the spectrum across the width of the terminal's typewriter. However, W could be of any useful length subject only to time limitations and the memory size of the workspaces which was 50 kbytes for these calculations.

```
      ∇GEX[⎕]∇
    ∇ G←P GEX W;M;I;G̲
[1]   I←1,,G← 0 12 ⍴W←,W
[2]  L1:G←G,[1](((-¯3↑G̲),3↑G̲)⌹M),G̲←(0 0 0 1 1 1)⌹M←⍉P MARRAY W[I]
[3]   →((⍴W)≥I←I+1)/L1
[4]   DVSUM←+/((1↑⍴W),3)↑3⌽G
[5]   →0
    ∇
      ∇MARRAY[⎕]∇
    ∇ M←P MARRAY W;A;MOFF
[1]   M←(A,-MOFF),[1](MOFF←DIAGM(¯3↑P)-W),A←(-P[1]×(⍳3)∘.≠⍳3)+DIAGM(2×P[1])+÷P[2 3 4]
    ∇
      ∇DIAGM[⎕]∇
    ∇ Z←DIAGM V;N
[1]   Z←(N,N)⍴((N×N)⍴1,(N←⍴V)⍴0)\V
    ∇
```

Figure A1. Reproduction of the spectrum function, GEX, written in APL language

The resolution of typewriter graphics is limited (*see* Figure 1). Nevertheless, resolution was sufficient for qualitative comparisons, and the results were immediately available. The G_{ex} function was included in a function that allowed modification of parameters simply and produced graphs of one or a number of the spectra that can be constructed out of the elements of $\boldsymbol{u}$ and $d\boldsymbol{u}/d\omega$. For comparison with experimental spectra, the frequency variable W was converted to the equivalent field that would obtain at a microwave frequency of 9.5 GHz.

Acknowledgments

We gratefully acknowledge conversations with L. H. Piette and his encouragement and help in working out this application of the spin label technique. We are most grateful for conversations with P. Devaux and J. C. Lang about exchange effects.

Literature Cited

1. Jolly, M., *Surface Colloid Sci.* (1972) **5** ,1.
2. Gains, G. L., "Insoluble Monolayers at Liquid-Gas Interfaces," (Interscience) J. Wiley, New York, 1966.
3. Mannheimer, R. J., Schechter, R. S., *J. Colloid Interface Sci.* (1970) **32,** 212.
4. Goodrich, F. C., Allen, L. H., *J. Colloid Interface Sci.* (1972) **40,** 329.
5. Poskanzer, A., Goodrick, F. C., Birdi, K. S., Allen, L. H., "Abstract 136, Colloid Division," Spring Meeting, ACS, 1974.
6. Blank, M., Britten, J., *J. Colloid Sci.* (1965) **20,** 789.
7. Cooper, E. R., Mann, J. A., *J. Phys. Chem.* (1973) **77,** 3024.
8. Povich, M. J., Mann, J. A., *J. Phys. Chem.* (1973) **77,** 3020.
9. McConnel, H. M., McFarland, B. G., *Quart. Rev. Biophys.* (1970) **3,** 91.
10. Muus, L. T., Atkins, P. W., Plenum Press, London, 1972.
11. Atkins, P. W., "Electron Spin Resonance—Volume I," Norman, R. O. C., Ed., The Chemical Society, London, 1973.
12. Snipes, W., Keith, A., *Research/Development* (1970) **21,** No. **2,** 22.
13. McGregor, T. R., Cruz, W., Fenander, C. I., Mann, J. A., ADV. CHEM. SOC. (1975) **144,** 308.
14. Sackman, E., Trauble, H., *J. Amer. Chem. Soc.* (1972) **94,** 4482.
15. *Ibid.*, p. 4492.
16. *Ibid.*, p. 4499.
17. Devaux, P., Scandella, C. G., McConnell, H. M., *J. Mag. Res.* (1973) **9,** 474.
18. Iverson, K. E., "A Programming Language," Wiley, New York, 1962.
19. See the several Users Manuals, e.g., IBM publication GH20-09061, for the modern description of this language based on Iverson's work.
20. Hirschfelder, J. O., Curtis, C. F., Bird, R. B., "Molecular Theory of Gases and Liquids," Wiley, New York, N.Y., 1964.
21. Lang, John C., private communication, 1974.
22. Eastman, M. P., Bruno, G. V., Freed, J. H., *J. Chem. Phys.* (1970) **52,** 2511.

RECEIVED November 6, 1974. This work was supported by a grant from PRF No. 6689-AC5, 6.

26

Monolayers of Dimethylsiloxane-Containing Block Copolymers

G. L. GAINES, JR.

General Electric Corporate Research and Development,
Schenectady, N. Y. 12301

Monolayers of styrene and poly(2,6-diphenyl)phenylene ether–dimethylsiloxane block copolymers have π–A characteristics similar to those of bisphenol A carbonate–siloxane copolymers. The π–A curves are sigmoidal, and the areas at low surface pressure are proportional to siloxane content, indicating that the non-spreading organic blocks occupy no area at the interface in the spread films. On the basis of more detailed examination of 3 bisphenol-A carbonate–dimethylsiloxane copolymer monolayers which are stable to 16–18 dynes/cm, a configuration having the organic block immersed in the aqueous phase is proposed. Methyl methacrylate–siloxane block copolymers, both of whose components are independently spreadable, behave differently; their monolayer characteristics, and those of physical mixtures of methacrylate and siloxane homopolymers, indicate a gross deviation from additivity of film areas of the two separate components.

The surface activity of block copolymers containing dimethylsiloxane units as one component has received considerable attention. Silicone–polyether block copolymers (*1*, *2*, *3*) have found commercial application, especially as surfactants in polyurethane foam manufacture. Silicone–polycarbonate (*4*, *5*), –polystyrene (*6*, *7*), –polyamide (*8*), –polymethyl methacrylate (*9*), and –polyphenylene ether (*10*) block copolymers all have surface-modifying effects, especially as additives in other polymeric systems. The behavior of several dimethylsiloxane–bisphenol A carbonate block copolymers spread at the air–water interface was described in a previous report from this laboratory (*11*). Noll *et al.* (*12*) have described the characteristics of spread films of some polyether–siloxane block co-

Table I. Surface Pressure Decay of DMS–BPAC Films

After Compression to 13 dynes/cm

Sample	*Average Block* DP *DMS*	*BPAC*	$\Delta\pi$, *dynes/cm decrease in 60 sec*
III	10	3.4	<0.1
1236	20	3.3	<0.2
923	20	3.9	<0.2
1000-B	60	3.6	<0.2
924	20	5	0.3
944	40	10	2.1
1428	100	16	1.2

Sample III, Compress and Hold at Constant Area for 15 min at Each Point

Area, m^2/mg	π, *dynes/cm*	$\Delta\pi$, *dynes/cm*
0.674	2.2	<0.1
.593	5.1	<0.1
.460	10.1	−0.1
.412	12.2	−0.2
.199	27.1 (initial)	−5.3 in 4 min

polymers. This paper presents additional information on polycarbonate block copolymer monolayers and compares them with films of polystyrene, polyphenylene ether, and polymethyl methacrylate block copolymers with polydimethylsiloxane.

Experimental

Films were spread from purified *n*-hexane, benzene, or chloroform, depending on the polymer solubility. The subphase in all cases was distilled water. Surface pressure or potential–area characteristics were measured using apparatus and procedures previously described (*11, 13*). In replicate measurements, π–A characteristics were reproducible within ±3%. All measurements were made at room temperature (24 ± 1°C).

Results and Discussion

Bisphenol A Carbonate–Dimethylsiloxane Block Copolymers. As reported previously (*11*), the surface pressure–area (π–A) curves for these materials are sigmoidal, with a sharp rise in pressure at low areas, but considerable variation in the stability of films compressed to surface pressures greater than 10 dynes/cm had been noted. On the basis of initial compression curves of 20 different samples, it seems that instability—rapid pressure decay on holding the film at constant area after compression to ~13 dynes/cm—is invariably observed when the average polycarbonate block length is greater than 5 BPA units. Table I illustrates typical results.

For further study, three samples which did not exhibit pressure decay after compression to 13 dynes/cm were selected. The stability limit for films of these materials was established by incremental compression, holding the monolayer at constant area for up to 15 min as surface pressure was progressively increased. In the second part of Table I, a typical experiment is detailed. By this procedure, these monolayers were each stable to a surface pressure of 16–18 dynes/cm. There was no hysteresis in compression–expansion cycles (at 2–10% of the total film area per minute) as long as this surface pressure was not exceeded. On the other hand, in a continuous compression experiment (at such rates), surface pressures of 27–32 dynes/cm could be developed before obvious evidence of film collapse (*e.g.*, an increase in compressibility, or bending over of the π–A curve). Figure 1 shows the π–A curves for these three samples, and characterization data are given in Table II.

It was noted previously (*11*) that the BPAC units of these copolymers occupy essentially no area at the interface when the films are

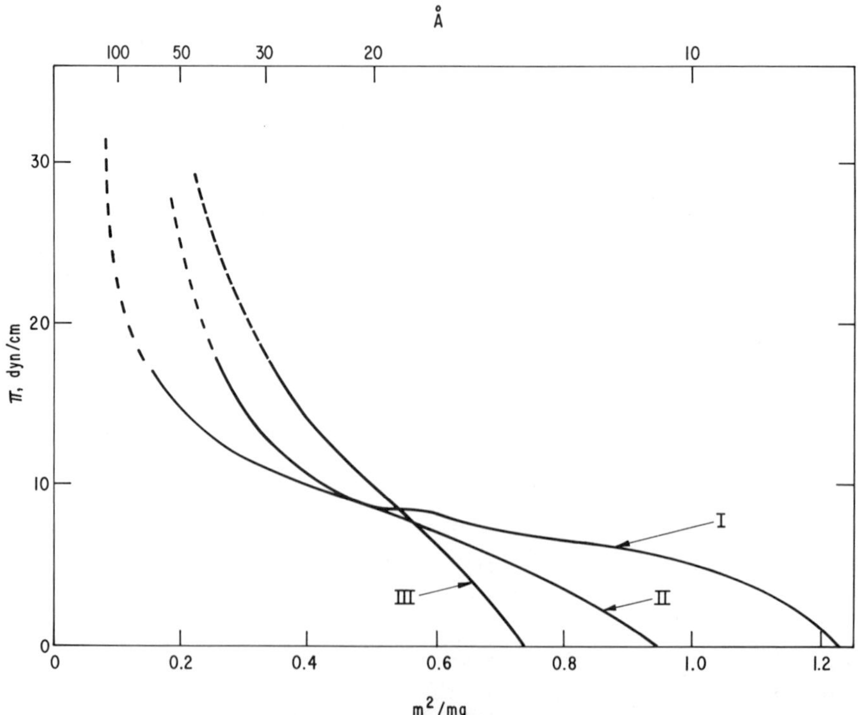

Figure 1. Surface pressure–area characteristics of DMS–BPAC block copolymers (cf. Table II) on water. Solid lines are region of stable surface pressure, dashed lines represent region where pressure decays with time. Equivalent film thickness (for density of 1.0 g/ml) indicated on axis at top.

Table II. Properties of DMS–BPAC Copolymers Forming Stable Monolayers

Sample	*Wt %* *DMS*	*Average Block* DP *DMS*	*BPAC*	*Repeat Unit* MW	$\overline{M}_n$ *(Osmotic)*	*Density, g/ml*
I	81.9	40	2.6	3620	79,000	1.04
II	63.1	20	3.4	2350	79,200	1.07
III	46.5	10	3.4	1600	36,200	1.13

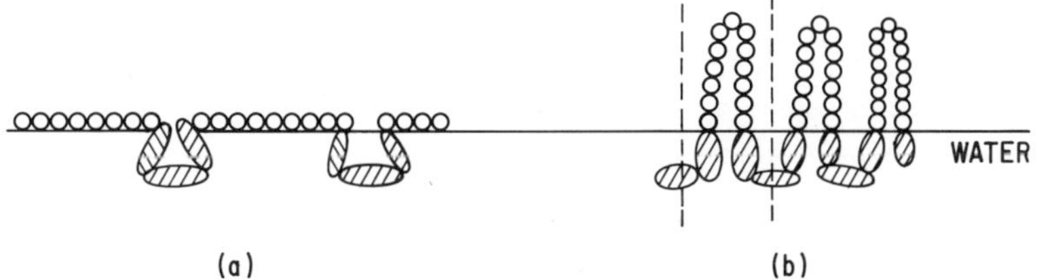

Figure 2. Schematic of structure of block copolymer films. Open circles are DMS segments, shaded lozenges are BPAC segments. (a) At low surface pressures; (b) at maximum compression.

spread at low pressure. (While no specific configuration was described in that report, a diagram suggested that the BPAC segments projected upward from the surface.) We have now established that bisphenol A monomer is slightly soluble in water—rough estimates suggest that the solubility is of the order of 0.01% at room temperature. Therefore, we now favor a configuration in which the BPAC segments are immersed in the aqueous subphase (Figure 2). Since the solubility of polymers decreases as the degree of polymerization rises, this may provide a rationale for the reduced film stability of copolymers with larger BPAC blocks. Additional support for the concept may be adduced from calculations utilizing the minimum stable film areas and thicknesses estimated from these areas and the polymer density. The average repeat unit molecular weight (*i.e.*, the sum of the average DMS block and BPAC block weights —values listed in Table II) permits calculation of an average repeat unit cross-section at the minimum film area. These areas (*cf.* Table III) are similar for all three samples (93 ± 5 A^2), and on the basis of measurements on molecular models they appear too small to admit a configuration with both DMS and BPAC segments folded upward from the water surface. The length of the average repeat unit can also be estimated from the model dimensions (14 A for a BPAC monomer and 2.8 A for a DMS monomer); these values are given in Table III. If we assume that the thickness of the film at the minimum area is equal to half of the repeat length, less the square root of the minimum repeat unit cross

Table III. **Monolayer Characteristics of DMS–BPAC Copolymers**

Sample	*Min Area* m^2/mg	*Thickness,* A	*Cross-section,* A^2	*Repeat Length,* A	τ_{calc}, A
I	0.15	65	90	148	64
II	.25	37	98	104	42
III	.34	27	89	75	29

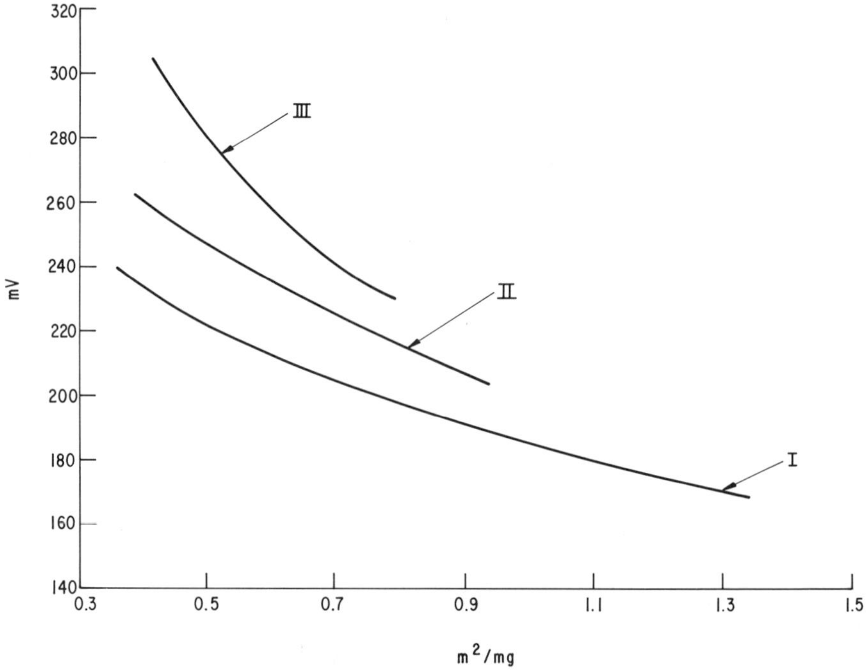

Figure 3. Surface potential, ΔV, vs. film area for DMS–BPAC block copolymer layers. Numbers refer to polymers in Table II.

sectional area, the values of τ_{calc} given in Table III are obtained. These are in good agreement with the experimental thickness values.

Surface potentials were also measured for these three materials; results are shown in Figure 3. The surface potentials of polydimethylsiloxane homopolymer monolayers have been reported by several workers (*14, 15, 16, 17, 18*); while there are slight quantitative differences, all the reported values lie between 135 and 200 mV, and in every study it was found that the product $A \cdot \Delta V$ (which may be taken as proportional to the monomer unit contribution to the measured potential) decreased as the film was compressed. The absolute values of ΔV for the block copolymers are larger than for the PDMS homopolymer. The change of $A \cdot \Delta V$ on compression is similar but appears to be progressively smaller as the BPAC content increases. While it is not obvious how these effects should be interpreted in terms of film structure, they do not appear to offer any contradictions to the structure postulated above.

Films of Other Copolymers with Insoluble, Non-Spreading Blocks. Samples of styrene–dimethylsiloxane and poly(2,6-diphenyl)phenylene ether–dimethylsiloxane block copolymers were also examined as spread films. The styrene–siloxane copolymers included AB, ABA, and repeating block copolymer types. In these cases, as with the polycarbonate, the organic homopolymers do not form monolayers when we try to spread volatile solvent solutions on water. The characteristics of the copolymer spread films, however, were similar to those of the BPAC–DMS copolymers. In all cases, sigmoidal π–A curves were obtained, and surface pressures above 10 dynes/cm were unstable. (All of the samples examined had organic blocks of 15 or more monomer units.) A typical curve, for a styrene–dimethylsiloxane repeating block copolymer (*19*), is shown in Figure 4.

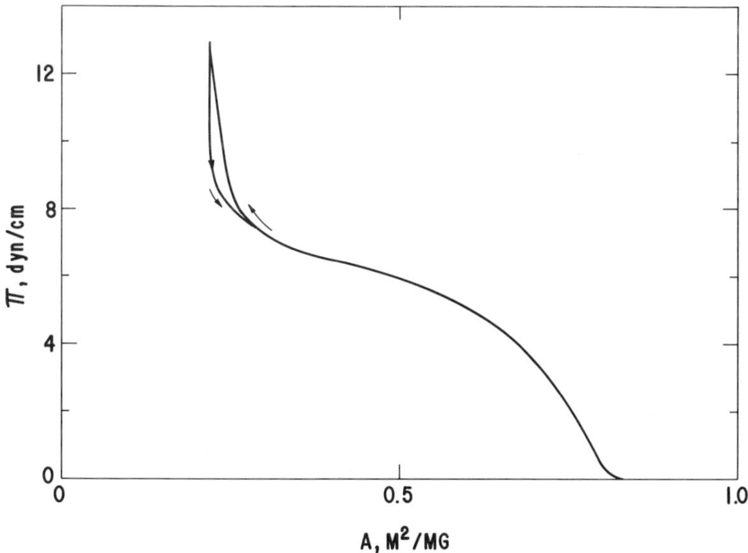

Figure 4. Surface pressure–area curve for $(DMS_{40}$*–styrene*$_{50}$*–*$DMS_{40})_m$ *block copolymer (54% dimethylsiloxane;* cf. *Ref.* 19). *Pressure drop at low area occurs in 30 sec, but no hysteresis was observed on subsequent expansion of film below 7 dynes/cm.*

As for the BPAC copolymers, the area at low surface pressure was in all cases proportional to the siloxane content of the copolymer (Figure 5): this suggests that for these copolymers, too, the siloxane chains spread at low pressures but the organic blocks occupy negligible area at the interface. Areas at high surface pressure did not correlate well with

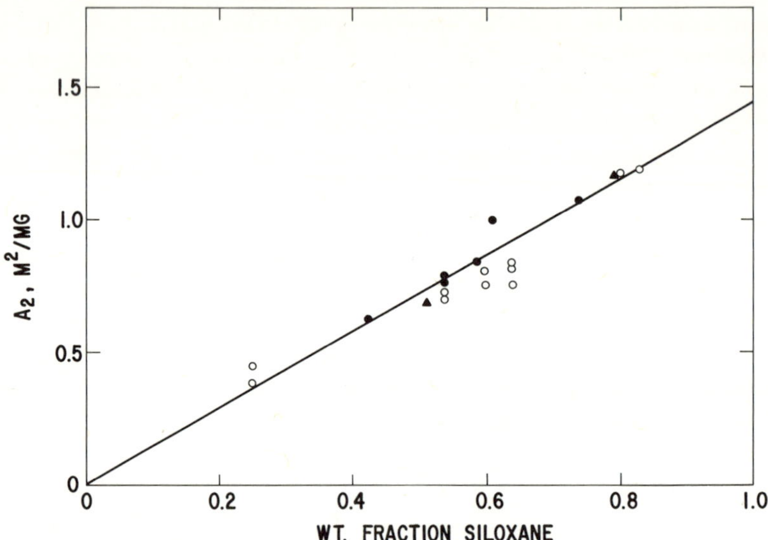

Figure 5. Monolayer area at $\pi = 2$ dynes/cm, vs. *weight fraction siloxane.* ○ *BPAC copolymers (from Ref.* 11); ● *Styrene–DMS copolymers;* ▲ *Phenylene ether–DMS copolymers.*

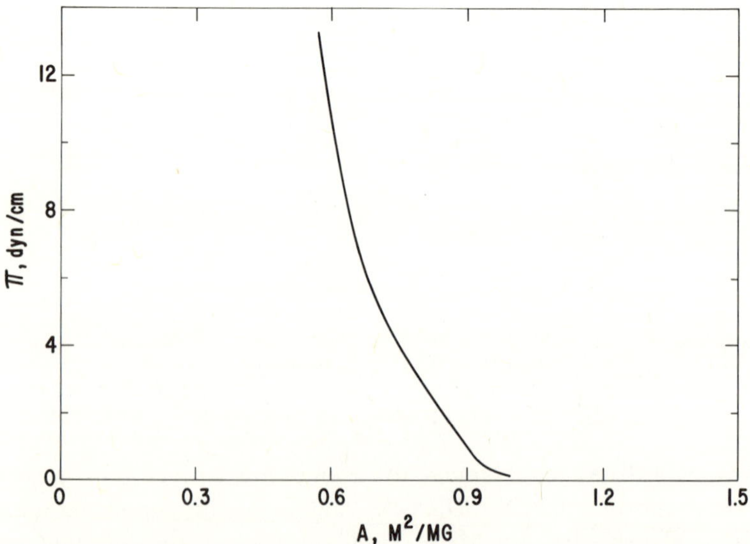

Figure 6. Surface pressure–area curve for MMA–DMS block copolymer (38-1-HI, Ref. 9) *containing 44.6% dimethylsiloxane*

composition, but in view of the rapid pressure decay observed with most of the films, this may not be surprising, and detailed calculations were not attempted.

Methyl Methacrylate–Dimethylsiloxane Block Copolymers (*9*). Films of these materials, which are of the AB type, and contain blocks both of which independently form insoluble monolayers on water, behave quite differently from those with non-spreading blocks. Monolayers of three samples, having dimethylsiloxane contents ranging from 25 to 58% were examined; their π–A curves were nearly identical. The curve for the intermediate material, containing 44.6 wt % dimethylsiloxane (corresponding to a structure $(MMA)_{200}$–$(DMS)_{217}$), is shown in Figure 6. The surface pressure was stable at least to the upper limit of this measurement, and no hysteresis was observed on compression and expansion in this range.

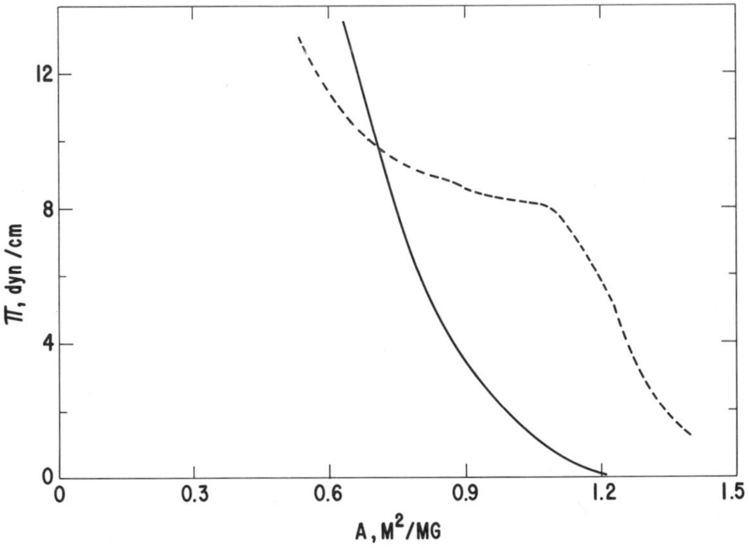

Figure 7. Surface pressure–area curve for mixture of dimethylsiloxane and methyl methacrylate homopolymers, 42.2% siloxane. Solid curve: observed for mixed film. Dashed curve: calculated for additivity of single component monolayers.

For comparison with this material, a mixture of the homopolymers was examined. Portions of an anionically polymerized polymethyl methacrylate ($\overline{M}_n = 39{,}200$) and a polydimethylsiloxane ($\overline{M}_n = 29{,}000$) were combined to yield a mixture containing 42.2 wt % siloxane; the mixture was spread to form a monolayer. Figure 7 shows the π–A curve of this mixed film, along with the curve calculated by adding, in the appropriate proportion, the π–A curves measured for the two homopolymers.

Significant interaction in mixed monolayers containing poly(methyl methacrylate) (as evidenced by non-additivity of film areas) was noted in certain cases by Wu and Huntsberger (*20*). However, there have been a number of reports (*21, 22*) which suggest that the monolayer characteristics of this polymer depend appreciably on both molecular structure details, such as tacticity, and experimental procedure, notably compression rate. Accordingly, a more detailed examination of this effect is beyond the scope of the present work.

Acknowledgment

I am indebted to J. W. Dean, P. C. Juliano, K. W. Krantz, and H. A. Vaughn for generously providing samples, and to D. G. LeGrand for helpful discussion.

Literature Cited

1. Kanner, B., Reid, W. G., Peterson, I. H., *Ind. Eng. Chem., Prod. Res. Develop.* (1967) **6**, 88.
2. Kendrick, T. C., Kingston, B. M., Lloyd, N. C., Owen, M. J., *J. Colloid Interface Sci.* (1967), **24**, 135, 141.
3. Owen, M. J., Kendrick, T. C., *J. Colloid Interface Sci.* (1968) **27**, 46.
4. LeGrand, D. G., Gaines, G. L., Jr., *Amer. Chem. Soc., Div. Polymer Chem., Preprints* **11**, 442 (1970).
5. LeGrand, D. G., Gaines, G. L., Jr., U.S. Patent **3,686,355** (1972).
6. Owen, M. J., Kendrick, T. C., *Macromolecules* (1970) **3**, 458.
7. German Offen. **1,915,789** (1969); *Chem. Abstr.* (1969) **71**, 125476g.
8. Owen, M. J., Thompson, J., *Brit. Polym. J.* (1972) **4**, 297.
9. Juliano, P. C., Floryan, D. E., Hand, R. W., Karttunen, D. D., "Block and Graft Copolymers," J. J. Burke and V. Weiss, Eds., p. 61, Syracuse University Press, 1973.
10. Clark, R. F., Krantz, K. W., U.S. Patent **3,696,137** (1972).
11. Gaines, G. L., Jr., *J. Polymer Sci., Pt. C* (1971) **34**, 115.
12. Noll, W., Sucher, C., de Montigny, A., *Koll. Z. Z. Polymere* (1973) **251**, 643.
13. Gaines, G. L., Jr., "Insoluble Monolayers at Liquid-Gas Interfaces," Chap. 3, Wiley-Interscience, New York, 1966.
14. Fox, H. W., Taylor, P. W., Zisman, W. A., *Ind. Eng. Chem.* (1947) **39**, 1401.
15. Newing, M. J., *Trans. Faraday Soc.* (1950) **46**, 755.
16. Willis, R. F., *J. Colloid Interface Sci.* (1971) **35**, 1.
17. Bernett, M. K., Zisman, W. A., *Macromolecules* (1971) **4**, 47.
18. Arslanov, V. V., Ogarev, V. A., *Dokl. Akad. Nauk SSSR* (1971) **196**, 1105.
19. Dean, J. W., *Polymer Letters* (1970) **8**, 677.
20. Wu, S., Huntsberger, J. R., *J. Colloid Interface Sci.* (1969) **29**, 138.
21. Sutherland, J. E., Miller, M. L., *Polymer Letters* (1969) **7**, 871.
22. Sutherland, J. E., Miller, M. L., *J. Colloid Interface Sci.* (1970) **32**, 181.

RECEIVED October 3, 1974. Work supported in part by the Office of Saline Water, U.S. Department of the Interior.

27

Macromolecular Conformation in the Two-Dimensional State

G. GABRIELLI and M. PUGGELLI

Institute of Physical Chemistry, Via G. Capponi 9, Florence, Italy

Experiments were done on monomolecular films of poly-β-benzyl-L-aspartate (PβBA) at the water/air interface obtained from a spreading solvent with and without pyridine and poly-γ-benzyl-L-glutamate (PγBG) obtained from a spreading solvent of chloroform on a support of HCl or dichloroacetic acid solutions. The results show that the α-helical monomolecular films of polymer PβBA are obtained when a solvent devoid of pyridine is spread; β-helical form is obtained from solvents containing a high percentage of pyridine. For monolayers of PγBG there is evidence of the α-helixes on the HCl support and the random coil on the dichloroacetic acid support.

By comparing experimental spreading isotherms and their thermodynamic values with theoretical ones (*1, 2, 3, 4, 5, 6, 7*), we can obtain parameters of the distribution and energies of macromolecules at an interface (*8, 9, 10, 11, 12*). The macromolecular configuration in the bulk phases (*13*) and the type of interface (*11, 14, 15, 16, 17, 18, 19, 20*) are important in determining the distribution and polymeric conformation in the two-dimensional state. Thus, by comparing the configurations in the three-dimensional and two-dimensional phases, we can define the role of the interface in modifying the configuration found in the bulk phase.

This work was done to obtain various macromolecular forms in the two-dimensional state for the same polymer. Since monomolecular films at the water–air interface are often treated theoretically (*21*) as two-dimensional solutions, the most convenient comparison appeared to be between two- and three-dimensional solutions. We studied two polypeptidic polymers which have conformations that can be modified in solution by various solvents:

(a) Two-dimensional films of PβBA at the water–air interface, obtained from various spreading solvents on the same liquid substrate.

(b) Two-dimensional films of PγBG at the water–air interface, obtained from the same spreading solvent on various liquid substrates.

The macromolecular forms at the interface were characterized by determining the spreading isotherms at various temperatures and by the technique of multiple internal reflection (MIR).

Experimental

The polymers PβBA (mean molecular weight, 4200) and PγBG (mean molecular weight, 100,000) were supplied by Miles Yeda, Ltd., Israel. The solvents, high purity chloroform and dichloroacetic acid, were supplied by Carlo Erba, Milan, Italy; the pyridine for chromatography was supplied by Riedel de Haën, Hannover. The support solutions were twice-distilled water, purified by activated charcoal, high purity hydrochloric acid (Riedel de Haën), and dichloroacetic acid (Carlo Erba).

Surface pressure, measured by an apparatus already described (*8, 9, 10*), was ±0.05 dyne/cm; the surface area was 0.02 m^2/mg. All isotherms were constructed by points after each area reached a constant surface pressure to guarantee that surface equilibrium was attained. All isotherms were obtained at various surface concentrations to ensure repro-

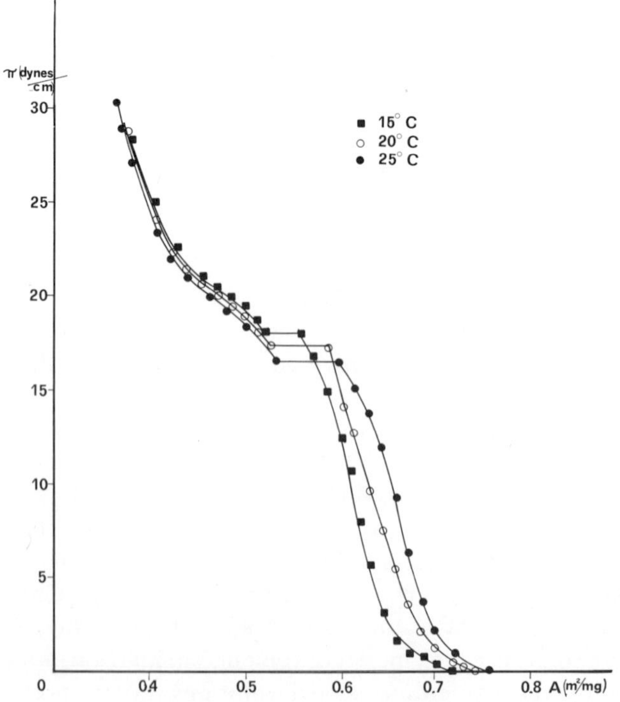

Figure 1. Spreading isotherms of PβBA at 15°, 20°, and 25°C obtained from solvent I

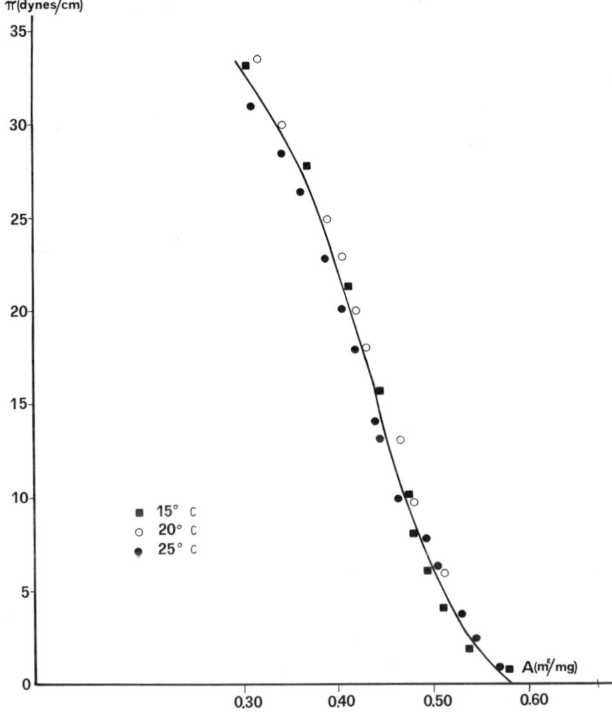

Figure 2. Spreading isotherms of PβBA at 15°, 20°, and 25°C obtained from solvent II

ducibility and perfect spreading. Special care was taken in forming PβBA monolayers, starting from pyridine-rich spreading solutions. To avoid introducing drops into the lower phase, the spreading was done slowly, with the microsyringe held nearly parallel to the surface. The spreading solutions, prepared for solubilizing the polymer in the solvent or mixture of solvents were used 24 hr after preparation.

IR spectra were recorded with a Perkin–Elmer 225 spectrophotometer with accessories for MIR from Wilks Scientific Co., model 50 with a germanium prism, 52.5 × 20, 1 mm thick. To transfer polymer films from the liquid substrate to the germanium plate, the Langmuir–Blodgett technique was used (*22, 23*); oleic acid was the piston oil, and the withdrawal rate was 1 mm/min. The germanium plate was washed with a detergent (Tide), carefully rinsed with water, acetone, and chloroform, and dried for 3 hr *in vacuo*. Cleaning was always controlled by the IR spectrum of reflection on the germanium plate only.

Results and Discussion

PβBA Monolayers. All PβBA monolayers were obtained on twice-distilled water, starting from two different spreading solvents. Spreading solvent I, composed of chloroform containing 0.8 vol % dichloroacetic acid, ensured complete solubilization of the polymer; spreading solvent II was composed of solvent I and 80 vol % pyridine. Figure 1 shows the spreading isotherms at 15°, 20°, and 25°C for PβBA monolayers obtained from solvent I. Figure 2 shows the monolayers obtained from solvent II. Since the π–A plot (Figure 2) was nearly independent of temperature, the line shows the mean values, and the symbols refer to the various temperatures.

For spreading solvent I (Figure 1), all the isotherms have a surface pressure "plateau" above 10 dynes/cm; thus, the curves are composed of two parts. For pressures lower than the plateau, we derived from π–A experimental values the corresponding equations of the two-dimensional state πA–π. From their coefficients, using an accepted procedure (*8, 9, 10, 11*), we derived the following parameters:

(1) B_1, the coefficient of the first degree term in the two-dimensional state equation, deduced theoretically by M. L. Huggins (*7*).

(2) z', the coordination number of the two-dimensional pseudo-lattice of Singer's theory (*1*).

(3) f_m, the partial submersion factor in the Frish and Simha theory (*4*).

(4) η^2/z, which, in the treatment by Motomura and Matuura (*6*), represents the interaction energy between macromolecular segments at the interface.

These parameters are shown together at the limiting area A_0 (obtained by extrapolating the straight portion of the high pressure range of the π–A isotherms) in Table I.

Table I shows that (a) the values of A_0 agree satisfactorily with those found by others (*24, 25, 26*), but they do not constitute a valid criterion

Table I. Parameters of PβBA in the Two-Dimensional State

Temperature, °C	A_o *(m²/mg)*	B_1	z'	f_m	η^2/z[a]
Pyridine-free Spreading Solutions					
15	0.650	0.997	2.22	0.81	−31.5
20	0.680	1.000	2.22	0.81	−46.5
25	0.698	1.010	2.22	0.81	−16.0
Spreading Solutions Containing 80% Pyridine					
15–25	0.535	1.008	2.23	0.83	−87.0

[a] The η^2/z values are in cal/mole of monomeric unit.

Table II. Thermodynamic Spreading Values of PβBA from Pyridine-Free Spreading Solutions

A *(m²/mg)*	ΔS_s *ergs/cm² K*	ΔH_s *ergs/cm²*
0.72	+0.10	+ 28.7
0.70	+0.16	+ 45.6
0.68	+0.38	+108.8
0.66	+0.68	+193.7
0.64	+0.82	+231.9

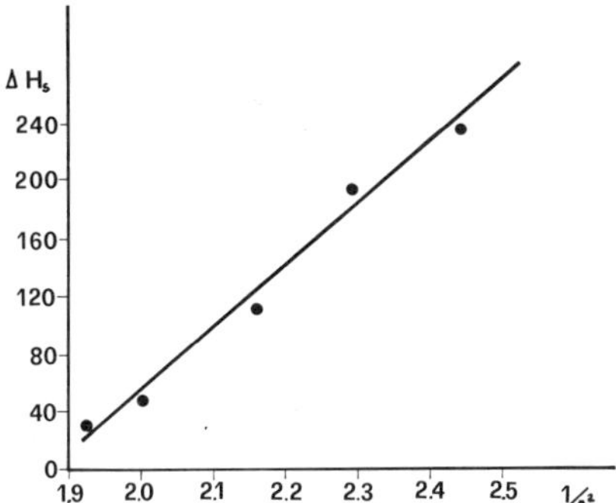

Figure 3. ΔH_s vs. $1/A^2$ *for PβBA polymer*

for distinguishing and characterizing the macromolecular form present at the interface; (b) the values of z' and consequently of ω flexibility (*19*) [even considering the estimates made in their deduction (*8, 9, 10*)] agree with a configuration at the two-dimensional helix which is not very flexible. Of course, even these parameters do not indicate a definite macromolecular conformation at the interface; (c) the values of f_m show that nearly 20% of the macromolecules are submerged in the aqueous substrate, typical of most of the polypeptidic type polymers; (d) the negative values of η^2/z indicate (*8, 9, 10*) that the main surface energies are attractive.

We also calculated the thermodynamic spreading functions [enthalpy (ΔH_s) and entropy (ΔS_s)] to further our knowledge of polymeric distribution at the interface (*9, 10, 11, 27, 28, 29*). Their values at 20°C for five surface areas are shown in Table II. Since the spreading functions are substantial, the polymer–interface system (*27, 28, 29*) is not athermic. In particular, the positive sign of the enthalpy means that in the monolayers the attraction energies between macromolecular segments are prevalent if other enthalpic contributions are constant. This result is confirmed by the fact that the most important of these (caused by submersion) should be negative if one considers the hydrophobic groups in the macromolecules. The nearly linear dependence of ΔH_s on $1/A^2$ (*see* Figure 3) further confirms that the spreading enthalpy depends largely on the attraction energy between macromolecular segments.

The above discussion suggests a two-dimensional spiral on the surface, with attractive energy between polymeric segments, which agrees satisfactorily with the α-helical form. This has been demonstrated by others for this and other macromolecules (*25, 26, 30, 31, 32, 33*). To confirm further the α-helical form on the surface, we recorded multiple reflection IR spectra at surface pressures both less than and greater than the plateau pressure (Figure 4, part 1). The band characteristic of the

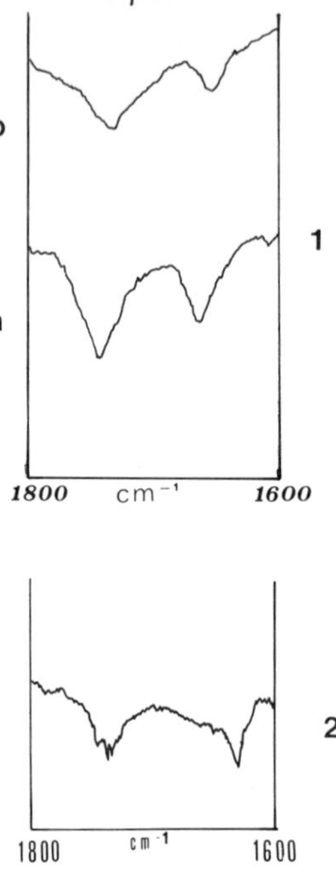

Figure 4. (1) MIR spectra of surface films of PβBA from solvent I transferred on germanium prism (a) at surface pressure less than the plateau pressure, (b) at surface pressure greater than plateau pressure (2) MIR spectra of surface films of PβBA from solvent II transferred on germanium prism at high surface pressure

β form is missing, and the bands of the α or coil form are present (*32*). Since the preceding data show the existence of a macromolecular form more rigid than the coil with attractive energy between the segments at the interface, the presence of the α-helical form appears to be confirmed on the surface as the bulk phase (*34*).

The pressure plateau with solvent I is attributed to either a first-order transition of the two-dimensional phase or to collapse. With polypeptidic monolayers, surface pressure plateau has often been interpreted (*30, 31, 33*) as a collapse and, more precisely, as a passage of the α form from the monolayer to the bilayer. We observed after every compression an obvious decrease of surface pressure with time which was characteristic of collapse. The identity of the IR spectra of monolayers transferred at pressures lower and greater than the plateau pressure may confirm that the plateau is not the result of a phase transition connected with a change in the macromolecular conformation. Even the rather low, negative surface entropy of about -0.15 erg/cm^{2}K at the plateau agrees with the observation of others for bilayer formation (*30, 31*). Finally, if we consider the collapse at equilibrium and calculate the transition enthalpy from the classic formula $\ln \pi_1/\pi_2 = -\Delta H/R[1/T_1 - 1/T_2]$, we obtain nearly 60 ergs/cm^2. This agrees favorably with the work required to remove the unit area of monolayer from the water surface as calculated by Malcolm (*30, 31*) for other polymers. Therefore, the pressure plateau may be caused by collapse, and the curves of the isotherms at pressures greater than this must be compression curves of collapsed monolayers which are not represented by equations which are valid only for monomolecular films.

Thus α-helical monolayers can be obtained from solvent I on the surface; in this case, the stable conformation in the three-dimensional spreading phase is not modified by the interface (*35, 36*). Therefore the polymer–support interactions are not as great as the polymer–polymer interactions, as confirmed by the small variation of the first degree coefficient of the equations of the two-dimensional state with temperature (Table I).

With regard to the monolayers from solvent II, particular care must be taken in spreading. Because these isotherms at various surface concentrations have high reproducibility, the difference between the curves obtained with solvent I and these are not attributed to loss of the polymer from the interface to the substrate. Even in this case we derived the two-dimensional state equation which corresponded to the mean isotherm and, therefore, is valid for the three temperatures. From it we deduced the same parameters described earlier for spreading solvent I (Table I).

The limiting area of about 18 A^2/monomeric unit is less than that of the α form; this agrees with the findings of others (*33*) for polymethyl glutamate attributed to the β form. Naturally, this is not a criterion for characterizing the macromolecular form found on the surface. The values of z' and ω are the same size as those for spreading solutions I and may only signify that at the interface a much more flexible macromolecular form, like the random coil, will not be found; however, these values cannot be used to distinguish the two helical forms. The value of f_m shows that the macromolecule is submerged to the same degree as an α form. The negative value of η^2/z indicates that the form at the interface is characterized by inter- or intramolecular attractive energies and therefore is probably not the random-coil type.

Since these spreading isotherms are independent of temperature, the coefficient of the first degree term is independent of temperature. Besides confirming the presence on the surface of a form different from that obtained with solvent I, the interaction energies between polymeric segments are smaller than in the preceding case. From earlier observations and the value of the first degree coefficient of the two-dimensional state equation (Table I) which is equal to or less than that of the α form (*37*), it seems improbable that the random coil form exists on the surface. A helical form more extended than the α one—possibly a β form—appears to be more sound.

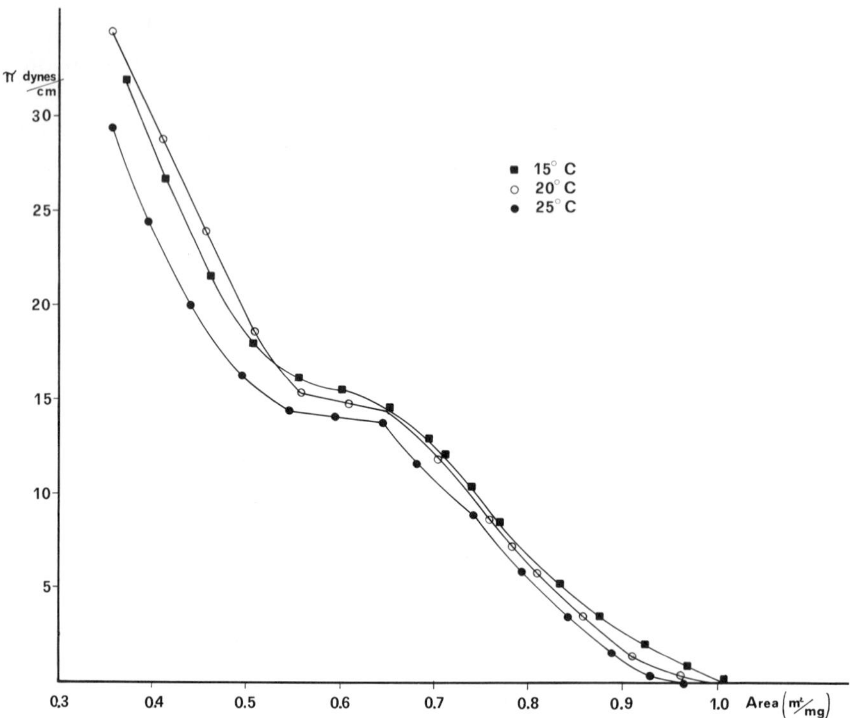

Figure 5. Spreading isotherms of PγBG at 15°, 20°, and 25°C on support I

A further, though indirect, confirmation of the existence of the form can be found by comparing the spreading enthalpies of the two forms. In fact, taking the difference between the enthalpy of the form obtained from solvent I and that from solvent II, taking as surface areas the limiting areas of the two forms, and considering the corresponding pressures, one obtains about 3500 cal/mole of monomeric unit; this agrees satisfactorily with the transition enthalpy $\alpha \rightarrow \beta$ in the bidimensional phase for other compounds (*38*), taking into account the diversity of the hydrophobic chains.

Finally, the multiple reflection IR spectra, obtained on the monomolecular films from solvent II and transferred at constant pressure in the zone of condensed film, are shown in Figure 4, part 2. The band at 1630 cm^{-1} is characteristic of the β form (*33, 39*). Thus, the surface contains two-dimensional films of β helixes in the case of pyridine-rich spreading solvents. This agrees with the results of others for different polymers (*33*). The β helixes for the same polymer in monolayer are present (*40*) because the sample had a low molecular weight.

Monolayers of PγBG. All the PγBG monolayers were obtained using chloroform as the spreading solvent on two different supports: support I was a solution of 0.001*N* HCl in twice-distilled water; support II was a solution of 0.25*M* dichloroacetic acid (DCA) in twice-distilled water. This concentration was the minimum strength necessary to achieve spreading isotherms of PγBG which were independent of DCA concentration in the support. Figure 5 shows the spreading isotherms on support I at 15°, 20°, and 25°C. Figure 6 shows the same isotherms on support II.

MONOLAYERS ON SUPPORT I. As Figure 5 shows, all the isotherms have a sharp plateau in surface pressure and therefore are made up of two parts—one less than and one greater than the plateau. For the areas below the plateau, we calculated the two-dimensional state equation πA–π from experimental data; from this equation, we calculated the same parameters that were used for the other polymers. These parameters are shown in Table III. According to Table III:

(1) The value of A_0, taking into account the experimental conditions to get the π–A curves, agrees satisfactorily with that found by others and is generally attributed to the α form although it does not constitute a definite criterion for characterizing the macromolecular form found at the interface.

(2) The values of z' and ω would indicate the presence of a rigid form even though one must always remember the approximations made in their deduction. These parameters for PγBG have lower values than

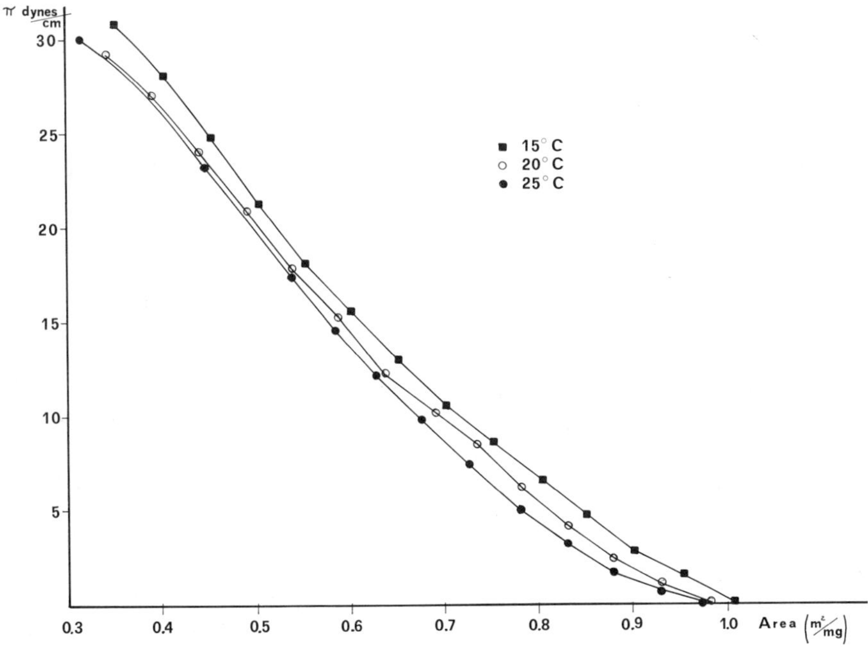

Figure 6. Spreading isotherms of PγBG at 15°, 20°, and 25°C on support II

Table III. Parameters of PγBG in the Two-Dimensional State

Temperature, °C	A_o *(m²/mg)*	B_1	z'	f_m	η^2/z [a]
		Subphase I			
15	0.880	1.082	2.009	0.81	−1430
20	0.866	1.041	2.009	0.81	−1438
25	0.858	1.019	2.008	0.77	−1350
		Subphase II			
15	0.833	1.146	2.010	0.86	+ 414
20	0.792	1.140	2.010	0.85	+1054
25	0.775	1.140	2.010	0.85	+1072

[a] Values of η^2/z are in cal/mole of monomeric unit.

Table IV. Thermodynamic Spreading Values of PγBG

A *(m²/mg)*	ΔS_s *(ergs/cm² K)*		ΔH_s *(ergs/cm²)*	
	Subphase I	*Subphase II*	*Subphase I*	*Subphase II*
0.687	−0.160	−0.22	−58.3	−75.0
0.742	−0.150	−0.24	−52.9	−73.0
0.756	−0.165	−0.23	−56.6	−75.0
0.783	−0.165	−0.24	−55.2	−76.4
0.825	−0.160	−0.24	−52.0	−64.5

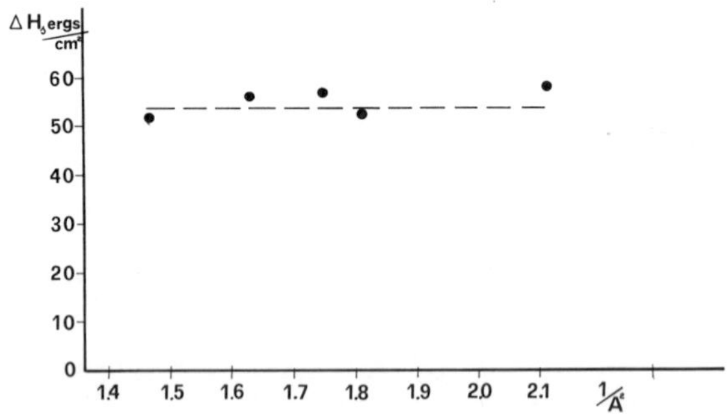

Figure 7. ΔH_s vs. *1/A² for PγBG polymer*

those for PβBA. Besides the different rigidity, this arises from the difference in molecular weight (*8, 9, 10*).

The values of f_m indicate that in this case the macromolecule is partially submerged in the substrate. Since the η^2/z values are all negative, they indicate energies of attraction between the macromolcular segments at the interface, characteristic of rigid forms like an α helix. Values of η^2/z for this polymer are higher than those for PβBA because of the longer hydrophobic chain of monomeric unity.

We also calculated the thermodynamic spreading functions (Table IV) for five surface area values. In this case, the polymer–interface system cannot be considered athermic, but ΔH_s is negative even though it is positive for PβBA. However, we calculated a total ΔH_s which includes all enthalpic contributions to the spreading, particularly those caused by submersion in the aqueous substrate. For PγBG, the submersion process could be facilitated by the presence of H^+ ions of the support acid, thus resulting in a higher corresponding negative enthalpic contribution. The independence of ΔH_s with respect to $1/A^2$ (Figure 7) further confirms that such ΔH_s is not derived mainly from the attractive and repulsive energies of interaction between macromolecular segments; therefore, the negative sign does not exclude the presence of a form characterized by cohesive energies between macromolecular segments. Finally, the multiple reflection IR spectra retaken on monolayers transferred above and below the plateau (Figure 8) show the spectrum characteristic of the α form or coil as found by others (*32, 33*).

Therefore, excluding the presence of the β form and considering the values of z' and ω, the negative sign of η^2/z, the variation with temperature of coefficient B_1, and the presence of plateau characteristic of the

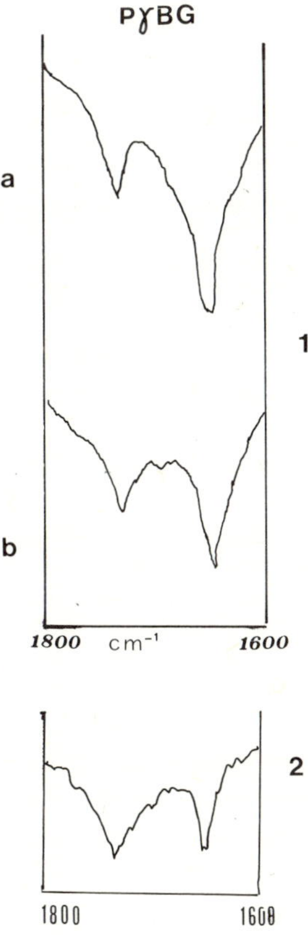

Figure 8. (1) MIR spectra of surface films of PγBG on support I transferred on germanium prism (a) at surface pressure less than the plateau pressure, (b) at surface pressure greater than plateau pressure (2) MIR spectra of surface films of PγBG on support II transferred on germanium prism at high surface pressure

α form, the monomolecular films of PγBG on 0.001N HCl are probably made up of the α helix which is stable in the spreading solution; this was found by others (*26, 30, 33*) and with different methods. In addition, starting with the initial area of plateau, the surface pressure decreases with time after every compression.

The multiple reflection IR spectra are identical for films transferred above or below the plateau pressure (Figure 8). The surface entropies corresponding to the plateau pressure of -0.15 erg/cm^2 K and the value of ΔH of transition (monomolecular film → bimolecular film), calculated at about 70 ergs/cm^2, agree well with the corresponding values necessary for the α form to pass from the mono- to the bilayer (*30, 31*). Therefore, the plateau represents "collapse," and the isotherm curves at pressures greater than the collapse pressure pertain to films which are not homogeneous monolayers.

Table V. Thermodynamic Values of Transition

A (*m²/mg*)	ΔS_{trans}, eu	ΔH_{trans}, *cal/mole of monomeric unit*
0.687	2.6	601
0.742	3.8	980
0.756	2.6	729
0.783	3.1	870
0.825	3.4	972
Average values	3.0	830

FILM ON SUPPORT II. The isotherms in this case do not have a plateau in surface pressure at any temperature (Figure 6). We derived from experimental values the πA–π equation and the parameters shown in Table IV, from which we note:

(1) The values of A_o are less than those found for support I; this may be the first clue to the presence of a macromolecular form different from the α form on the surface.

(2) The values of z' and ω are higher and thus indicate a more flexible form than one on support I.

(3) The f_m factor is of the same order and indicates that even here the macromolecules are partially submerged.

(4) The η^2/z factor is positive; this means that at a difference of that observed for the form on support I, the predominant energies among the macromolecular segments are repulsive; this is characteristic of forms less orderly than the helixes.

(5) Even the value of B_1, which is higher than the corresponding ones for the α form, indicates the more random form of the helix (*37*).

The higher negative values of the spreading enthalpy (Table V) further confirm the presence of a macromolecular form that has attractive energies lower than those of the stable form on support I; the negative enthalpic contribution of submersion is of about the same order on the two supports.

The IR spectra for monomolecular films transferred at a pressure in the range of condensed films are identical to those obtained for support I (Figure 8). Thus the β form is not present, but either the α form or the random coil is present though not distinguishable on the basis of these infrared spectra alone as noted by others (*32, 33, 41*).

Since the macromolecular form must be less orderly and different from the α form, the random-coil form seems plausible. We calculated the α-helix, random coil transition enthalpy from the difference of the spreading enthalpies of the forms on supports I and II (Table V); the average is 830 cal/mole of monomeric unit. Table V also shows the transition entropies calculated as the difference between the spreading entropies of the two forms; the average is 3.0 eu. The ΔH and the single values agree well with those found by others using different methods (*42, 43, 44, 45, 46, 47, 48, 49, 50, 51, 52, 53*) in the bulk phase.

This result confirms the existence of the coil form in the DCA support and shows that in this case the support (the interface) can modify the stable form in the three-dimensional spreading solution; further, it allows us to consider a new method for measuring the heat of helix coil transition. This method is analogous to that used in the bulk phase, also based on the difference in heat of the solution by Giacometti and Kagemoto (*50, 51, 52*), it validates the comparison between two-dimensional and tridimensional solutions. Even the value of ΔS transition agrees well with that found in the bulk phase (*45*).

Conclusions

(1) For PβBA the macromolecular conformation in the two-dimensional phase is modified by varying the spreading solvent. Thus it is possible to obtain two different macromolecular forms, at the interface, that modify their conformation only in the bulk phase.

(2) For PγBG the macromolecular conformation is modified if DCA is in the support. Thus it is possible to obtain different macromolecular forms on the surface that modify the conformation only in the two-dimensional state.

These results are particularly interesting not only in preparing various methods for obtaining conformations of the same macromolecules at an interface but also because they indicate the possibility of comparing energies of macromolecules in the two-dimensional phase with those in the three-dimensional phase.

Literature Cited

1. Singer, S. J., *J. Chem. Phys.* (1948) **16**, 872.
2. Saraga, L. T. M., Prigogine, I., *Mem. Serv. Chim. Etat (Paris)* (1953) **38**, 109.

3. Davies, J. T., Llopis, J., *Proc. Roy. Soc. Ser. A* (London) (1955) **227**, 537.
4. Frisch, H. L., Simha, R., *J. Chem. Phys.* (1957) **27**, 702.
5. Kawai, T., *J. Polymer Sci.* (1959) **35**, 401.
6. Motomura, K., Matuura, R., *J. Colloid Sci.* (1963) **18**, 12.
7. Huggins, M. L., *Makromol. Chem.* (1965) **87**, 119.
8. Gabrielli, G., Puggelli, M., Ferroni, E., *J. Colloid Interface Sci.* (1970) **32**, 242.
9. Gabrielli, G., Puggelli, M., Ferroni, E., *J. Colloid Interface Sci.* (1970) **33**, 133.
10. Gabrielli, G., Puggelli, M., Ferroni, E., *J. Colloid Interface Sci.* (1971) **36**, 401.
11. Gabrielli, G., Puggelli, M., *J. Colloid Interface Sci.* (1971) **35**, 460.
12. Birdi, K. S., Gabrielli, G., Puggelli, M., *Kolloid-Z. Z. Polym.* (1972) **250**, 591.
13. Rosoff, M., in "Surface Chemistry and Polymers," Marcel Dekker, Ed., pp. 44-54, 1969.
14. Crisp, D., *J. Colloid Sci.* (1946) **1**, 49.
15. Hotta, H., *Bull. Chem. Soc. Japan* (1953) **26**, 386.
16. Isemura, T., Hotta, H., Miwa, T., *Bull. Chem. Soc. Japan* (1953) **26**, 380.
17. Hotta, H., *J. Colloid Sci.* (1954) **9**, 504.
18. Shick, M., *J. Polymer Sci.* (1957) **25**, 465.
19. Davies, J. T., Rideal, E. K., "Interfacial Phenomena," pp. 246-247, Academic Press, New York, 1963.
20. Gabrielli, G., Puggelli, M., *J. Colloid Interface Sci.* (1973) **45**, 217.
21. Huggins, M. L., *Kolloid-Z. Z. Polym.* (1973) **251**, 449.
22. Blondgett, K. B., *J. Amer. Chem. Soc.* (1935) **57**, 1007.
23. Blondgett, K. B., Langmuir, I., *Phys. Rev.* (1937) **51** ,964.
24. Ikeda, S., Isemura, T., *Bull. Chem. Soc. Japan* (1961) **34**, 416.
25. Malcolm, B. R., *Nature* (1968) **219**, 929.
26. Malcolm, B. R., *J. Polymer Sci.* (1971) **34**, 87.
27. Llopis, J., Rebollo, D. V., *J. Colloid Sci.* (1956) **11**, 543.
28. Llopis, J., Subirana, J. A., *J. Colloid Sci.* (1961) **16**, 618.
29. Llopis, J., Subirana, J. A., *Proc. Intern. Congr. Surface Activity, 3rd, Koln*, Verlag der Universitat-sdruckerei (1960) **II**, 149.
30. Malcolm, B. R., *Polymer* (1966) **7**, 595.
31. Malcolm, B. R., *Proc. Roy. Soc. Ser. A* (1968) **305**, 363.
32. Loeb, G. I., *J. Colloid Interface Sci.* (1968) **26** ,236.
33. Loeb, G. I., Baier, R. E., *J. Colloid Interface Sci.* (1968) **27**, 38.
34. Bradbury, E. M., Downie, A. R., Elliot, A., Hanby, W. E., *Proc. Roy. Soc. Ser. A* (1960) **259**, 110.
35. Deboeck, Y., Jaffé, J., *Intern. Kongr. grenzflächen.*, Carl Hanser Verlag, München (1973) **II** (1), 305.
36. Birdi, K. S., Fasman, G. D., *J. Polym. Sci. Part C* (1973) **42**, 1099.
37. Jaffé, J., Ruysschaert, J. M., Heca, W., *Biochim. Biophys. Acta* (1970) **207**, 11.
38. Glazer, J., Alexander, A. E., *Trans. Faraday Soc.* (1951) **47**, 401.
39. Loeb, G. I., *J. Colloid Interface Sci.* (1969) **31**, 572.
40. Malcolm, B. R., *Biopolymers* (1970) **9**, 911.
41. Masuda, Y., Miyazawa, T., *Makromol. Chem.* (1967) **103**, 261.
42. Zimm, B. H., Doty, P., Iso, K., *Proc. Natl. Acad. Sci. U.S.* (1959) **45**, 1601.
43. Zimm, B. H., Bragg, J. K., *J. Chem. Phys.* (1959) **31**, 526.
44. Lifson, S., Roig, A., *J. Chem. Phys.* (1961) **34**, 1963.
45. Applequist, J., *J. Chem. Phys.* (1963) **38**, 934.
46. Ackermann, T., Rüterjans, H., *Z. Phys. Chem.* (1964) **41**, 116.
47. Karasz, F. E., O'Reilly, J. M., Bair, H. E., *Nature* (1964) **202**, 693.
48. Karasz, F. E., O'Reilly, J. M., Bair, H. E., *Biopolymers* (1965) **3**, 241.
49. Karasz, F. E., O'Reilly, J. M., *Biopolymers* (1966) **4**, 1015.

50. Giocometti, G., Turolla, A., *Z. Phys. Chem.* (1966) **51,** 108.
51. Giacometti, G., Turolla, A., Boni, R., *Biopolymers* (1968) **6,** 441.
52. Kagemoto, Akihiro, *Biopolymers* (1968) **6,** 1753.
53. Karasz, F. E., Gajnos, G. E., *J. Phys. Chem.* (1973) **77,** 1139.

RECEIVED September 23, 1974.

INDEX

D

E

T

U

V

The text of this book is set in 10 point Caledonia with two points of leading. The chapter numerals are set in 30 point Garamond; the chapter titles are set in 18 point Garamond Bold.

The book is printed offset on Danforth 550 Machine Blue White text, 50-pound. The cover is Joanna Book Binding blue linen.

Jacket design by Norman Favin.
Editing and production by Mary W. Rakow.

The book was composed by the Mills-Frizell-Evans Co., Baltimore, Md., printed and bound by The Maple Press Co., York, Pa.